KNOWLEDGE SHARING IN THE INTEGRATED ENTERPRISE

IFIP – The International Federation for Information Processing

IFIP was founded in 1960 under the auspices of UNESCO, following the First World Computer Congress held in Paris the previous year. An umbrella organization for societies working in information processing, IFIP's aim is two-fold: to support information processing within its member countries and to encourage technology transfer to developing nations. As its mission statement clearly states,

> *IFIP's mission is to be the leading, truly international, apolitical organization which encourages and assists in the development, exploitation and application of information technology for the benefit of all people.*

IFIP is a non-profitmaking organization, run almost solely by 2500 volunteers. It operates through a number of technical committees, which organize events and publications. IFIP's events range from an international congress to local seminars, but the most important are:

- The IFIP World Computer Congress, held every second year;
- Open conferences;
- Working conferences.

The flagship event is the IFIP World Computer Congress, at which both invited and contributed papers are presented. Contributed papers are rigorously refereed and the rejection rate is high.

As with the Congress, participation in the open conferences is open to all and papers may be invited or submitted. Again, submitted papers are stringently refereed.

The working conferences are structured differently. They are usually run by a working group and attendance is small and by invitation only. Their purpose is to create an atmosphere conducive to innovation and development. Refereeing is less rigorous and papers are subjected to extensive group discussion.

Publications arising from IFIP events vary. The papers presented at the IFIP World Computer Congress and at open conferences are published as conference proceedings, while the results of the working conferences are often published as collections of selected and edited papers.

Any national society whose primary activity is in information may apply to become a full member of IFIP, although full membership is restricted to one society per country. Full members are entitled to vote at the annual General Assembly, National societies preferring a less committed involvement may apply for associate or corresponding membership. Associate members enjoy the same benefits as full members, but without voting rights. Corresponding members are not represented in IFIP bodies. Affiliated membership is open to non-national societies, and individual and honorary membership schemes are also offered.

KNOWLEDGE SHARING IN THE INTEGRATED ENTERPRISE

Interoperability Strategies for the Enterprise Architect

Edited by

Peter Bernus
Griffith University
Australia

Mark Fox
University of Toronto
Canada

 Springer

Library of Congress Cataloging-in-Publication Data

A C.I.P. Catalogue record for this book is available from the Library of Congress.

Knowledge Sharing in the Integrated Enterprise, Edited by Peter Bernus and Mark Fox

p.cm. (The International Federation for Information Processing)

ISBN 978-1-4419-3893-0   e-ISBN 978-0-387-29766-8

Printed on acid-free paper.

Printed in the United States of America.

9 8 7 6 5 4 3 2 1
springeronline.com

CONTENTS

Preface to ICEIMT'04

Knowledge Sharing in the Integrated Enterprise – Interoperability Strategies for the Enterprise Architect

The ICEIMT series of conferences 1992, 1997 and 2002 was originally started as a strategic initiative of NIST and the European Union to review the state of the art in Enterprise Integration (EI) and to make recommendations to industry and research, creating roadmaps for EI research and product development. Pre-conference workshops had been organised with internationally recognised experts and reports of these workshops were presented at the conference.

Enterprise Integration has grown in the past ten years at a pace where there is an obvious need for a more frequent forum where these strategic discussions can be continued bringing together leading thinkers of industry, defence and research.

The IFIP (International Federation of Information Processing) Working Group 5.12 on Enterprise Integration (a majority of members being organisers of past ICEIMT conferences and invited experts to past workshops) has taken the responsibility to sponsor this more frequent reincarnation of ICEIMT. In addition, the INTEROP European Network of Excellence has been invited to present the results of their interoperability workshop series at the conference. As EI is an interdisciplinary field, the International Programme Committee includes important figures from industrial engineering and management, supply chain management, software engineering, systems engineering, artificial intelligence and computer science, CALS, and most importantly, representatives of tool developers. Members also include strategic leaders of Virtual Enterprise research and ongoing projects on interoperability.

A particular feature of EI, and interoperability within that area, is the prominent role of international and regional standardisation bodies as well as industry consortia. An important role of ICEIMT04 will be to conduct discussions about the strategic fit between the short and medium term steps that industry needs to take (which should enable the development of interoperable products and software systems), and the long term strategic considerations. Without the deep understanding of this issue industry may end up facing a new 'Y2K problem' in the years to come.

In the past five years is has become apparent that creating the technical conditions of interoperability must be supported by cultural, socio-economic and psychological conditions. The interoperability of our software tools crucially depends on the motivation of people who create them, their ability to learn as well as to communicate in order to create a mutually accepted common understanding. Thus this conference intends to also investigate interoperability from the point of view of communication between humans.

This last point inspired the title of ICEIMT'04: Knowledge Sharing in the Integrated Enterprise – Interoperability Strategies for the Enterprise Architect, because there is interoperability between humans (so they understand one another on the basis of commonly defined concepts) and interoperability between modelling tools (so they can exchange and interpret models in the same way).

Topics of this conference include:

- Enterprise Modelling (modelling languages, scope and detail of modelling, model development methodologies)
- Enterprise Reference Models (modularity, sharability, quality, scalability)
- Enterprise Architecture Frameworks (practice and theory), role of standardisation, relation to systems and engineering and software engineering
- Interoperability – present and future trends & standardisation
- Common ontologies (level of definition – logic, XML, etc –, competency questions, evolvability, standardisation)
- Enterprise Modelling Tools (functionality, interoperability, methodology support)
- Hot spots of interoperability
- New theories and techniques, interdisciplinary approaches
- Human understanding and communication as a condition of interoperability. Suitable social structures that create the motivation and the opportunity to achieve common understanding and consensus.

Papers of this conference have been peer reviewed through a double blind refereeing process and we would like to thank all members of the International Programme Committee who undertook this task.

Peter Bernus
Chair IPC ICEIMT'04

Mark Fox
Chair Organising Committee

Preface to DIISM'04

Manufacturing and Engineering in the Information Society: Responding to Global Challenges

Since the first DIISM working conference, which took place nearly 11 years ago, the world has seen drastic changes, including the renovation of manufacturing softwares. The conditions for engineering and manufacturing science have changed on a large scale, in terms of technology-enabled collaboration among the fields of design, engineering, production, usage, maintenance and recycling/disposal. These changes can be observed in rapidly growing fields such as supply chain management. On factory floors, new visions of co-existing human-machine production systems involve both knowledge management and multi-media technologies.

As a consequence of these changes, the importance of information infrastructures for manufacturing has stunningly increased. Information infrastructures play a key role in integrating diverse fields of manufacturing, engineering and management. This is in addition to its basic role as the information and communication platform for the production systems. Eventually, it should also serve the synthetic function of knowledge management, during the life cycles of both the production systems and their products, and for all stakeholders.

These proceedings is the compilation of those leading authors, who have contributed to the workshop 'Design of Information Infrastructure Systems for Manufacturing' (DIISM 2004) that was held at the University of Toronto from Oktober 10 - 11, 2004. Prominent experts from both academia and industries have presented significant results, examples and proposals. Their themes cover several necessary parts of the information infrastructure.

The workshop was sponsored by the International Federation of Information Processing (IFIP), through Working Groups 5.3 (Computer Aided Manufacturing) and 5.7 (Computer Applications in Production Management).

We sincerely thank all the authors, the program committee members and the participants for their contribution.

In conclusion, we strongly hope that these proceedings will be a useful source of information for the further development and foundation of the information infrastructure for engineering and manufacturing.

On behalf of the Organizing Committee

Jan Goossenaerts
Eindhoven University of Technology

ICEIMT 2004 Conference Chair

Mark S. Fox, University of Toronto, Canada

ICEIMT 2004 Programme Chair

Peter Bernus, Griffith University, Australia

ICEIMT 2004 Members of the International Programme Committee

Hamideh Afsasmanesh, UvA, The Netherlands

Marc Alba, DMR Consulting Group, Spain

Sam Bansal, Consultant, USA

Mihai Barbuceanu, U Toronto, Canada

Giuseppe Berio, University of Torino, Italy

Peter Bernus, Griffith University, Australia; IFIP TC5.12 Chair

Dennis Brandl, Sequencia

Carlos Bremer, EESC- University of Sao Paulo

Jim Browne, CIMRU, Ireland

Christopher Bussler, DERI Galway, Ireland

L. Camarinha-Matos, New University of Lisbon, Portugal

Yuliou Chen, Tshinghua University. PRC

David Chen, Universite de Bordeaux I, France

Niek. D. Preez, Uni of Stellenbosch, South Africa

Emmanuel delaHostria, Rockwell Automation, USA

Ziqiong Deng, Narvik Inst of Tech, Norway

Guy Doumeingts, GRAISOFT, Bordeaux

J.J.P. Ferreira, INESC Porto

Mark Fox, U Toronto, Canada

Yoshiro Fukuda, Hosei University, Japan

Marco Garetti, Politecnico Di Milano, Italy

Florin Gh. Filip, ICI-Inst, Romania

Jan Goossenaerts, Eindhoven Uni, the Netherlands

H.T. Goranson, Sirius-Beta, USA

Michael Gruninger, NIST, USA

Gérard Guilbert, EADS, Paris

Richard H. Weston, Loughborough University

Michael Huhns, University of South Carolina, USA

Matthias Jarke, Technical University Aachen, Germany

Brane Kalpic, ETI Co, Slovenia

Bernhard Katzy, CeTim / Uni BW Munich, Germany

ICEIMT 2004 Members of the International Programme Committee (cont'd)

Kurt Kosanke, CIMOSA Association, Germany
Robert L. Engwall, USA
Francisco Lario, University of Valencia, Spain
Meir Levi, Interfacing Technologies, Canada
Hong Li, Columbus, USA
Frank Lillehagen, COMPUTAS, Norway
Michiko Matsuda, Kanagawa Inst. of Techn, Japan
Arturo Molina, U Monterrey, Mexico
John Mylopoulos, University of Toronto, Canada
Richard Neal, NGM Project Office
Ovidiu Noran, Griffith University, Australia
Gus Olling, Daimler-Chrysler, USA; IFIP TC5 chair
Angel Ortiz, UP Valencia, Spain
Herve Panetto, Université Henri Poincaré Nancy, France
Michael Petit, University of Namur, Belgium
Carla Reyneri, Data Consult, Italy
Tom Rhodes, NIST, USA
August-Wilhelm Scheer, Universität des Saarlandes, Germany
Gunter Schmidt, Universität des Saarlandes, Germany
Guenter Schuh, University of St Gallen, Switzerland
Gérard Ségarra, Renault DIO-EGI, France
David Shorter, IT Focus, UK
Dirk Solte, FAW Ulm, Germany
Francois Vernadat, INRIA Rhone-Alpes, France & CEC, Luxembourg
Simon Y. Nof, Purdue Univ, USA
Martin Zelm, CIMOSA Association, Germany

ICEIMT Steering Committee

Peter Bernus, Griffith University, Australia
Mark S. Fox, University of Toronto, Canada
J.B.M. Goossenaerts, Eindhoven University of Technology, The Netherlands
Ted Goranson, USA
Mike Huhns, Univeristy of South Carolina University, USA
Kurt Kosanke, CIMOSA, Germany
Charles Petrie, Stanford University
Francois Vernadat, INRIA Rhone-Alpes, France & CEC, Luxembourg
Martin Zelm, CIMOSA, Germany

DIISM 2004 Conference Chair

Mark S. Fox, University of Toronto, Canada

DIISM 2004 Program Chair

Fumihiko Kimura , The University of Tokyo, Japan

DIISM 2004 Members of the International Programme Committee

Robert Alard, ETHZ, Switzerland
Rainer Anderl, University of Darmstadt, Germany
Eiji Arai, Osaka University, Japan
Hiroshi Arisawa, Yokohama National University, Japan
Peter Bernus, Griffith University, Australia
Jim Browne, CIMRU, Ireland
Luis Camarinha-Matos, New University of Lisbon, Portugal
Yuliou Chen, Tshinghua University, PR China
Mark Fox, University of Toronto, Canada
Florin Gh. Filip, ICI-Inst, Romania
Jan Goossenaerts, Eindhoven University of Technology, Netherlands
Dimitris Kiritsis, EPFL, Switzerland
Kurt Kosanke, CIMOSA Association, Germany
John J. Mills, University of Texas at Arlington, USA
Michiko Matsuda, Kanagawa Inst. of Techn, Japan
Kai Mertins, IPK, Germany
John Mo, CSIRO, Australia
Laszlo Monostori, MTA SZTAKI, Hungary
Odd Mycklebust, SINTEF, Norway
Laszlo Nemes, CSIRO, Australia
Gus Olling, Daimler-Chrysler, USA & IFIP TC5 chair
Masahiko Onosato, Osaka University, Japan
Herve Panetto, Université Henri Poincaré Nancy, France
Henk Jan Pels, Eindhoven University of Technology, Netherlands
Keith Popplewell, University of Coventry, UK
Paul Schoensleben, ETHZ, Switzerland
Keiichi Shirase, Kobe University, Japan
David Shorter, IT Focus, UK
Johan Stahre, Chalmers University, Sweden
Marco Taish, Politecnico Di Milano, Italy
Klaus-Dieter Thoben, University of Bremen, Germany
Fred van Houten, University of Twente, Netherlands
Paul Valckenaers, KULeuven, Belgium

DIISM 2004 Members of the International Programme Committee (Cont'd)

Part I

ICEIMT 04

Knowledge Sharing in the Integrated Enterprise –Interoperability Strategies for the Enterprise Architect

Part I

CHAPTER ...

Knowledge Sharing in the Integrated Enterprise: Interoperability Strategies for the user, the Architect ...

1. A 'Standards' Foundation for Interoperability

Richard A. Martin

Convener ISO TC 184/SC 5/WG 1 Email: tinwisle@bloomington.in.us

Participants of ISO TC184/SC 5/WG 1 will present a series of papers that address the group's work and our thoughts on the direction we feel appropriate for the establishment of new international standards in manufacturing automation. The focus of WG1 is on architecture and modelling aspects in support of the automation standards objectives of TC184. This paper sets the stage, so to speak, upon which current and future group efforts will play out.

1. INTRODUCTION

Members of ISO TC184/SC 5/WG 1 are presenting a series of papers that address the group's work and our thoughts on the direction we feel appropriate for the establishment of new international standards in manufacturing automation. The focus of WG1 is on 'architecture and modelling' aspects in support of the automation standards objectives of TC184. To set the stage, this paper describes the backdrop that frames our current work, identifies a few key terms of our dialog (including a note of caution), introduces the actors in leading roles, and presents an overview of past performances now published as international standards. Upon this stage, Kurt Kosanke will address current draft documents, David Chen will address efforts related to our interoperability standard objectives, and David Shorter will address the topic of meta-modelling as a means to achieve our modeller and model view objectives.

2. BACKDROP

Central to WG1, and many other groups, is the effort to bring standardization that supports integration and interoperability to manufacturing enterprises. Today we are far from achieving the levels of interoperability among manufacturing system components that many believe are essential to significant improvement in manufacturing efficiency (IDEAS, 2003). We continue the exchange of capital and labor to reduce cost and increase productivity per unit of expense, and we improve the communication channels that are now essential to production systems. However, our dynamic response to changes in strategy, tactics, and operational needs continues to be limited by the paucity of interoperability between systems, and between components within systems (National, 2001).

The extent to which we are successful in component and system interoperability is expressed in the current international standards and de-facto industry standards that define the extent of information exchange in use today. Having emerged from the automation of tasks and the adoption of information management as a key factor in modern manufacturing, the need for interoperability of the kind we seek is rather new. Reliance upon human mediated interoperation is no longer sufficient.

3. DIALOG TERMS

3.1 Unified, integrated, interoperable

Systems and components thereof interact in different ways ranging along a continuum from isolated action to complete interoperability. When all connections between components are direct, almost in a physical sense, we can say that the components of a system are unified. A model of this system is a unified model and model components have essentially the same conceptual representation although distinctions in levels of detail resulting from decomposition, and of properties emerging from aggregation, remain.

When a component connection becomes indirect, i.e., a transformation from one representational form or view to another occurs, and system behaviour results from specific knowledge about the means to transfer information, products, and processes, then we can say that the system is integrated. The models of this system, often with distinct conceptual representations, form an integrated system model wherein individual components interact using fixed representations known by other components a-priori.

When component connections become malleable or ad-hoc in their manifestation, then system behaviour must move from static descriptions to incorporate dynamic features that enable interoperability. This situation allows one component, or agent as it is often called, to act as if it were another component while maintaining its own distinct features. Interoperable components interact effectively because they know about effective communication.

These same distinctions, unified, integrated, and interoperable can be used to classify the relationships between systems as well. Systems integration is now the standard of practice and the area of interest to most practitioners. In fact, the vast majority of our standards effort to date has targeted enablement of integration. But interoperability, especially in a heterogeneous setting like a supply chain, goes beyond our methodologies for integration and offers new challenges for system and enterprise understanding. WG1 is pursuing the codification of that understanding into new international standards.

3.2 The 'resource' example

Since standards, both international and de-facto, are developed by working groups, each standard bears a perspective on word choice and meaning that represents an agreement among those approving adoption of the standard. And even then, we tend to allow wide latitude in word use. Take, for example, the use of the term 'resource' that is commonly found in our manufacturing standards, and focus on just one sub-committee – SC5 of TC 184 (Kosanke, 2004). Within SC5 some groups consider 'resource' to include material consumed by manufacturing processes as well as the capital and human resources required to conduct those processes. Other groups, like our WG1, restrict 'resource' to non-consumables. Some even advocate including processes as a deployable resource. All are valid uses of the term but one must be aware of the usage context.

To be interoperable, components and systems must correctly interpret words used as labels and data in an appropriate context. While resolving this aspect of interoperability is beyond the charge of WG1, we are constantly reminded of its importance to our efforts.

4. ACTORS

WG1 is one of several working groups in SC5 developing standards for manufacturing automation. A complete listing of ISO Technical Committees is found at http://www.iso.ch where TC184 is charged with 'Industrial automation systems and integration'. SC5 is now responsible for six working groups and has working group collaboration with TC 184/SC 4 'Industrial data' (SC 5, 2004).

In addition to the collaborations between ISO committees and sub-committees, ISO partners with other international bodies to promulgate standards of common interest. ISO TC184/SC 5 and IEC TC65 are working together at the boundary between automation control systems and production management systems that encompass the information exchange content necessary to direct and report manufacturing operation and control (ISO 62264-1, 2003).

WG1 is working closely with CEN TC310/WG 1 (International, 2001) to produce two standards that are the subject of Kurt Kosanke's presentation and we expect to receive substantive material from other European efforts including those detailed by David Chen in his presentation.

5. PAST PERFORMANCES

5.1 Describing industrial data

The development of international standards is an evolutionary process that mimics the evolution of industrial practice as supported by academic and industrial research. One of the more successful standardization efforts toward integration began in 1979 and continues to his day with the efforts of TC 184/SC 4. At that time NIST (National Institute of Standards and Technology, USA) began work in establishing standards for the exchange of engineering drawing elements, beginning with IGES (Goldstein, 1998), that has evolved through several iterations into ISO 10303 and its many application protocol (AP) parts (Kemmerer, 1999). Today ISO 10303, better know as STEP (STandard for the Exchange of Product model data) by many practitioners, is a robust foundation for the exchange of information about product components and, increasingly, system attributes codified as data elements. ISO 10303 continues its evolution with new APs and revisions to established parts.

A recent study commissioned by NIST concludes that the STEP standard accounts for an annual two hundred million dollar benefit for adopting industries (Gallaher, 2002). One key factor in the success of STEP related to that savings is the enablement of information migration between product and process versions. This reuse of data through changes in operations comprises half of the standards benefit to industry.

One feature of ISO 10303 is the EXPRESS language (ISO 10303-11, 1994) and its graphical extension subset that enables the programmatic description of primitives identified in the standard. In a manner similar in concept to ISO 10303, the new PSL language standard (ISO 18629-1, 2004) seeks to emulate the success of STEP.

5.2 Describing industrial processes

PSL (Process Specification Language) and its extension parts target the exchange of process descriptions among process modelling and enablement tools. Note that these two language standards, EXPRESS and PSL, go beyond the format definition of descriptive information exchange, e.g., EDI, to allow a more flexible resolution of rule based semantic exchange for well defined situations.

A distinguishing characteristic of PSL is its origin as a joint effort between the data centric charge of ISO TC 184/SC 4 and the process centric charge of ISO TC 184/SC 5. SC5 collaboration with SC4 also involves a multi-part standard for 'Industrial manufacturing management data' known as MANDATE (ISO 15531-1, 2004).

5.3 SC5 Integration standards

ISO TC 184/SC 5 is producing a series of standards devoted to integration and interoperability:

- component to component information exchange protocols under the 'Open System application integration frameworks' multi-part standard (ISO 15745-1, 2003),
- the establishment of 'Manufacturing software capability profiles' (ISO 16000-1, 2002),
- and recently a Technical Report on Common Automation Device Profile Guidelines' (IEC/TR 62390, 2004) was approved.

These standards codify existing industry practice and focus industrial efforts on common feature support. These are detailed descriptive standards that can be utilized to enable integration and to support interoperability.

5.4 WG1 integration standards

At the other end of the spectrum is ISO 14258 (ISO 14258, 1998) that describes concepts and rules for enterprise models. This WG1 produced standard provides an overview of the issues that must be considered when modelling in the context of enterprises. It establishes system theory as the basis for modelling and introduces the primary concepts of modelling that include: life-cycle phases, recursion and iteration, distinctions between structure and behaviour, views, and basic notions of interoperability.

Upon this conceptual foundation, ISO 15704 (ISO 15704, 2000) constructs a more detailed model representation and adds concepts for life history, and model genericity. This standard also begins the elaboration of methodologies to support enterprise modelling. A significant feature of ISO 15704 is its informative Annex A that presents the GERAM (Generalised Enterprise Reference Architecture and Methodologies) developed by the IFIP/IFAC Task Force on Architectures for Enterprise Integration. Currently we are amending ISO 15704 to add user centric views, Economic View and a Decision View, as informative annexes. ISO 15704 identifies the structural features available for further development of model and system interoperability.

6. ON WITH THE SHOW

All of these standards support the interactions necessary to construct unified manufacturing operations and enhance integration among systems of differing origin. But the difficult tasks of dynamic interoperation are yet to be addressed in a standard way. These past efforts lay a solid foundation and begin to articulate the system and component features necessary to achieve robust interoperability. We invite your support for international standards and our efforts. Should you wish to participate, please contact the author.

The presentation of Kurt Kosanke will describe in more detail two standards now in preparation that continue our articulation of enterprise representation through models.

Acknowledgments

The author wishes to thank all participants of ISO TC 184/SC 5 and in particular those who attend our WG1 meetings.

REFERENCES

Gallaher, M., O'Connor, A., Phelps, T. (2002) Economic Impact Assessment of the International Standard for the Exchange of Product Model Data (STEP) in Transportation Equipment Industries. RTI International and National Institute of Standards and Technology, Gaithersburg, MD

Goldstein, B., Kemmerer, S., Parks, C. (1998) A Brief History of Early Product Data Exchange Standards – NISTIR 6221. National Institute of Standards and Technology, Gaithersburg, MD

IDEAS Roadmaps (2003) Deliverable D2.3 The Goals, challenges, and vision (2nd version) (M8), Information Society Technologies, European Commission, Available at www.ideas-roadmap.net

IEC/TR 62390 (2004) Common Automation Device Profile Guideline. IEC, Geneva

International Organization for Standardization and European Committee for Standardization, (2001) Agreement on Technical Co-operation between ISO and CEN (Vienna Agreement). Version 3.3

ISO 10303-11 (1994) Industrial automation systems and integration – Product data representation and exchange - Part 11: Description methods: The EXPRESS language reference manual. ISO, Geneva

ISO 14258 (1998) Industrial automation systems – Concepts and rules for enterprise models. ISO, Geneva

ISO 15704 (2000) Industrial automation systems – Requirements for enterprise-reference architectures and methodologies. ISO, Vienna

ISO 15531-1 (2004) Industrial automation systems and integration – Industrial manufacturing management data – Part 1: General overview. ISO, Geneva

ISO 15745-1 (2003) Industrial automation systems and integration – Open systems application integration framework – Part 1: Generic reference description. ISO, Geneva

ISO 16000-1 (2002) Industrial automation systems and integration – Manufacturing
    software capability profiling for interoperability – Part 1: Framework. ISO,
    Geneva
ISO 18629-1 (2004) DIS Industrial automation systems and integration – Process
    specification language – Part1: Overview and basic principles. ISO, Geneva
ISO 62264-1 (2003) Enterprise-control system integration – Part 1: Models and
    terminology. ISO, Geneva
Kemmerer, S (1999) STEP The Grand Experience NIST Special Publication 939.
    Manufacturing Engineering Laboratory, National Institute of Standards and
    Technology, Gaithersburg, MD
Kosanke, K., Martin, R., Ozakca, M., (2004) ISO TC 184 SC5 - Vocabulary
    Consistency Study Group Report, SC 5 Plenary Meeting, 2004-April, AFNOR,
    St. Denis, France. Available at http://forums.nema.org/~iso_tc184_sc5_wg1
National Coalition for Advanced Manufacturing, (2001) Exploiting E-
    Manufacturing: Interoperability of Software Systems Used by U.S.
    Manufacturers. Washington D.C.
SC 5 Secretary (2004) Meeting Report SC 5 Plenary Meeting 2004-04-22/23,
    AFNOR, St. Denis, France. Available at http://forums.nema.org/~iso_tc184_sc5

2.   Standards in Enterprise Inter- and Intra-Organisational Integration

Kurt Kosanke
CIMOSA Association
kosanke@t-online.de

World-wide collaboration and co-operation of enterprises of all sizes increases the need for standards supporting operational interoperability in the global environment. Such standards are concerned with the communication aspects of information and communication technology (ICT), like communication protocols as well as the syntax and semantics of the communication content. Communicating parties have to have the same understanding of the meaning of the exchanged information and trust both the communication itself and the validity of its content. Focus of the paper is on business process modelling and its standardisation in the area of enterprise inter- and intra-organisational integration. Relations to the subject of interoperability are discussed.

Keywords: business process modelling, enterprise integration, enterprise engineering, standardisation.

1. INTRODUCTION

Business in today's global environment requires the exchange of physical products and parts and even more importantly the exchange of the related business information between co-operating organisations. The latter is true for such an operation itself, but to an even larger extent for the decision making processes during the establishment of the cooperating enterprise with market opportunity exploration and co-operation planning and implementation. The need for a knowledge base to be used for organisational interoperation and decision support on all levels of the enterprise operation is recognised as an urgent need in both business and academia (Kosanke *et al* 2002).

Building and maintaining the enterprise knowledge base and enabling its efficient exploitation for decision support are major tasks of enterprise engineering. Enterprise models are capable of capturing all the information and knowledge relevant for the enterprise operation (Vernadat 1996). Business processes and their activities identify the required and produced information as inputs and outputs. Since business processes may be defined for any type of enterprise operation, including management-oriented activities, their models will identify and capture all relevant enterprise planning knowledge as well and thereby complementing the operational knowledge of the enterprise. Process-oriented and model-based enterprise engineering and integration will be a significant contributor to the needed support for enterprise interoperation, provided it can become an easy to use and commonly accepted decision-support tool for the organisation.

Today's challenges concern the identification of relevant information, easy access across organisational boundaries and its intelligent use. To assure interoperation between organisations and their people the exchanged items have to be commonly understood by both people and the supporting ICT and have to be useable by the receiving parties without extensive needs for costly and time consuming preprocessing and adaptation. Therefore, we have to distinguish between two types of interoperability issues: the understanding by people and by the ICT environment.

To achieve common understanding does not mean to establish an Esperanto like all-encompassing business language, but to provide a commonly agreed language base onto which the different professional dialects can be mapped.

The other major interoperability problem encountered in the area of information exchange is due to the use of ICT systems with incompatible representation of information syntax and semantics. Many solutions have been proposed to improve interoperability using both unifying and federating approaches (Petrie, 1992). However, ICT unification is again not to be understood as a semantically unified universe. Only the universe of discourse of the interchange between participating enterprise models and their corresponding processes has to be founded on a common base. A common semantic modelling base will be sufficient to reduce the needs for unification for only those items or objects which have to be exchanged – the inputs and outputs of co-operating models and business processes, respectively.

Standards-based business process modelling will play an important role in creating this needed ease of use and common acceptance of the technology (Clement 1997, Kosanke, 1997). Only with a common representation of process models and its global industry acceptance will the exchange of models and their interoperability become common practice, and only then will decision support for creation, operation and discontinuation for the new types of enterprise organisation become reality.

Focussing on semantic unification, the European ESPRIT project AMICE developed CIMOSA[1], an enterprise modelling framework including an explicit modelling language (AMICE 1993). The European standards organisation developed standards on enterprise modelling based mainly on the AMICE work (CEN/CENELEC 1991, CEN 1995).

These standards have been further developed by CEN jointly with ISO leading to revisions of the original standards (CEN/ISO 2002, 2003). The revisions have been guided by GERAM (GERAM 2000), the work of the IFAC/IFIP Task Force (Bernus, et al 1996), which in turn has been the base for the ISO standard IS 15704 on requirements for enterprise architectures (ISO 2000).

In the following the key features and the expected use of the two CEN/ISO standards supporting enterprise integration are presented. The basic principles of GERAM and ISO 15704 are introduced as well.

2.   GERAM AND ISO 15704

The IFAC/IFIP Task Force developed GERAM (Generalised Enterprise Reference Architecture and Methodologies), which is the result of the consolidation of three initiatives: CIMOSA, GRAI-GIM[2] and PERA[3]. GERAM provides a framework for

---

[1] CIMOSA = Computer Integrated Manufacturing – Open System Architecture
[2] GRAI/GIM = Graphes de Résultats et Activités Interalliés/ GRAI-IDEF0-Merise
[3] PERA = Purdue Reference Architecture

enterprise integration which identifies a set of relevant components and their relations to each other. These components group specific concepts selected from the three initiatives underlying GERAM. The most important concepts are: life cycle and life history, enterprise entities, modelling language and an integrated enterprise model with user oriented model views.

GERAM has been the base for the standard ISO 15704 Requirements for enterprise reference architectures and methodologies. This standard defines several key principles: applicability to any enterprise, need for enterprise mission/objectives definition, separation of mission fulfilment and control, focus on business processes, and modular enterprise design. Based on GERAM, the standard places the concepts used in methodologies and reference architectures such as ARIS[4], CIMOSA, GRAI/GIM, IEM[5], PERA and ENV 40003 within an encompassing conceptual framework. It states the concepts and components that shall be identified in enterprise reference architectures, which are to support both enterprise design and enterprise operation. Included are the framework for enterprise modelling and modelling languages supporting model-based methodologies a concept and a component further defined in the two standards CEN/ISO 19439 and 19440 described below.

3. CEN/ISO 19439

This standard on Enterprise Integration – Framework for Enterprise Modelling describes the modelling framework that fulfils the requirements stated in ISO 15704. The work is based on the Generalised Enterprise Reference Architecture (GERAM) proposed by the IFAC/IFIP Task Force and recognises earlier work in ENV 40003 (CEN/CENELEC 1991). The framework is described as a three dimensional structure consisting of a life cycle dimension with seven life cycle phases, a view dimension with a minimum set of four model views and a genericity dimension with three levels of genericity (Figure 1).

4. MODEL LIFE CYCLE

The phases of the life cycle dimension identify the main activities to be carried out in an enterprise modelling task, but they do not imply a particular sequence to be followed during the modelling process. Especially the life cycle phases may be applicable with different work sequences (top-down, bot-tom-up) for modelling tasks like business process re-engineering or business process improvement.

The domain identification phase allows to identify the enterprise domain to be modelled – either a part or all of an enterprise – and its relations to the environment; especially the sources and sinks of its inputs and outputs. Relevant domain concepts like domain mission, strategies, operational concepts will be defined in the following phase. Operational requirements, the succeeding system design and its implementation are subject of the next three phases. The released model will be used in the operational phase to support the operation in model-based decision processes and in model-based operation monitoring and control. Any needed end-of-life

---

[4] ARIS = ARchitecture for integrated Information Systems.
[5] IEM = Integrated Enterprise Modelling

activities like recycling of material, retraining of personnel or reselling of equipment may be described in the final life cycle phase.

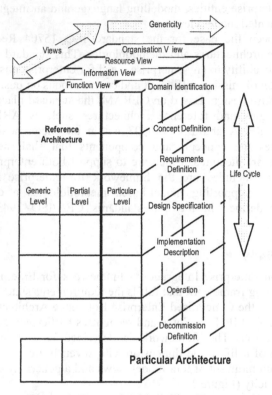

Figure 1: Enterprise Modelling Framework

5. MODEL VIEW

Enterprise models representing enterprise domains of any significance will become rather complex and not very easy to comprehend by both the modellers and the expected model users. Therefore, the framework provides the concept of model view enabling the representation of sub-models, which allow to show only those parts of the model that are relevant for the particular decision making, monitoring or control task. The model view concept is applicable throughout the life cycle.

The four model views identified in the standard are: function view, information view, resource view and organisation view. Other model views e.g. product view or economic view may be derived from the model by providing the appropriate filtering functionality in the modelling tools.

The function view allows to represent the business processes of the domain and the corresponding enterprise activities. All necessary domain, process and activity inputs and outputs as well as the process control flow (the process dynamic behaviour) will be defined in this view. Inputs and outputs for processes will be mainly material and products, whereas activity inputs and outputs define resources, constraints and related messages as well. However, the representation of the function view may be reduced to process structures (static representation or activity modelling) or activity nets (dynamic representation or behavioural modelling) both with or without selected inputs and outputs.

Enterprise objects and their relations will be represented in the information view. This information model holds all the information needed (inputs) and created (outputs) by business processes and enterprise activities during model execution. Enterprise Objects may be any object identified in the enterprise, e.g. products, raw material, resources, organization entities. To reduce complexity enterprise object of types resource and organization will be presented in their own model views. The inputs and outputs usually use only some parts of the enterprise objects. These parts are named object views.

The two subsets of enterprise objects, enterprise resources and organisational entities are represented in the resource view and in the organisation view, respectively. All enterprise resources: people, equipment and software are represented in the resource view. The organisation view shows the enterprise organisation with their objects being people with their roles, departments, divisions, etc. This allows the modeller to identify the responsibilities for all the enterprise assets presented in the three other views – processes, information and resources. Again object views identify those particular parts of the enterprise objects in the resource and organisation view, which are used for the description of the resource and organisation inputs and outputs in the particular model.

6. MODEL GENERICITY

The third dimension of the framework represents the concept of genericity identifying three levels: generic, partial and particular where less generic levels are specialisations of the more generic ones. The generic level holds the generic modelling language constructs applicable in the different modelling phases. Reference, or partial, models, which have been created using the generic modelling language(s) identified in the generic level, are contained in the partial level. Both, modelling language constructs and partial models are used to create the particular model of the enterprise under consideration.

7. CEN/ISO 19440

The standard on *Language Constructs for Enterprise Modelling* fulfils the requirements for a modelling language also stated in ISO 15704 and supports the framework for enterprise modelling described in CEN/ISO 19439 with its life cycle phases, model views and genericity levels.

The standard is based on ENV 12204 (CEN 1995) and defines a set of fourteen language constructs for enterprise modelling (see Figure 2). Models generated using constructs in accordance with the modelling framework will be computer processable and ultimately enable the model-based operation of an enterprise.

The standard contains definitions and descriptions – the latter also in the form of templates – of the set of constructs for the modelling of enterprises. Figure 2 shows nine core constructs and four additional constructs, which are specialisations of one of the core constructs (enterprise object) or even specialisations of a specialisation

(Resource – Functional Entity). Also indicated in Figure 2 are the relations to the four model views, which are supported by the particular modelling language constructs.

8. FUNCTION VIEW

The operational processes and the associated activities are represented in the function view. Four core constructs are used to model the functional aspects of the enterprise.

Starting with the domain construct, which identifies the <u>domain</u> scope, its goals, missions and objectives as well as the relations to the environment with domain inputs and outputs and their sources and sinks.

From the relations between inputs and outputs the needed transformation function – the main *business processes* of the domain – can be identified. These processes can be further decomposed into lower level processes until the desired level of granularity for the intended use of the model is reached. Process dynamics will be described by behavioural rule sets, which are defined as part of the business process construct.

The lowest level of decomposition is the level of <u>enterprise activities</u>, which usually would be rep-resented as a network of activities linked by the control flow defined using the business process behavioural rule sets. Enterprise activities transform inputs into outputs according to activity control/constrains information and employ resources to carry out the activity. All needed and produced inputs and outputs are identified as object views and are defined for each of the activities participating in the particular process.

The business process dynamics is controlled by behavioural rules and <u>events</u>. The latter are generated either by the modelled environment or by enterprise activities in the course of processing. Events start identified business processes through their identified enterprise activities and may provide additional process information as well.

9. INFORMATION VIEW

Enterprise objects and their relations are represented in the information view in the form of an information model. Two core constructs are defined for modelling of the information.

The <u>enterprise objects</u> are organised as a set of high level objects, which in general have lower level sub-objects. Different enterprise objects and sub-objects have relations to other objects in the same or other views.

A special set of sub-objects are the enterprise activity inputs and outputs, which only used in the function view. These sub-objects are selected views on particular enterprise objects and are named <u>object views</u>. Object views are of temporal nature; they only exist during the model execution time.

Three different enterprise object specialisations are defined in the standard: product, order and resource. These language constructs provide means to identify specific aspects relating to these enterprise sub-objects.

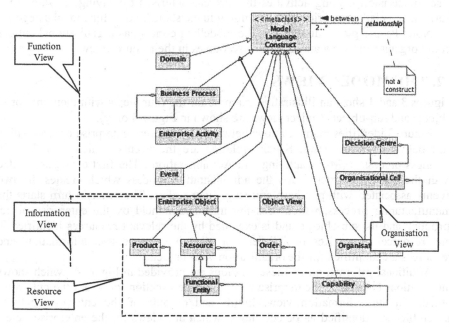

Figure 2: The set of enterprise modeling language constructs (EN/ISO 19440)

10. RESOURCE VIEW

The Resource View represents the enterprise resources, which can be organised into higher-level structures as well. These structures may represent the some of the organisational structure of the enterprise like shop floor, assembly line, etc.

In addition to the specialisations of the enterprise object – resource and functional entity – a core construct – capability – is provided as well. Whereas the resource construct (and its specialisation, functional entity) is used to describe general aspects of resources, the capability construct captures both the required (by the enterprise activity) and the provided (by the resource) capabilities. Functional entities are resources, which are capable of sending, receiving, processing and storing information.

11. ORGANISATION VIEW

The organisational entities of the enterprise and their relations are represented in the organisation view. It allows to identify authorisations and responsibilities for processes, information, resources and organisational entities.

Three different core constructs are defined in the standard: organisation unit, organisation cell and decision centre. The first two allow to model the organisation, using the organisational unit construct to describe the organisation relevant aspects of people and the organisation cell for describing organisational groupings like departments, divisions, etc.

The third construct enables the representation of the enterprise decision making structure that identifies the relations between different decision makers. It identifies a set of decision making activities that are characterised by having the same time horizon and planning period, and belonging to the same kind of functional category.

Note: People play a dual role in the modelling concept as organisational entities in the organisation view and as human resources in the resource view.

12. THE MODEL VIEWS

Figures 3 and 4 show an illustrative example for the four views with their enterprise objects and sub-objects (object views are shown in Figure 4 only).

Figure 3 identifies main relations between the different enterprise objects within and between model views. Special relation are the events, which start the two business processes (Manufacturing and Administration). The first event, associated with the customer order, starts the administrative process which creates the two events associated with purchase and shop floor orders. The latter in turn starts the manufacturing process, which uses the information held by the enterprise object 'product' and its sub-objects and is executed by the relevant resources identified in the resource view. Responsibilities for planning, processes, information and resources are identified for the organisation view objects.

Additional information on these relations are provided in Figure 4, which shows the relations between the enterprise activity in the function view and the enterprise objects in the information view. Inputs and outputs of the enterprise activity 'assemble' are identified as object views, which are views on the enterprise object 'product' and its sub-object 'part', respectively.

Similar diagrams are shown in CEN/ISO 2002 for the relations between function view and resource view and the organisation view and all other views.

13. ISSUES

Terminology is still a major problem in standardisation. A particular issue in the two standards described in this paper is the view concept. This concept is used in the sense of filtering the contents or presentation of specific aspects of the particular model by means of enterprise model views as well as presenting sets of selected attributes of enterprise objects by means of enterprise object views. This means the same concept is used in a rather similar way in its two applications in the standards. However, using the term view without its different qualifiers leads to misunderstandings. But to find a meaningful new term for either one of the two uses of the term view seems to be rather difficult. Therefore, it is essential to understand the meaning of the two qualifiers to be able to have meaningful discussions about the model content with co-workers.

Definitions
- *View:* visual aspect or appearance (Collins Dictionary 1987) perspective, aspect (WordNet 1.7.1 2001)
- *Enterprise model view:* a selective perception or representation of an enterprise model that emphasizes some particular aspect and disregards others. (ISO/CEN 19439 2002)
- *Model view:* a shortened form of, and an alternative phrasing for, 'enterprise model view' (ISO/CEN 19440 2003)

- *Enterprise object view:* <construct> a construct that represents a collection of attributes of an Enterprise Object. The collection is defined by a selection of attributes and/or a constraint on these attributes (ISO/CEN 19440 2003)
- *Object view:* <construct> a shortened form of, and the usual alternative phrasing for, 'Enterprise Object View'.(ISO/CEN 19440 2003)

These definitions imply for the enterprise model views that they will reduce the complexity of the particular model for both the modeller and the model user. However, to do this in a useful way the model views have to retain their links to the underlying complex model and thereby allow model manipulations of the model contents via modifications of the individual views.

Similarly enterprise object views will help to reduce the amount of information to be identified in the particular enterprise model. Only those attributes, which are needed in the course of model execution will be selected from the relevant enterprise objects and will be used as enterprise object views to define the business process and enterprise activity inputs and outputs.

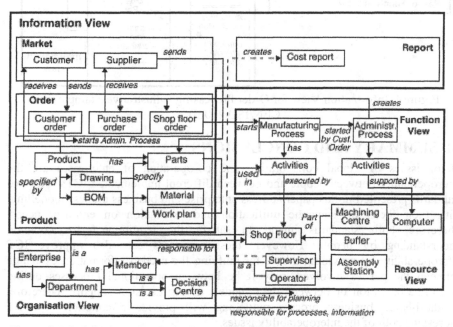

Figure 3: Model Views for Order Processing (illustrative example from CEN/ISO 2002)

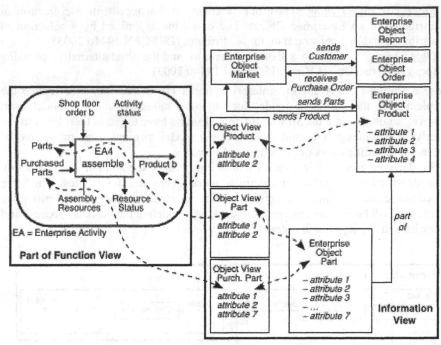

Figure 4. Information View for Order Processing ((illustrative example from
CEN/ISO 2002))

14. SUMMARY AND CONCLUSIONS

Major issues of global collaboration and co-operation of enterprises are the
interoperability between people and between different implemented ICTs. The two
standards presented in this paper address enterprise and business process modelling,
with their focus on semantic unification and orientation on end-user needs.
Therefore, both standards intend to satisfy the requirement for common
understanding by people. However, these standards can also improve ICT
interoperability by providing a base for unifying the needed information exchange
between the parties involved, may it be between operational processes during
enterprise operation or between their models during decision support. Therefore,
standard-based business process modelling will provide a common base for
addressing both of the interoperability issues.

In ISO and CEN the work is progressing in joint projects that will lead to
additional international standards for business process modelling and its application
in enterprises. More work is still required especially on the human-related aspects
like model representation to the user, representation of human roles, skills and their
organisational authorities and responsibilities. In addition standardisation is required
in the area of business co-operations as well.

Standardisation for enterprise integration is considered an important subject.
However, the current state of standardisation is not yet sufficient to allow easy
implementation at the operational level. Many of the standards are still on the

conceptional level and more detail is required to make them truly useable in the operation. Work is required in areas of languages and supporting modules, especially for the business process model creation and execution. To enable cross-organisational decision support especially the subject of 'common' semantics has to be addressed. ISO/CEN 19439 modelling constructs are defined using a meta-model and accompanying text (sufficient to define an intuitive semantics as well as to define a model repository database). However, the capture of finer points of these meanings requires even more detailed formal semantics. Ontologies will play an important role in this area as well as in the area of semantic unification (Akkermans 2003).

REFERENCES

Akkermans, H. Ontologies and their Role in Knowledge Management and EBusiness Modelling, in K. Kosanke *et al* (Eds.) Enterprise Engineering and Integration: Building International Consensus; Proc. ICEIMT'02 (Intern. Conference on Enterprise Integration and Modelling Technology), Kluwer Academic Publishers, pp 71-82.

AMICE (1993), ESPRIT Consortium (Eds.), CIMOSA: Open Systems Architecture, 2nd revised and extended edition, Springer-Verlag.

Bernus, P. Nemes, L. Williams, T.J. (1996), Architectures for Enterprise Integration, Chapman & Hall.

CEN/CENELEC (1991), ENV 40003 Computer Integrated Manufacturing - Systems Architecture - Framework for Enterprise Modelling.

CEN, (1995), ENV 12204 Advanced Manufacturing Technology - Systems Architecture - Constructs for Enterprise Modelling, CEN TC 310/WG1.

CEN-ISO (2002), DIS 19439 Framework for Enterprise Modelling, CEN TC 310/WG1 and ISO TC 184/SC5/WG1.

CEN-ISO (2003), CD 19440 Constructs for Enterprise Modelling, CEN TC 310/WG1 and ISO TC 184/SC5/WG1.

Clement, P. (1997), A Framework for Standards which support the Virtual Enterprise, in Kosanke, K. & Nell, J.G. (Eds.) Enterprise Engineering and Integration: Building International Consensus Pro-ceedings of ICEIMT'97 (Intern. Conference on Enterprise Integration and Modelling Technology), Springer-Verlag, pp 603-612.

Collins Dictionary (1987), The Collins Dictionary and Thesaurus, William Collins and Sons & Co Ltd. Glasgow and London.

GERAM (2000), GERAM 1.6.3: Generalised Enterprise Reference Architecture and Methodology, in ISO 15704

ISO (2000), IS 15704 Requirements for Enterprise Reference Architecture and Methodologies, TC 184/SC5/WG1.

Kosanke, K. (1997), Enterprise Integration and Standardisation, in Kosanke, K. & Nell, J.G. (Eds.). Enterprise Engineering and Integration: Building International Consensus Proceedings of ICEIMT'97 (Intern. Conference on Enterprise Integration and Modelling Technology), Springer-Verlag, pp 613-623.

Kosanke, K. Jochem, R. Nell, J.G. & Ortiz Bas, A. (Eds.) (2002). Enterprise Engineering and Integra-tion: Building International Consensus; Proceedings of

ICEIMT'02 (Intern. Conference on Enter-prise Integration and Modelling Technology), Kluwer Academic Publisher.

Petrie, C.J. (Ed.) (1992). Enterprise Integration Modelling Proceedings of the First International Con-ference, The MIT Press.

Vernadat, F.B. (1996), Enterprise Modelling and Integration, Principles and Applications; Chapman and Hall.

WordNet 1.7.1 (2001) WordNet Browser, Princeton University Cognitive Science Lab

3. Integrating Enterprise Model Views through Alignment of Metamodels

David Shorter, IT Focus
Convenor, CEN TC310 WG1[6]
david.itfocus@zen.co.uk

Standards have been developed and are still developing for enterprise modelling frameworks and modelling constructs. Recently they have made increasing although informal use of UML to metamodel and hence to clarify the concepts involved. However, the relationships between those concepts (for example, positioning the modelling constructs within a framework) have not been defined to the degree of precision required. This paper describes an approach as a work-in-progress, which proposes to use metamodelling to ground the concepts within the framework, and so to resolve difficult issues such as federated views on an enterprise model. In the longer term it should also provide benefits through alignment with Object Management Group (OMG) developments, especially the Model Driven Architecture, MDA™.

1. INTRODUCTION

Two major strands can be identified in standards-related work on enterprise integration. They are:
- architectures and conceptual frameworks for the design of an enterprise, and
- concepts from which to develop enterprise models as the basis of enterprise integration.

In the manufacturing domain (although also applicable more widely), the following standards are currently available or due in the near future:
- a general framework (IS 15704)[7] for the assessment of methodologies and reference architectures,
- a more specific conceptual framework for enterprise integration (ENV 40003), which focuses on model-based integration – this standard is being superseded by (prEN/FDIS 19439), which has been aligned with the (IS 15704) framework, and
- constructs for enterprise modelling (ENV 12204), which is being superseded by (prEN/DIS 19440).

(prEN/DIS 19440) defines a modelling language *construct* as an "element of a modelling language designed to represent in a structured way the various information about common properties of a collection of enterprise entities". This definition specialises but aligns with that of (ISO 10303) which defines a construct as "a generic object class or template structure, which models a basic concept independently of its use". Given this 'object-orientation', the drafting process for

---

[6] Following a meeting of ISO TC184 SC5 WG1 (12-14/10/2004), this paper incorporates revisions to clarify the structure and purpose of the metamodel for [prEN/DIS 19440].
[7] References are listed in Section 8.

(prEN/DIS 19440) naturally made extensive use of the class notation of the Unified Modelling Language™, UML™ [8], to represent constructs and the relationships between them within an integrated metamodel.

The reasons for a model-based approach to specifying the constructs are well-argued in (Flater 2002): "But as the emphasis has shifted increasingly towards conceptual design, there has been an increasing role for specifications that deal with 'meta-level' concepts, or that relate to the mapping from these concepts down to the implementation level. These specifications do not fit naturally within the requirements of standards for software component interfaces ... yet the need for them is undisputed." Additionally "UML models capture more information at the concept level and more of the intent of the designer. This makes them more effective tools of communication to assist (standards developers[9]) in understanding each other and to assist users of standards in understanding the intent".

UML has achieved wide industry acceptance but is still evolving. The version used for the (prEN/DIS 19440) work was UML 1.4. The current specification is UML 1.5, which has added action semantics – not relevant for class modelling. However, UML is evolving to UML v2.0, which will have a more rigorous specification (achieved by metamodelling UML itself, see below) and which will integrate with the Model Driven Architecture, MDA™ 10 to provide platform independence and to enable composable models.

The (MOF 2002) specification explains that "the UML and MOF are based on a conceptual layered metamodel architecture, where elements in a given conceptual layer describe elements in the next layer down. For example,

- the MOF metametamodel is the language used to define the UML metamodel,
- the UML metamodel is the language used to define UML models, and
- a UML model is a language that defines aspects of a ... system[11].

"Thus, the UML metamodel can be described an 'instance-of' the MOF metametamodel, and a UML model can be described as an 'instance-of' the UML metamodel. However, these entities need not necessarily exist in the same domain of representation types and values. This approach is sometimes referred to as *loose metamodelling*."

Using UML in developing (prEN/DIS 19440) has been helpful – however the integrated metamodel of the constructs is not a normative part of the standard, and there is no publicly available version[12]. Also the way in which the Constructs[13] are to be deployed within the Framework is described only in illustrative terms. As

---

[8] http://www.uml.org/
[9] Author's insertion
[10] http://www.omg.org/mda/
[11] The [MOF 2002] specification says "of a computer system" but that is seen as over-restrictive.
[12] This metamodel uses UML notation but – for reasons of consistency checking with construct templates – contains several relational attributes which would normally be regarded as redundant since the relationships between constructs are shown explicitly.
[13] Hereafter 'Constructs' will be used as an abbreviation for "constructs as defined in [prEN/DIS 19440]", and 'Framework' for "[prEN/FDIS 19439] Framework"

explained in the next section, this is a particular problem for *views* on a model composed of *Constructs*.

This paper will argue that it is possible and desirable to:

- use metamodelling as a basis for reasoning about problematic issues such as views on an enterprise model,
- represent the concepts of the (prEN/FDIS 19439) framework in a metamodel,
- provide a 'mapping' mechanism, called Framework-Construct-Mapping or **FCM**, which will allow standardizers and modellers to refine relationships between the Framework and the Constructs, and
- base both the framework metamodel and the FCM on the UML meta-language (as is already largely the case for (prEN/DIS 19440)), and so enable integration within the MDA.

2. THE ISSUE OF VIEWS

A common theme for the model-related content of the Framework and Architecture standards (ENV 40003), (IS 15704), (prEN/FDIS 19439), and other frameworks such as the Zachman Framework for Enterprise Architecture[14] is the notion of *views* on a model. In (prEN/FDIS 19439), Enterprise Model Views are defined as "a selective perception or representation of an enterprise model that emphasizes some particular aspect and disregards others". This definition is well-aligned with the (OMG MDA Guide 2003) definition of *view* as "a viewpoint model or view of a system (that) is a representation of that system from the perspective of a chosen viewpoint", where *viewpoint* is defined as "a technique for abstraction using a selected set of architectural concepts and structuring rules, in order to focus on particular concerns within that system".

The motivation for such views is that in any real-world complex enterprise, there are multiple concerns to be modelled (e.g. functional, informational, economic, decisional) involving a combination of concern-specific and more general concepts, and that views allow the modeller and model-user to concentrate on only those details that are relevant for the purpose. However, the reconciliation of independently developed concern-specific models would be a problematic, if not impossible, exercise. The approach adopted in (prEN/FDIS 19439) is therefore to postulate an 'integrated enterprise model'[15], and to regard these concern-specific models (Enterprise Model Views) as representing major aspects of that integrated model and corresponding to a view on the enterprise itself. This corresponds to the 'federated development' approach (Whitman *et al* 1998) as described later and is also analogous to the 'Single Model Principle' (Paige and Ostroff 2002) which has been proposed to address consistency of UML modelling and the UML itself.

Just what the minimum set of views is to be, or whether there should be a minimum set, has been a matter of some debate. In an early standard (ISO 14258) two such views were defined – Function and Information. (ENV 40003 and IS 15704) later extended these to Function, Information, Resource and Organisation, but the

---

[14] http://www.zifa.com/

[15] [IS 15704] distinguishes between enterprise-reference architectures and methodologies that are model-based, and those that are not.

possibility of other views being required was accepted. More recently (prEN/DIS 19440) has been drafted to include a construct (Decision Centre) to support a Decision View, and it is anticipated that a future revision of (ISO 14258) will also include an Economic View.

Model consistency requires that changes made in any one view be propagated to the integrated model, and to any other view that is affected by that change. The issue is discussed in (Whitman *et al* 1998) which distinguishes between:

- a master view, requiring all relevant information to be entered in a single view ("found by several to be difficult if not impossible");
- the driving approach, where a view with the largest content is populated first (e.g. the Function View for CIMOSA) and then other views are populated from that information; and
- the federated approach "which allows the user to populate each view as information becomes available in that view (with the advantage of allowing) the addition of knowledge in the view most conducive to that form of knowledge".

However, this requires that whichever "approach (is used, it needs to) ensure the consistency between views" (*ibid*). The authors also point out that "this method is highly tool-dependent' and "the rigor of the tool capability in ensuring the proper mapping between views is critical to the success of this method..."

A central issue with the *views on an integrated model* approach is therefore how to guarantee that views (and changes made to the content of a view) are consistent. (prEN/DIS 19440) partly addresses this by the principle of representing the integrated model for any specific enterprise in terms of constructs which are themselves described in an integrated metamodel. However, other mechanisms are also needed to manage changes in the model content.

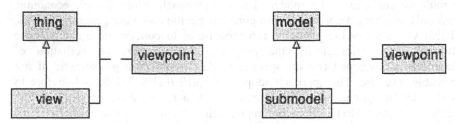

Figure 1 – Views and viewpoints

At its most general, a view is a selective perception from a particular viewpoint, so it is a selector of something – a mapping from a thing to a view on that thing, e.g. from a model to a submodel. (See Figure 1.)

A view is also a selective encapsulation of a modelled object's attributes or state (or those of a collection of modelled objects). It involves selection corresponding to the viewpoint, and encapsulation because it identifies and contains the information (state) relevant to that selected view.

This is not the same as the (Gamma *et al* 1997) *State* pattern because that externalises the state as an abstract class, which has alternative subclass implementations. Views are not alternatives, because several views can exist simultaneously of the same modelled object(s) and some object attributes may be present in different views.

However a view is valid only if the content of the view is not changed independently of the modelled object(s) visible in the view. Views are therefore more akin to an *Observer* pattern (or *Model-View-Controller*), because any change to the content of a view has to cause a corresponding change in the state of the enterprise object or enterprise objects being viewed.

An implementation model[16], adapting the (Gamma *et al* 1997) Observer pattern, might appear as Figure 2[17].

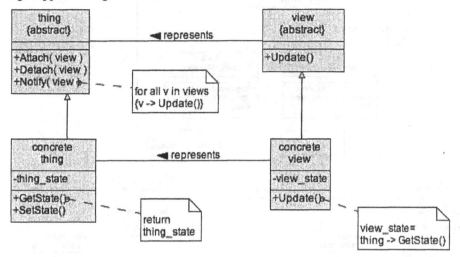

Figure 2 – Implementation model for views (*observer* pattern)

Something additional is needed to record each view's subset of the modelled object(s) state, for example as in Figure 3, modifying the approach to include aspects of interest as suggested by (Gamma *et al* 1997) in "specifying modifications to interested parties explicitly".

One issue is what is to be notified from the changed object to the interested party, because this will limit the extent of unification of the underlying model. For example, propagating attribute changes seems straightforward, but what of propagating, say, constraints on a modelled object? Or specialisations thereof?

3. LIMITATIONS OF ENTERPRISE OBJECTS AND OBJECT VIEWS IN (PREN/DIS 19440)

In (prEN/DIS 19440), an Enterprise Object (EO) is defined as "a construct that represents a piece of information in the domain of the enterprise (and) that describes

---

[16] Which might support for example a software tool to present view on models generated from prEN/DIS 19440 constructs

[17] The terms *thing* and *view* have been chosen to avoid confusion with (Enterprise) Object and Object View – the latter having a distinct and restricted meaning as described in the following section.

a generalized or a real or an abstract entity, which can be conceptualized as being a whole", and an Object View (OV) as "a construct that represents a collection of attributes of an Enterprise Object".

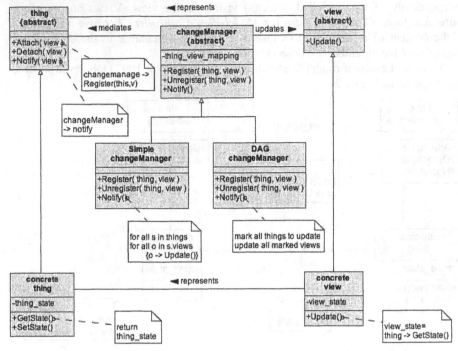

Figure 3 – Implementation mode for views (*interested parties* pattern)

However, while an EO can in principle describe any physical or informational entity, in practice (prEN/DIS 19440) restricts its usage to those entities that are inputs or outputs of transforming processes such as Enterprise Activities. The current template for an EO does not allow the EO to represent some other construct – for example, to be a description of a Business Process. And OVs are essentially low-level mechanisms providing a transient subset of information for selected EOs – they are not intended for and do not provide a general mechanism for obtaining a view on an integrated model[18]. OVs are geared to the limited objective of linking Resources to Enterprise Activities, as reflected in the Figure 4, which is generated from the integrated metamodel that underpins (prEN/DIS 19440), and showing all OV-related relationships:

Consider the situation where it is desired to analyse 'value added' in a chain of processes. This would mean extracting financial information from the processes involved – one could aggregate the costs of the involved resources, but that is only part of the story because it does not include the added value that only the process 'knows about'. Since it is not possible in (prEN/DIS 19440) to obtain an Object

---

[18] The terminology for Enterprise Object and Object View is unfortunate – CEN TC310 WG1 tried to find terms which did not imply a more general usage but was unsuccessful.

View of a Business Process or Enterprise Activity, it is not possible to build such a 'value-added' view.

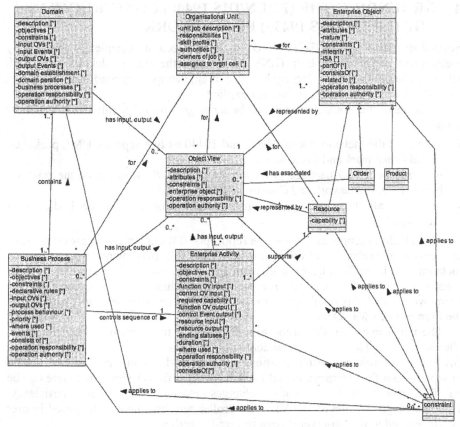

Figure 4 – Object Views and related constructs

Consider also the case where it is required to develop a performance model of processes (this example is suggested by (Salvato 2002)). One possibility is to introduce additional classes for performance and performance indicators, both of which might be associated with Business Process – but it is not clear whether these need to be defined as specialisations of some existing construct, or whether they can be simply attributes of business process. In the former case, they could be treated independently of the process itself, for example, as performance measures that can be associated with several different, but similar, types of business processes.

There is, therefore, a need for a more general 'modelled object' view mechanism, to allow selected attributes of any modelled object or collection of objects to be represented as a named collection. One way of handling this[19] in any revision of (prEN/DIS 19440) would be to define a 'Construct View' as a named

---

[19] Not considered further in this paper.

collection of selected attributes from a collection of constructs. Object View would then be a specialisation of Construct View.

4. GROUNDING THE (PREN/DIS 19440) CONSTRUCTS IN THE (PREN/FDIS 19439) FRAMEWORK

Several members of CEN TC310 WG1 have criticized the complexity of the early metamodels of the concepts in (ENV 12204) and the later models of (prEN/DIS 19440). The problem arose largely because of an attempt to ground these models in (prEN/DIS 19439) without making that explicit.

To manage this complexity, and to better align with other initiatives, it is now proposed:

- to present the metamodels of 19439 and 19440 as two separate UML packages, called *Framework* and *Constructs*,
- to introduce a new package provisionally entitled *FCM* to define the relations between the Framework and Constructs packages, and
- to define all three packages as metamodels expressed in the UML meta-language[20].

The term FCM is proposed because the (concepts of the) Framework provides meta-concepts for the (concepts of the) Constructs; however, there are no direct entity <–> meta-entity relationships between the concepts of the two models. For example, the construct Resource is not an instance or specialisation of a concept in the Framework – and in particular, an Object View is not an instance or specialisation of the Framework's Model View.[21]

The initial impact of this proposal will be on the treatment of Enterprise Model Views.   The   argument   runs   as   follows: Given that different modelling situations will require different views, there is a need to separate out the enterprise model itself from the different views that there can be on that model, and to control changes to both to ensure consistency. This is analogous to the Model-View-Controller paradigm now widely used in user interfaces, and to the Observer pattern referred to earlier.

An early MDA working paper (MDA 2001) provides some guidance on how this might be done. "UML provides an important modelling element that is relevant for separating viewpoints, levels of abstraction, and refinements in the context of MDA – the UML package, a UML construct for grouping model elements. A package can import other packages, making elements from the imported package available for its use. Packages help understand the MDA since the models being interrelated for integration are typically in different packages. Models of a system from two different viewpoints unrelated by refinement (or models of two distinct interfaces of a component, or models of two different systems to be integrated) would be defined in two separate packages.

---

[20] Current thinking is that if the packages are legal UML v1.X or later v2.0, then it is not necessary (and would provide little benefit) to define these packages as stereotyped extensions of MOF base classes.

[21] This clarification is derived from the author's interpretation of [Robertson 2004].

"The interrelationships between the two would be defined in a model correspondence ... defining the integration of those two viewpoints or systems at the current level of abstraction." The latest MDA proposals have replaced *model correspondence* by concepts such as mappings, markings and transformations, but the principle remains the same.

This suggests, therefore, defining (within the FCM) a structure for views that holds visibility attributes for each construct in the different views[22] – this is a construct-visibility map, and corresponds to the mappings used in MDA. There is an issue here about non-construct concepts, for example behavioural rules and associations. How is the visibility for these to be handled? Perhaps they should be Visually Representable (VR) *if and only if* any referenced construct is visible? And an association (or constraint?) is VR *if and only if* all its end-points are VR?

Extending metamodelling to (prEN/FDIS 19439) and introducing a new FCM package is seen as the easiest way to formally resolve the meaning of Framework Views on a Construct-derived model23. The alternative would be to explain this in a separate Annex as in (prEN/DIS 19440) or in a separate document as in (Zelm 2001). However, this would not provide an accessible, extensible and maintainable structure for the description of additional views, including those required by a modeller for particular and possible transient purposes. Note that a modeller view may be concerned with something yet to be designed, for which no appropriate view exists beforehand.

The FCM package might also be extended to the (prEN/FDIS 19439) dimension of Life Cycle Phase, so defining and constraining what constructs are visually representable in each phase and what attributes and constraints are visible for modification in each phase. This possibility is not further addressed in this paper, but (prEN/DIS 19440) provides some guidance for this in the A2.x, B2.x sections of the templates for each construct.

5. THE BENEFITS OF FORMALISING STANDARDS AND STANDARDS PROPOSALS THROUGH METAMODELLING

Some informal class models were constructed for (ENV 40003), but solely for developing understanding of concepts during the drafting process. A subsequent analysis by (Petit *et al* 2000) as a contribution to work on drafting prEN/DIS 19440

---

[22] This would need to accommodate both pre-defined and user-defined views.
[23] Extending the existing [prEN/DIS 19440] UML model by introducing new classes derived from the MOF base classes has been proposed by [Salvato 2002]. The specific proposal was to specialise *modelElement* into new abstract classes of *functionElement*, *informationElement* etc, and then to define construct classes as specialisations of those. However that would introduce additional complexity because of the need for multiple inheritance to handle Enterprise Activity and Enterprise Object (both of which are used in more than one View), and it is not clear that it would be sufficient or manageable in supporting extension to new Views (modeller- or even user-defined).

identified several problems and ambiguities in the intended grounding Framework of (ENV 40003) by using the Telos Knowledge Representation Language[24] .

The drafting of the (prEN/DIS 19440) made extensive use of UML modelling techniques, in particular by developing a metamodel of all the constructs in one consolidated metamodel using a UML modelling tool, MagicDraw,™ 25 from No Magic, Inc. One facility provided by the tool (also available in similar tools) was that of defining a new diagram containing a single modelled construct and then adding automatically only those constructs and other modelled elements that were directly associated (through inheritance or association26) with that selected construct. This was a major benefit in checking the consistency of the normative text. Additionally, a concordance program was used to check consistent usage of terms.

While these procedures greatly increased confidence in the rigour of the draft, they are not to be seen as any form of guarantee that the latest draft of (prEN/DIS 19440) is completely error free[27]. While a metamodel can ensure consistency of the things expressed in that metamodel, it does not solve all problems[28] – and manual checking is still necessary against the normative text. Where feasible, metamodelling should be complemented by formal analysis as in the previous Telos analysis of (ENV 40003).

(Holz 2003) argues that basing the formal description of the constructs on the MOF metamodel may (to be proven) assist with the transformation of models based on metamodel formalisms, which are in turn derived from a common metamodel. Previously (Pannetto *et al* 2000) had proposed an approach to generalising two UML fragments, justified by similar semantic description (and presumably a similar usage or context). The IST UEML project found that common concepts can be defined by demonstrable equivalence in a (number of interesting) well-defined context(s) or under some explicit hypotheses (UEML 2003). However, interoperability with other enterprise modelling languages may well require new abstractions to be introduced – for example (UEML 2003) also found that a new class of ExchangeObject was found necessary to represent a common abstraction for process input/outputs in three different modelling languages. An extension mechanism is therefore needed to allow modellers to introduce new generalizations and specialisations, both for classes and (probably) for associations (the latter is a facility to be provided in UML2.0).

---

[24] See http://www.cs.toronto.edu/~jm/2507S/Notes04/Telos.pdf for overview and further references

[25] http://www.magicdraw.com/

[26] This is the reason for representing associations explicitly in additional to the relational attributes captured in the templates.

[27] It is known that in the informative Annex, an association of Event with Enterprise Activity (EA) should show both EA generates Event, and Event initiates EA – however the normative text is correct.

[28] For example, a detailed analysis of the metamodel for UML1.x [Fuentes 2003] identified 450 errors in terms of non-accessibility of elements, empty/duplicate names and derived associations.

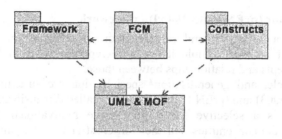

Figure 5 – Packaging the standards

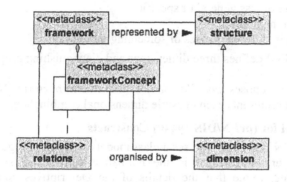

Figure 6 – Organising a framework

6. REVISED CLASS MODELS

6.1 General approach

Figure 5 shows the metamodelling-based approach that is proposed for the alignment of the (prEN/FDIS 19439) Framework and (prEN/DIS 19440) Constructs.

- UML & MOF are essentially givens (meaning that they already exists and will be maintained by the OMG) – however, some care will be needed to reflect any changes made by the OMG as UML 2.0 develops. The other three packages are at very different stages of development:
- The integrated metamodel for (prEN/DIS 19440) Constructs is well developed as an internal document but will need maintenance to accommodate any changes made to (prEN/DIS 19440) after the on-going enquiry and to delete redundant relationships or attributes that were introduced in order to check template consistency. Stereotyping the constructs, associations etc. in terms of MOF base classes would also need additional work if that were seen as necessary.
- Some internal class models were constructed during the drafting of (prEN/FDIS 19439) but these need consolidation and further work.
- The class diagram for the FCM package is still work to be done.

6.2 Class model for (prEN/FDIS 19439) Framework

A framework is a collection and representation of concepts and the relations between them. At its most simple it could be represented as in Figure 6 as a collection of concepts and relationships between those.

View, life cycle, and generalisation/ specialisation are structuring principles which in (ENV 40003) and (prEN/FDIS 19439) are called dimensions.

A dimension is a selective focus from some motivational (stake holder) viewpoint on the domain entities and their different states. It is also a classifier, having some property that enables classifications to be distinguished, for example:

- by being more or less abstract or concrete
- by being more or less general or specific
- by simple enumeration
- by occurring in different phases of something's life cycle.

(prEN/FDIS 19439) defines three dimensions and distinguishing properties for each, as in Figure 7[29].

Figure 7 also proposes *modellingStance* (not a concept in (prEN/FDIS 19439)) for the tuple that represents each possible dimensional combination.

6.3 Class model for (prEN/DIS 19440) Constructs

Annex C of (prEN/FDIS 19440) uses a class model as shown in Figure 8 to describe how constructs are represented in the normative part of the standard using a common template. (Note that the details of the Descriptives and Relationships containers are in general different for each construct.)

A high-level representation of the thirteen constructs is shown in Figure 9 (suppressing attributes and relationships between them).

The Annex then models the constructs in more detail from the viewpoint of each of the model views, each derived from a single integrated metamodel. Figure 10 shows the Function View of the constructs

6.4 A provisional model for mapping constructs into the framework

The approach proposed and illustrated in Figure 11 is to develop an FCM package that conceptually would act as a ViewManager, and as a container for defining what constructs would be used in the existing and new Model Views, further qualified by Enterprise Model Phase and Genericity.

Note that this is not intended as a basis for an implementation (although any such implementation could make use of the *Observer* and *Interested Party* patterns described earlier). Further work is needed in CEN TC310 WG1 and elsewhere to confirm the validity or otherwise of this approach, and to design the details of the FCM package. In particular, the issues of how to handle additional and user-defined modelViews, and detailed constraints and visibility rules for specific attributes are yet to be considered.

---

[29] The issue of how to handle additional viewNames, e.g. decision, economic, user-defined is not addressed in this Figure.

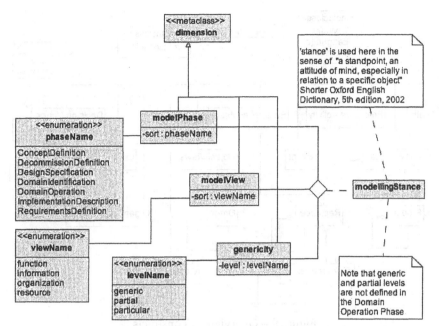

Figure 7 – The dimensions of the (prEN/FDIS 19439) framework

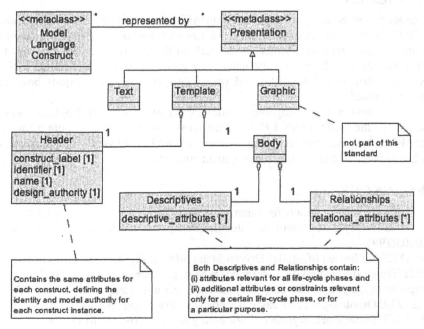

Figure 8 – Template for constructs

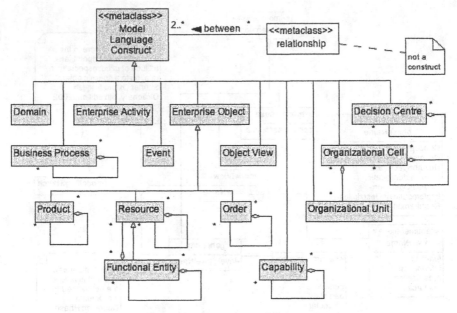

Figure 9 – Overview of constructs

7. CONCLUSIONS

This paper proposes an approach for maintaining the (separately developed and/or modified) *views on an integrated enterprise model* through a change propagation mechanism, based on more formal metamodels of the constructs and framework for enterprise modelling. It sets out the need for a generalization of the current Object View mechanism in (ENV12204) and (prEN/DIS 19440), and suggests how this might be achieved.

Lastly, it argues that using metamodelling to ground the (prEN/DIS 19440) Constructs in the (prEN/FDIS 19439) Framework would provide a route by which standards can be developed to manage current and additional views on an enterprise model, within a federated model development context.

REFERENCES

ENV 12204 (1996) Constructs for Enterprise Modelling, CEN ENV 12204:1996

ENV 40003 (1991) CIM systems architecture framework for modelling, CEN ENV 40003:1991

Flater, D (2002) Impact of Model-Driven Standards. NIST, in Proceedings of the 35th Hawaii International Conference on System Sciences, January 2002, or http://www.omg.org/mda/mda_files/mda_1.7_cleanformat.pdf

Fuentes,J.M, Quintana, V., Llorens,J., Génova,G., Prieto-Díaz,R. (2003) Errors in the UML Metamodel. *Software Engineering Notes*, 28(6). ACM Digital Library 1-13

E Gamma, R Helm, R Johnson, J Vlissides (1997) Design Patterns CD, Elements of reusable Object Oriented Software. Addison Wesley, 1997, ISBN 0-201-63498-8

Holz (2003) E Holz, a metamodel approach for the combination of Models in multiple languages, Modelling, Simulation, and Optimization. 308-313

IS 15704 (2000) Requirements for enterprise-reference architectures and methodologies, ISO IS 15704:2000

ISO 10303 (1994) STEP, Standards for the exchange of Product Model Data, 1994

ISO 14258 (1998) Industrial Automation Systems – Concepts and rules for enterprise models, ISO 14258:1998

MDA (2001) Model Driven Architecture (MDA) Document number ormsc/2001-07-01, Architecture Board ORMSC1 July 9, 2001

MOF (2002) OMG-Meta Object Facility, v1.4, April 2002, © OMG, available via http://www.omg.org/technology/documents/formal/mof.htm

OMG MDA Guide (2003) MDA Guide Version 1.0.1, © 2003 OMG, Document Number: omg/2003-06-01, 12th June 2003

Paige R., Ostroff J (2002) The Single Model Principle, R Paige and J Ostroff, *Journal of Object Technology*, 1(5) 63-81

H Pannetto, F Mayer, P Lhoste (2000) Unified Modelling Language for metamodelling: towards constructs definitions, Proc AIS ,2000 conference, Bordeaux, France, ISBN 960-530-050-8

Petit,M., Ferier,L., Heymans,P. (2000), Some hints for clarification of CEN ENV 12204, Contribution to CEN workshop on Evolution in Enterprise Engineering and Integration, 2000

prEN/FDIS 19439 (2004) Enterprise Integration – Framework for Enterprise Modelling, issued for parallel EN/IS ballot on 2004-09-16

prEN/DIS 19440 (2004) Enterprise Integration – Constructs for Enterprise Modelling, issued for parallel ENQ/DIS ballot on 2004-06-10

Robertson, E (2004) E Robertson, Explicitly Modelling Metadata, private communication

Salvato, G (2002) ENV 12204 metamodel, Contribution to CEN TC310 WG1, (ENV-UML Paper vs2.doc) approx. February 2002

UEML (2003) Common and Non Common concepts referring to the Scenario and GRAI, IEM, EEML metamodels, UEML working paper, IST-2001034229, February 2003

Whitman,L., Huff,B., Presley,A. (1998) Issues encountered between model Views. Flexible Manufacturing and Integrated Manufacturing Conf. Proc. 1998, Begell House, Inc.

Zelm,M. (2001) Representation of Modelling Constructs. Contribution to CEN TC310 WG1, 12/5/01

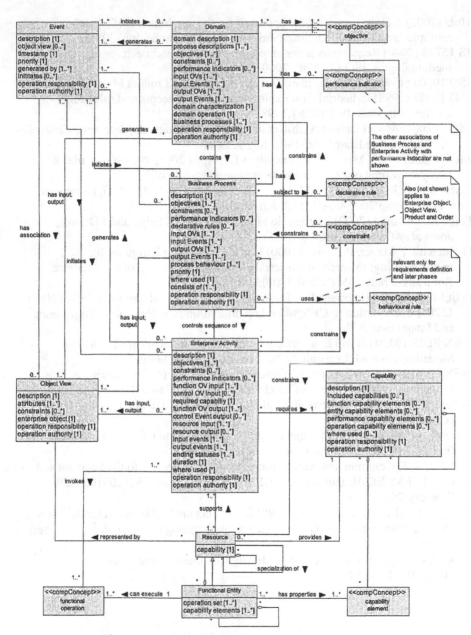

Figure 10 – Use of Constructs in the Function View

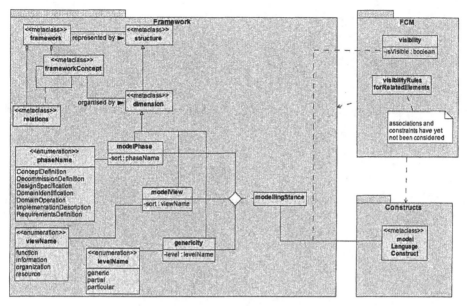

Figure 11 – Role of FCM package

4.   Semantic Distance and Enterprise Integration

H T Goranson
Sirius-Beta and Old Dominion University

1. INTRODUCTION

Work toward enterprise integration is easily justified as the core science of the engineering discipline that drives world economies by empowering infrastructure. Basics of collaboration and the resulting work in industry depend on the ability to convey meaning in a trustworthy manner. In 1990, the major research sponsors in the U. S. and European Union formed a partnership to define a research agenda for the underlying sciences of enterprise integration. That collaborative exercise was repeated twice at five years apart since as the International Conference on Enterprise Integration Technology (ICEIMT). ICEIMT has recently transitioned into the hands of the community.

In 1992, the international workshops and associated book codified the discipline of enterprise integration and directly contributed to unified approaches such as enterprise resource planning (Petrie, 1992). The 1997 exercise was a landmark in recognizing the economic advantages of opportunistic integration in the form of virtual enterprises. A conclusion was that prior integration strategies based on centralization and homogeneity were hampering business flexibility. The science behind enterprise integration shifted from standard interfaces to ontologies (Kosanke, Nell, 1997).

The 2002 activity noted the reality of many competing ontologies with the costs and difficulties of harmonizing them (Kosanke, Nell, Jochem, Ortega Bas, 2003). A concern emerged to consider context. Often integration is measured as a matter of exhaustive possibility: two diverse methods or representations are said to be integratable if every possible condition and context permits complete semantic conveyance. But the real virtual enterprise situation is that partners need to integrate in a specific context consisting of processes that will present only a few of all the possible conditions.

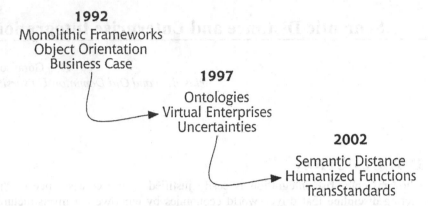

1992
Monolithic Frameworks
Object Orientation
Business Case

1997
Ontologies
Virtual Enterprises
Uncertainties

2002
Semantic Distance
Humanized Functions
TransStandards

Figure 1: ICEIMT Results

In such cases, it may be possible that the integration as a whole is imperfect, but is "close enough;" either it is perfect in a limited context, or it is imperfect but a single message easily repairable, or it is imperfect but the consequences are tolerable. The notion of "semantic distance" was developed to cover the notion of "how close is close enough."

The U. S. National Institute of Standards and Technology (NIST) had independently identified this need in the course of developing support of ontology standards. In November of 2003, they – with the aid of several European projects — hosted a several day international workshop on the topic to determine best approaches. A variety of disciplines and viewpoints were represented, with the workshop identifying a number of challenges. The concept of semantic distance is likely to play a major role in some way in the future of virtual enterprise integration and incidentally the semantic web (and other applications). But it is too early to guess in exactly what form, as there are all sorts of market, other economic and political forces at work.

This paper represents one proposal for addressing the need for a measure of semantic distance. As it happens, the term "virtual enterprise" has been significantly watered down by many from its original intent. Today, people use it for uninteresting cases: distributed but stable aggregations of firms (even supply chains!), or firms that band together for coordinated marketing of their ordinary services. In this paper, we use the original intent: opportunistic, often temporary aggregations of mostly small and medium-sized firms who come together to address or create an opportunity. A key part of the notion is that the integration is sufficiently tight that partners may radically adapt their processes as a result of requirements of the system. They may even have been identified as partners because they are judged to be capable of doing something that they currently do not, and may never have thought of. The virtual enterprise is dynamic in the sense that it forms and dissolves but also in the more interesting behavior that it evolves when operating.

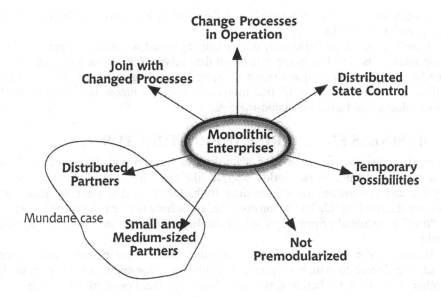

Figure 2: Features of Advanced Virtual Enterprises

2. HIGH LEVEL ONTOLOGICAL DISTANCE

The science behind enterprise management suffers from a wide variety of theories and philosophies variously applied to the purpose of design and management. This fact affects both the root problem (we have fundamental ontological mismatches within the enterprise), and it also complicates the problem of shaping a solution (we have many differing theories of just what constitutes and operates an enterprise and particularly a virtual enterprise). Under Advanced Research Project Agency (ARPA) tasking and the guidance of the Defense Manufacturing Board, we (Goranson, 1999) developed a parsing of the enterprise intended to:

- Identify the fundamental ontological domains (which correspond in some respect to different functions and theories within the enterprise),
- Provide for an easy mapping of tools and philosophies from similar breakdowns that played significant roles in the marketplace and academy, and
- Provide a basis for a rigorous study of metrics within and about the virtual enterprise.

That decomposition divides the problem space first into "infrastructure," then "metrics."

Infrastructure describes the "medium" in which an enterprise operates. This includes the various types of rules and constraints that apply to it as well as its kinds and sources of energy. This is all of the stuff of the environment, including the underlying laws and "physics," plus the material of which the enterprise is made. Some of the infrastructure is man-made (like telephones and some business rules) but other elements are "natural" (like the laws of physics and most psychology of group dynamics). This parsing of the environment is independent of its representations, and can be equated to differences in high level ontologies (world views) and therefore distance.

Metrics concern the basic stuff of the language used when an enterprise and its components reason and communicate about themselves. We use the term in a richer, broader sense than mere quantitative measures, intending instead to focus on the notions of "value" and "effect," that motivate activity, compose the intent of much communication and advise decision-making.

3. BUSINESS ENTERPRISE INFRASTRUCTURES

Enterprise infrastructure is divided according to fundamental differences in how their worlds operate. Some worlds operate like the "real" world and are tied to physics and the impression of absolute truth. Other worlds are man-made, for instance the legal world. There, for instance, something is true if it can be shown to be "true" by artificial principles of submissability even if it is not so in the physical world.

Because these infrastructures are something that we can perceive and reason about, the degree to which they can formally and unambiguously be defined is another discriminator. Therefore, we have three large families of infrastructures:

- those that can be explicitly described and also conform to the laws of natural physics;
- those which can be explicitly described but do not conform to natural physics; and
- those that have neither quality – that is they neither conform to physics nor can be explicitly modeled.

Each has further breakdowns of discrete ontologies as listed shortly below. The integration problem in an enterprise is of two orders: integrating across infrastructures that are in the same domain but use different terms (like the shipping departments of two companies), and between infrastructures that live in different worlds (like the goals of a legal department and the operations on a manufacturing floor).

The reason we spend so much time on these divisions is to provide an ontological framework for the distance metrics. Similar parsings have been performed for other enterprises, for example combat and terrorist enterprises.

4. PHYSICALLY-BASED AND EXPLICABLE INFRASTRUCTURES: PHYSICAL LAWS: BASIC PROPERTIES OF CONTAINMENT, GRAVITY, MOTION AND SO ON

Physical Activities: concerns the actions associated with physical operations of manufacturing, conversion of material and assembly. This is differentiated from the above by adding human intent.

Logistics: supports principles associated with presence, location and movement. This differs from the above two: it captures intent but the basic ontology is driven by the environment rather than the action.

Most process modeling (especially those associated with enterprise resource planning) only addresses the above infrastructures with some annotations from business rules.

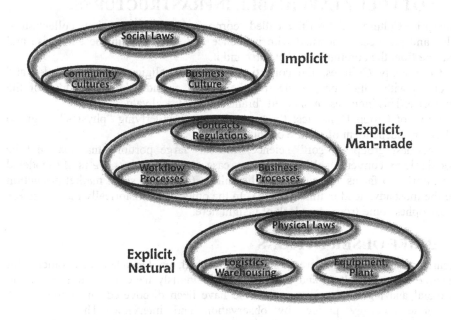

Figure 3: Key (Ontological) Enterprise Infrastructures

5. NON-PHYSICALLY BASED BUT EXPLICABLE INFRASTRUCTURES:

Business Rules: supports the actions that define and drive how the organization operates as a business. Included here are responsibility and control dependencies and most processes associated with trust.

Financial Rules: concerns the world defined by the reward structure, denominated in value metrics and associated currency. In some cases, this infrastructure splits into two siblings: the financial models associated with internal operations and the (often quite different) accounting rules associated with the reporting for the financial infrastructure that finances the enterprise from the capital ecology that surrounds it.

Legal Systems: this is the ontology concerned with contracts, liabilities and responsibilities, and societal constraints that are codified. This is the least "logical" of the three. In countries with a British colonial heritage this ontology has unusual ontological properties as a result of dynamic "case law." The rest of the civilized world uses more explicit "code" whereas some regions have individual, capricious ontologies as result of despotism.

Since the above group consists entirely of man-made "rules," one can say that every element is modeled in some way by the "maker" of the process/infrastructure. Both this and the previous group have formal standard ontology efforts underway in each of the discrete areas at various levels of maturity and formalism.

6. NOT FULLY EXPLICABLE INFRASTRUCTURES:

Enterprise Culture: what is often called "corporate culture," the unique collection of rules and practices concerned with influence and status within the enterprise and discrete from the communities that surround it.

Community Cultures: the collection of ethnic, religious and civil rules and practices with which people identify themselves as individuals "outside" of the enterprise. This includes engineered "brand" and political values.

Laws of Group Dynamics: these are the basic underlying "physics" of group behavior, independent of culture or enterprise.

This last group is "soft" science, and has large portions that may not be modelable by conventional logics. In any case, these behaviors *are* rarely modeled and poorly, so far as computable predictability. (Tools for stock market prediction are the most advanced in this domain.) On the other hand, historically most business catastrophes come from some lack of insight here.

7. BRIEF OBSERVATIONS

Clearly, some ontologies are more closely linked, or dependent than others: for instance business culture and business rules obviously have a dependency, as do financial and legal infrastructures. These have been discovered under the ARPA enterprise ontology project by observation and interview. The ontological dependencies are an essential tool in formalizing discrete ontologies that minimize problems between infrastructure and between simulations and reality.There is much to say about this ARPA effort. The original impetus was to guide ontological research to aid in metrics for integration. (The approach is outlined in the next section.) Since then, ontologies have become a focus for several large communities: as the basis of the "semantic web," as a key component in engineering intelligent agents, software engineering and simulation of complex systems. Ontologies continue to be the center of the newly revived (and huge) discipline of enterprise engineering for business enterprises and particularly advanced virtual enterprises.

One result is worth mentioning: one would guess that successful enterprises would be those that do well in all of these infrastructures and that lack of excellence in any one would drag the whole system down. Extensive case studies (Dove, 1995) have discovered the unintuitive result that this is not so. There does appear to be a threshold of incompetence in each infrastructure, but once beyond that, simple competence in most is adequate so long as one or two of the others have special strengths. For instance, if your corporate culture is particularly strong, you can bridge problems in poor management of business rules and legal issues.

We should note that this breakdown of infrastructures is for the ontology level only and is not intended to replace any paradigm used in the actual representation of models or formalisms: the ontological issues are independent of modeling paradigms such as: actors; actions; events; relations; dependencies; constraints, behaviors, interactions or what have you.

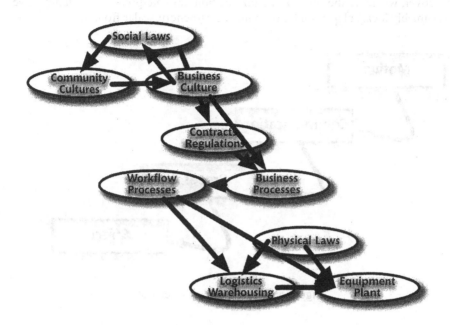

Figure 4: Infrastructure Linkages

8. METRICS AND SEMANTIC DISTANCE

Ontological foundations are an essential part of the solution to a large general class of problems, but researchers now understand that some better tools are needed concerning the semantics of the communications within and between the ontological domains we identified above as "infrastructures." Two results are notable: a focus on metrics and the previously mentioned research agenda in "semantic distance."

8.1 Metrics

All sorts of messages are conveyed within an enterprise. Fortunately, all of these are unlike communications in the open world in that they have a generally explicable purpose. Any reasonable approach to the semantics of collaboration needs to focus on the core semantics of the enterprise. For historical reasons we call that subset of the semantics the "metrics" subset, but we intend it in a larger sense than a scalar measure like dollars or quality.

The reasoning behind this is simple: we want to reason about the effectiveness of communication within a situation that includes ontological context. The semantics of effectiveness reside in those metrics. Indeed, they constitute a metasemantics of sorts, information that one can employ when evaluating information. Moreover, the metrics are often embedded in the communications themselves, or motivate them.

Instead of a number, we propose that metrics are semantic entities and that a combination of several metrics in a given context can be characterized algebraically or geometrically in some manner that conveys "fittedness" or "closeness."

Moreover, whatever the form of the information, an enterprise will certainly have a (presumably local) algorithm for deriving a cost/benefit scalar from it.

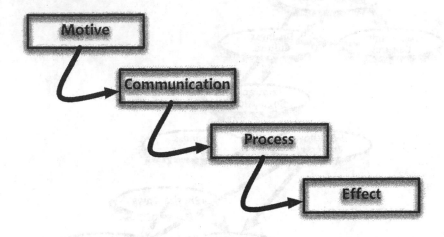

Figure 5: Four Levels of "Metrics"

8.2 Semantic Distance

The second fundamental element of the approach is brand new. In the past, we crudely assumed that the infrastructure had only two states of effectiveness: either communication was perfect or it was not. In the case where it was not, fatal problems could occur so the infrastructure was not to be trusted.

We now know otherwise. After all, in the real world communication among all the various ontological domains is seldom perfect. People negotiate to clarify meaning until it is decided that they understand well enough to do what they need to do.

We need a notion of "semantic distance" (or "fittedness"). If we were reasoning about semantics effectively, we would able to tell things like (given a communication between two different representation systems in a specific context):

(1) This is perfect (the information sent is precisely as understood), or
(2) This is not perfect, but it is good enough for the use intended, or
(3) This is not good enough, but it is "close" and worth the trouble of clarifying this one time, or
(4) This is not good enough, but it is "close" and it is reasonable to change things permanently, or
(5) This is not good enough, and it is "close," and things will or could go wrong, but the consequences are manageable or recoverable and probably tolerable, or
(6) This is not good enough, and it is "close," and things will or could go wrong, and the consequences are potentially catastrophic, or
(7) This is too far apart to be easily fixed, regardless of the extent of consequences.

The key elements of the problem appear to be:

- A method of "zooming" from very inexpensive high level abstraction to elementary details. The high level perspectives will allow identification of potential mismatches in semantics.
- Formalisms to characterize context, application and consequences without requiring a complete and/or certain model of the immediate world.
- Expressions to usefully report and reason about "fittedness."

Leading approaches to these challenges are (respectively) situation theory (Devlin, 1997; Barwise, 1989), some techniques in reasoning under uncertainty, and a synthesis of group and graph theories (Leyton, 1992).

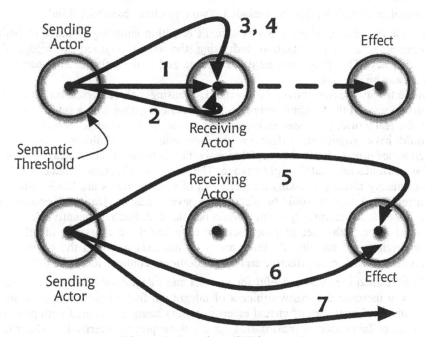

Figure 6: A Notion of "Distances"

The notion of distance is better suited to a normal form of "fittedness, " perhaps geometric (as in graph patterns) or topological. But there likely needs to be a facility at some point to use local methods with accounting practices to reduce the "geometric distance" to a cost-derived scalar. In that way, managers can "see" the cost to adapt or the cost of consequence. Nevertheless, this number would be a derived, flattened result.

9. TWO PROBLEM SPACES

The workshop identified two scenarios that likely would produce different tools:

- the "lab testbed" scenario where a tool is tested and certified against a number of peer tools in a wide set of characterized contexts

- the "field environment" scenario where an operating or newly formed virtual enterprise encounters a single, limited context and wishes to know how well it collaborates.

In the lab case, you have the luxury of time. You have the ability to test and discover failure by cheap observation. You almost certainly will have a well characterized set of scenarios (a. k. a. "a test suite") against which the effectiveness of semantic conveyance is tested. The distance characterizations are likely to consist of a spectrum of effectiveness against this collection of contexts.

The metric in this case is likely to include information such as:

- which of the infrastructure categories listed above the tested configuration falls in. (There will be a finer breakdown of ontology characterizations of course.)
- a characterization of the situations or contexts in which the condition holds.

Additional information might be included. Two types have been identified:

- in case (3) above, where the semantic fit is within shooting distance of being acceptable, a characterization indicating the effort required to bridge the inadequacy. This may even be a cost metric and is the only result expected to naturally be a scalar.
- In the first two cases above, the "semantic robustness characterization" is for the current state of the sending and receiving process, together with a set of contexts. In the real world, processes rarely remain the same. Any change, however small, could have significant effect on the semantic interoperability even if the semantics proper don't change. Obviously, that is because the contexts in which the semantics are "safe" might change. The semantic robustness characterization presumably already contains a description of what contexts are "safe" with the current semantics. It would be nice to also have a characterization of contexts in which certain semantic "growth" would be tolerated. Such a "negative distance" would report: "this set of processes not only has these measures of effective conveyance and additionally there are other contexts in which the conveyance can be expected to be effective and those additional contexts look like this."

The Lab Testbed scenario is useful for vendors and integrators who want to certify products or increase the trustworthiness of integration frameworks. But there are a large number of instances of virtual enterprise users being confronted with process-to-process collaboration scenarios that have not be precharacterized as described above.

These users will need the ability to determine semantic robustness on the fly, and may need additional tools to help correct an identified problem. In this case, many of the conveniences of the lab will be gone. Time is likely to be an issue. Probably, the most useful implementations would be iterative in that a very inexpensive process would be applied to identify a problem with successively more expensive and detailed iterations that drill into the semantics and context.

This use has been identified in other forums as the "self-organizing (or self-annealing or self-integrating) enterprise" (Kosanke & Nell, 1997).

The projected set of tools includes those of the testbed but adds some additional mechanisms to conduct conversations and support the layered zooming. Anticipated services might be:

- a means of identifying when a mismatch has occurred or is likely to. This could simply be a gross characterization that one or more of the processes involved haven't been evaluated in a Testbed mode (or, obviously, semantically harmonized). In this case, all semantic interactions are suspect.
- a lightweight language to support dialogue about the semantics involved. This might be called a "semantic interoperability language."
- a technique for quickly guessing contexts and semantic "anchor points" for a first, cheap evaluation to advise on whether further drilling is required.
- a process for guided drilling. The Testbed has the luxury of potentially exhaustive examinations of every pocket in every context. The field situation will instead only examine the instant context and the relevant subset of semantics. Identifying these may be non-trivial; it may be easier to follow-and-certify. However, there is a suspicion that guided anticipatory drilling is possible.
- a concurrent metric of cost of the process for incremental examining and certifying (or not). This might be tied to a "semantic benefit" metric.
- remaining tools and metrics as inherited from the "simpler" use scenario.

10. HOW THIS MIGHT WORK

Already, this topic has attracted attention and there are many suggested directions for solutions and research topics. We feel that the approach which characterizes ontology types by infrastructure and separately employs internal metrics (trust, effectiveness) as the basis for semantics of the external metric (semantic distance) is the way to go. It will require research in three areas to enhance the applicable formal tools.

11. A SEMANTIC INTEROPERABILITY LANGUAGE

We need a semantics to reason about semantics; it needs to include a logic to support formal reasoning over contexts and semantics. Ideally, it should support some sort of "zooming" from high level, cheap abstraction to thorny details. Fortunately, we have such a thing in situation theory, a system of logic originally developed by linguist mathematicians to formally manage the information from context (Barwise, 1989). Incidentally, it is suited for reasoning about semantics in general and has been used in "zooming" applications in the enterprise context (Devlin 1991, Devlin & Rosenberg, 1996).

The focus for activity in situation theory is the Center for the Study of Language and Information at Stanford University.

The first order of business is to extend the Situation Logic and Process Specification Language (PSL) to be friendlier to one another. PSL is a sufficiently formal framework for process-aware ontology dialog.

12. A METHOD OF CHARACTERIZING UNKNOWN CONTEXTS

This speaks only to the operational field environment; the test bed will have well-formed models of the test contexts and associated environments. The field environment is blessed with a simpler case in one regard; it has only one context. But that context is likely to be poorly understood and almost certainly unmodeled in important respects. One must reason over unknowns and uncertains, rather than forcing the enterprise to go through the extraordinarily expensive process of discovering and modeling their environment. Even many of the facts that will be known by someone may be too expensive to harvest.

We will require a grab bag of techniques for reasoning over uncertainty. The NIST workshop revealed that there is certainly no clear winner here and that a variety of theories will likely come to bear. Just what techniques are appropriate for which situations is a research topic, one in which our group has not yet invested.

Note that this supposes that modeling the environment can be orthogonally separated from models of the processes. This is routinely done in the business enterprise but is to be examined for other contexts. For instance, we have studied the combat enterprise (Goranson, 2004) and determined that the uncertainties span both worlds.

Almost certainly this will require further sponsorship in early exploration.

13. A ROBUST MEANS FOR MODELING AND OPERATING ON THE "DISTANCE"

Preference and tradition seem to converge on a graph or lattice expression for the actual form of the characterization we have been calling the "fitness metric." We believe it likely that such a thing can have a user friendly graphical expression using a structured, hyperlinked narrative. Toward that end, we are exploring tools such as Tinderbox and have established an expertise in outliner interfaces (Goranson, 2004).

However, we need a theory and algebra to manage the representations themselves apart from the logic — the semantic interoperability language — that generates them. This would in effect be a metamodeling method, geared toward two levels:

One level which maps to whatever the native semantics of the metric are. These are abstracted from the models and process codes involved and are one step removed from them. (As mentioned, it is a matter of practice and philosophy whether those models and process codes represent an abstraction from reality or constitute a part of the enterprise reality.) This level must have some correspondence between expression and content (between syntax and semantics if you will) to be able to support both the less abstract intuitive graphical user display (based on shape) and the higher level described below.

A second level which supports an algebra over distance models so that: history (context) is captured and also that supports a higher level of abstraction for semantic clustering by representation topology. By this clever means (infrastructure categories to distance shape-based groups via "core metrics" semantics) we can work with the clean and flexible mechanics of group theory. It is our belief that if we intend to have an ultimate algebra of semantics, this is the level one must seek.

We favor an emerging cognitive theory (Leyton, 1992) for this. It develops a rudimentary but workable system in the product model domain that has the links to intuitive shape perception, enterprise-sensitive models (albeit not process models), and higher level group-driven bundles for both simple calculations and metareasoning.

We intend to bring These tools from the product model side to the process semantics side, something that "follows the tide" in enterprise integration studies already.

14. A COLLECTION OF ACCESSIBLE META METAPHORS

No metric will survive in the business domain unless it is intuitively accessible to managers. We've already noted the requirement to map the complex representation of fittedness into a cost scalar using context-specific mappings. But a semantic distance characterization is a metametric, a metric of metrics. That's because we based our reduction of the system semantics to those elements that have effect, in other words those that affect basic metrics.

Managers will require an accessible metaphor for such "folding." Elsewhere we describe our proposal for such a metaphor, drawn from popular film (Goranson, 2000, 2003). As it happens, a great many popular movies employ sophisticated folding metaphors that are readily understandable to an ordinary viewer. The notion may seem a little strange, but no more than using sports or war metaphors.

These four areas are being tracked by our group at Old Dominion University. Further international workshops are planned and an on-line collaboration infrastructure has been established by NIST (Goranson, 2004).

15. CONCLUSION

The discipline of enterprise integration is maturing beyond the "one-religion" model and dealing with the real world situation faced by advanced virtual enterprises. We will have to deal with ontological mismatches that are imperfect but sufficiently effective. Some hard research topics are in front of us, but with enormous potential.

We are already committed to catalyzing the community and serving as a forum for firming up the research agenda, which at this point is wide open.

However, we have embarked on what we think may be the most promising directions, as described. Probably other approaches will be useful earlier but it appears to us that the community should be aiming high. All productivity gains since World War II can be attributed to improvements in the science underlying infrastructure. We can and must create revolutions for the next era.

In a related activity, the ICEIMT gathering has been taken over by the community as a more regular conference on advancing the science of Enterprise Integration and could serve to advance the agenda.

REFERENCES

Barwise, J. (1989). The Situation in Logic. Palo Alto: CSLI Press.
Barwise, J & Seligman, J. (1997) Information Flow, The Logic of Distributed
    Systems. Cambridge: Cambridge University Press.
Bernstein, M. (2004) Tinderbox. < http://www.eastgate.com/Tinderbox>

Devin, K. (1991). Logic and Information. Cambridge: Cambridge University Press.
Devlin, K & Rosenberg, D. (1996) Language at Work. Palo Alto: CSLI Press.
Dove, R. ed (1995). Agile Practice Reference Base. Bethelhem: Lehigh University.
Goranson, H T. (1999). The Agile Virtual Enterprise. Westport: Quorum,
Goranson, H T. (2003). Metaphoric Concepts for Scopable Enterprise Modeling.
    Norfolk: AERO/J9 report.
Goranson, H T. (2004). Counterterrorism Infrastructure Modeling. Norfolk:
    AERO/J9 report.
Goranson, H T. (2004). Semantic Distance Collaboration Group.
    http://interop.cim3.net/
Kosanke, K. & Nell, J G. ed (1997). Enterprise Engineering and Integration. New
    York: Springer-Verlag.
Kosanke, K. & Nell, J G., Jochem, R., Ortega Bas, A. ed (2003) Enterprise Inter and
    Intra Organizational Integration. Berlin: Kluwer.
Leyton, M. (1992) Symmetry, Causality, Mind. Cambridge: MIT Press.
Petrie, C J. ed (1992). Enterprise Integration Modeling. Cambridge: MIT Press

5. The Nature of Knowledge and its Sharing through Models

Peter Bernus,[1] Brane Kalpic[2]

1 Griffith University Email: P.Bernus@bigpond.com
2 ETI Elektroelement Email: Brane.Kalpic@eti.si

Enterprise Modelling has been repeatedly proposed as a way to share knowledge within and among companies. However, industry practitioners – especially in Small and Medium Enterprises – are slow to take up this practice, and models are usually only built to support the development of application programs, databases or other information technology artefacts, rather then for the broader purpose of knowledge sharing.

The article examines knowledge categories previously proposed in the literature and proposes an extension of previous work in order to better understand the nature of knowledge sharing processes and the role of models in these.

1. INTRODUCTION

In the literature, several different definitions of knowledge can be found. The Oxford English dictionary (1999) defines knowledge as the "facts, feelings, or experiences known by a person or group of people".

According to Baker *et al* (1997), knowledge is present in "ideas, judgements, talents, root causes, relationships, perspectives and concepts". Knowledge can be related to customers, products, processes, culture, skills, experiences and know-how.

Bender and Fish (2000) consider that knowledge originates in the head of an individual (the mental state of having ideas, facts, concepts, data and techniques, as recorded in an individual's memory) and is built on the basis of information transformed and enriched by personal experience, beliefs and values with decision and action-relevant meaning. Relevantly, therefore, knowledge formed by an individual could differ from knowledge possessed by another person receiving the same information.

Similarly to the above definition Baker *et al* (1997) define knowledge in the form of a simple formula:

(1) *Knowledge = Information + [Skills + Experience + Personal Capability]*

This simple equation must be interpreted to give knowledge a deeper meaning: knowledge is created from data which becomes information as interpreted and remembered by a person with given skills, experience, personal capabilities and previously developed mental models.

Knowledge gives a person the ability to use information to guide the actions of the person in a manner that is appropriate to the situation. It is noteworthy that this does not imply that the person is *aware* of this knowledge or that he/she can *explain* (externalise) it. These distinctions are important to consider when planning to discover what knowledge is available, or intending to establish knowledge transfer/sharing.

Reading equation (1) it seems to suggest that knowledge *equals* the sum of the listed components. However, the intention is clearly to suggest that knowledge is an *outcome* of a process performed by an individual, i.e. it is a function of the listed components, which gives equation (2).

(2) *Knowledge = f (Information, Skills, Previous Experience, Personal Capability)*

Still, equation (2) is not clear about the role of pre-existing knowledge in gaining new knowledge nor about the role of unlearning / transforming existing knowledge. Also neither equation explains what knowledge *is* – they only state that knowledge is created using these components. We would at least expect an equation that would have the pattern:

(3) $Knowledge_{t2} = f (Information\ t_{<t1-t2>}, ... , Knowledge_{t1}, ...)$

Thus such an equation would explain how information gained between times t1 and t2 transforms knowledge, depending on many factors, including knowledge possessed before time t1.

The authors believe that without improving the understanding of the nature of knowledge it would be difficult to pinpoint the role of models in gaining, capturing or sharing knowledge. Therefore this article sets out to investigate categories of knowledge (Section 2) and then identifies processes (Section 3) that transform knowledge in one category to knowledge in another category. Once such processes have been identified it is possible to identify those which can (or could) use models.

Note that the word 'models' here refers to a mathematical construct that can be used to represent a significant set of properties of some existing or proposed artefact, such that all relevant properties of the artefact can be derived by investigating the model rather then the artefact itself and no relevant properties can be derived from the model which are not properties of the artefact. Mathematical logic (model theory) actually calls such a mathematical construct a 'theory', rather then a 'model'. However, many other disciplines, including engineering, use the term 'model' for these mathematical constructs and this is the meaning adopted in this article. Thus an IDEF0 schema is an 'activity model' of some process, an IDEF1X schema is a 'model of some data', etc.

2. KNOWLEDGE CATEGORIES

Knowledge Management (KM) literature defines two main knowledge categories: explicit and tacit. Polanyi (1966) defines tacit knowledge as knowledge, which is implied, but is not actually documented, nevertheless the individual 'knows' it from experience, from other people, or from a combination of sources. Explicit knowledge is externally visible; it is documented tacit knowledge (Junnarkar and Brown, 1997).

Skryme and Amidon (1997) define explicit knowledge as formal, systematic and objective, and it is generally codified in words or numbers. Explicit knowledge can be acquired from a number of sources such as company-internal data, observing business processes, records of policies and procedures, as well as from external sources such as through intelligence gathering. Tacit knowledge is more intangible.

It resides in an individual's brain and forms the basis on which individuals make decisions and take action, but is not externalised in any form.

Polanyi (1958) also gives another detailed and substantial definition of knowledge categories. He sees tacit knowledge as a personal form of knowledge, which individuals can only obtain from direct experience in a given domain. Tacit knowledge is held in a non-verbal form, and therefore, the holder cannot provide a useful verbal explanation to another individual. Instead, tacit knowledge typically becomes embedded in, for example, routines and cultures. As opposed to this, explicit knowledge can be expressed in symbols and communicated to other individuals by use of these symbols.

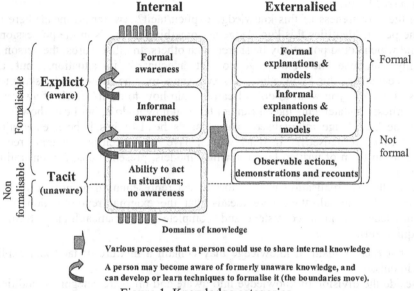

Figure 1. Knowledge categories

Bejierse (1999) states that explicit knowledge is characterised by its ability to be expressed as a word or number, in the form of hard data, scientific formulas, manuals, computer files, documents, patents and standardised procedures or universal works of reference that can easily be transferred and spread. Implicit (tacit) knowledge, on the other hand, is mainly people-bound and difficult to formalise and therefore difficult to transfer or spread. It is mainly located in people's 'hearts and heads'. Considering the above definitions, the authors give the following definitions:

(4)   *Explicit knowledge is knowledge, which can be articulated and written down. Therefore, such knowledge can (or could) be externalised and consequently shared and disseminated.*

(5)   *Tacit knowledge is subconscious, it is understood and used but it is not identified in a reflective, or aware, way[30]. Tacit knowledge is developed and derives*

---

[30] If a person geathers evidence that makes him/her aware of knowledge previously categorised as tacit then this knowledge becomes informal explicit knowledge.

from the practical environment; it is highly pragmatic and often specific to situations in which it has been developed.

Although tacit knowledge is not directly externalisable, it is sometimes possible to create externalisations[31] that may help someone else to acquire the same tacit knowledge. Tacit knowledge could be made up of insights, judgement, know-how, mental models, intuition and beliefs, and may be shared through direct conversation, telling of stories and sharing common experiences.

Definitions (4) and (5) give rise to a categorisation that can be used to make practically important differentiations between various categories of knowledge. The authors propose to divide knowledge into categories according to the following criteria (see Figure 1):

- Is there awareness of this knowledge explicit/tacit? Awareness means here that the person identifies this knowledge as something he/she is in the possession of and which could potentially be shared with others. In other words, the person not only can use the knowledge to act adequately in situations, but also conceptualises this knowledge. This awareness may be expressed by statements as "I can tell you what to do", "I can explain how to do it". Lack of awareness manifests is statements like "I can not tell you how to do it, but I can show".
- Is the knowledge internalised in a person's head or has it been externalised (internal/externalised)? In other words, have there been any external records made (in form of written text, drawings, models, presentations, demonstrations, etc.)?
- Does the externalisation have a formalised representation or not (formal/not-formal)? Formalisation here means that the external representation of the knowledge is in a consistent and complete mathematical/logical form (or equivalent).

Note that each domain of knowledge may contain a mixture of tacit and explicit constituents.

Beside the division of knowledge into aware and unaware categories, additional categorisation of knowledge, according to whether the knowledge *could* be externalised, into the category of formalisable and non-formalisable, may be added. While explicit knowledge can always be externalised (applying different processes, mechanisms and approaches) tacit knowledge could not be fully externalised, however there are parts that can be communicated through indirect externalised means. This externalisation could be achieved by a) indirect externalisation through conversation, telling of stories, sharing common experiences and other similar approaches, or b) thought an awareness-building process, where the unaware knowledge is transformed into an aware knowledge (even if not formal). A more detailed definition of knowledge processes and their relations to the postulated knowledge categories are presented in Section 3.

---

[31] I.e., these externalisations do not contain a record of the knowledge itself, rather they would contain information that another person could (under certain circumstances) use to construct the same knowedge combining it with his/her already possessed internal knowledge.

3. KNOWLEDGE PROCESS AND KNOWLEDGE RESOURCES

A comprehensive survey of the KM literature shows various knowledge management frameworks and KM activities. Some frameworks are composed of very low-level activities and in some frameworks it seems that elementary activities group into higher-level activities.

Nonaka and Takeuchi (1995) define four processes:

- *Internalisation* is the process in which an individual internalises explicit knowledge to create tacit knowledge. In Fig.1 this corresponds to turning externalised knowledge into internalised – Nonaka does not differentiate between formal and informal awareness.
- *Externalisation* is the process in which the person turns their tacit knowledge into explicit knowledge through documentation, verbalisation, etc. In Fig. 1 this process corresponds to turning internalised, formalisable knowledge into externalised knowledge and subsequently communicating it (internal → externalised).
- *Combination* is the process where new explicit knowledge is created through the combination of other explicit knowledge.
- *Socialisation* is the process of transferring tacit knowledge between individuals through observations and working with a mentor or a more skilled / knowledgeable individual. In Fig. 1 this corresponds to tacit knowledge → observable actions, etc.

Devenport and Prusak (1998) identify four knowledge processes: knowledge generation (creation and knowledge acquisition), knowledge codification (storing), knowledge transfer (sharing), and knowledge application (these processes can be represented as various transitions between knowledge categories in Figure 1).

Alavi and Marwick (1997) define six KM activities: a) acquisition, b) indexing, c) filtering, d) classification, cataloguing, and integrating, e) distributing, and f) application or knowledge usage, while Holsapple and Whinston (1987) indentfy more comprehensive KM process, composed of the following activities: a) procure, b) organise, c) store, d) maintain, e) analyse, f) create, g) present, h) distribute and i) apply. (Again, these processes can be represented as various transitions between knowledge categories in Figure 1.)

Holsapple and Joshi (2002) present four major categories of knowledge manipulation activities:

- *acquiring* activity, which identifies knowledge in the external environment (form external sources) and transforms it into a representation that can be internalised and used;
- *selecting* activity identifying needed knowledge within an organisation's existing resources; this activity is analogous to acquisition, except that it manipulates resources already available in the organisation;
- *internalising* involves incorporating or making the knowledge part of the organisation, and
- *using*, which represents an umbrella phrase for a) generation of new knowledge by processing of existing knowledge and b) externalising knowledge that makes knowledge available to the outside of the organisation.

These four processes are applicable to the organisation as an entity, rather then addressing knowledge processes from the point of view of an individual.

As a conclusion: organisations should be aware of the complete process of knowledge flow, looking at the flow between the organisation and the external world and the flow among individuals within (and outside) the organisation. This latter is an important case, because in many professional organisations individuals belong to various *communities*, and their links to these communities is equally important to them as the link to their own organisation.

3.1 Knowledge resources

Knowledge manipulation activities operate on knowledge resources (KR) to create value for an organisation. On the one hand, value generation depends on the availability and quality of knowledge resource, as well productive use of KR depends on the application of knowledge manipulation skills to execute knowledge manipulation activities.

Holsapple and Joshi (2002) developed a taxonomy of KR, categorising them into schematic and content resources. The taxonomy identifies four *schematic* resources and two *content* resources appearing in the form of participant's knowledge and artefacts. Both schema and content are essential parts of an organisation's knowledge resources.

Content knowledge is embodied in usable representations. The primary distinction between participant's knowledge and artefacts lies in the presence or absence of knowledge processing abilities. Participants have knowledge manipulation skills that allow them to process their own repositories of knowledge; artefacts have no such skills. An organisation's participant knowledge is affected by the arrival and departure of participants and by participant learning. As opposed to this, a knowledge artefact does not depend on a participant for its existence. Representing knowledge as an artefact involves embodiment of that knowledge in an object, thus positively affecting its ability to be transferred, shared, and preserved (in Figure 1 knowledge artefacts correspond to recorded externalised knowledge).

Schema knowledge is represented or conveyed in the working of an organisation. It manifests in the organisation's behaviours. Perceptions of schematic knowledge can be captured and embedded in artefacts or in participant's memories, but it exists independent of any participant or artefact. Schematic knowledge resources are interrelated and none can be identified in terms of others. Four schematic knowledge resources could be identified: a) culture (as the basic assumptions and beliefs that are shared by members of an organisation), b) infrastructure (the knowledge about the roles that have been defined for participants), c) purpose (defining an organisation's reason for existence), and d) strategy (defining what to do in order to achieve organisational purpose in an effective manner).

Note, that the above-described content knowledge is also referred to in contemporary management literature and can be named as 'individual knowledge'; while schema knowledge is identified as 'collective knowledge' and is closely related to the organisation's capability.

In addition to its own knowledge resources, an organisation can draw on its environment that holds potential sources of knowledge. Through contacts with its environment, an organisation can replenish its knowledge resources. The environmental sources do not actually belong to an organisation nor are they controlled by the organisation. When knowledge is acquired form an environment source, it becomes an organisational source.

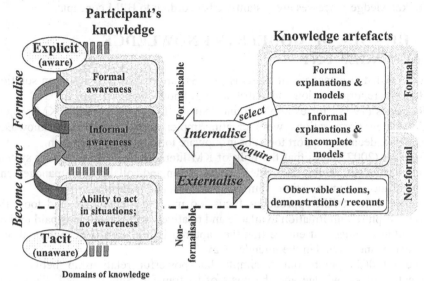

Figure 2. Knowledge process model

3.2 Knowledge process model

Considering the definitions of a) knowledge processes proposed by different authors (like Nonaka and Takeuchi (1995), and Holsapple and Joshi (2002)) and b) knowledge categories defined in the knowledge category model in Section 3.2), the authors further propose a knowledge process model, which identifies main internal and external knowledge processes and their relationships to knowledge categories.

This model defines two major categories of knowledge process: the knowledge *externalisation* process and the knowledge *internalisation* process.

The knowledge internalisation process, considers the source or environment from where that knowledge derives (originates) and applies two major mechanisms: a) the *selection* process internalises knowledge from inbound KR and b) the *acquisition* process acquires knowledge from external KR. However, a KR could appear in different forms as a) knowledge artefacts in formal or not-formal presentation and b) schema knowledge and knowledge present in data and information which has to be processed (in the form of observation of actions, demonstrations, recount and data and information processing) which is still to be turned into a usable and transferable form of knowledge.

Knowledge externalisation includes the articulation and codification of knowledge in the form of formal or not-formal knowledge. Formal, aware knowledge could be externalised by formal explanations and models, while informal knowledge can be externalised using informal explanations or incomplete models.

Beside the externalisation and internalisation processes, two other important participant-bounded processes can be identified – the awareness process and the formalisation process. The awareness process transforms the formalisable part of unaware knowledge into aware knowledge, while the formalisation process converts already aware knowledge into structured and formal form. Awareness and formalisation knowledge processes are discussed in more detail in Section 4.2, where knowledge processes are instantiated according to BPM concepts.

4. THE ROLE OF MODELS IN KNOWLEDGE MANAGEMENT

Many knowledge management systems (KMSs) are primarily focused on solutions for the capture, organisation and distribution of knowledge.

Rouggles (1998), for example, found that the four most common KM projects conducted by organisations were creating/implementing an intranet, knowledge repositories, decision support tools, or groupware to support collaboration.

Spender (2002) states that the bulk of KM literature is about computer systems and applications of 'enterprise-wide data collection and collaboration management', which enhance communication volume, timeliness, and precision.

Indeed, current KM approaches focus too much on techniques and tools that make the captured information available and relatively little attention is paid to those tools and techniques that ensure that the captured information is of high quality or that it can be interpreted in the intended way.

Teece (2002) points out a simple but powerful relationship between the codification of knowledge and the costs of its transfer. Simply stated: the more a given item of knowledge or experience has been codified (formalised in the terminology of Figure 1), the more economically it can be transferred.

Uncodified knowledge is slow and costly to transmit. Ambiguities abound and can be overcome only when communication takes place in face-to-face situations. Errors of interpretation can be corrected by a prompt use of personal feedback.

The transmission of codified knowledge, on the other hand, does not necessarily require face-to-face contact and can often be carried out by mainly impersonal means. Messages are better structured and less ambiguous if they can be transferred in codified form.

Based on the presented features of business process modelling (and in the broader sense enterprise modelling) and the issues in knowledge capturing and shearing, BPM is not only important for process engineering but also as an approach that allows the transformation of informal knowledge into formal knowledge, and that facilitates externalisation, sharing and subsequent knowledge internalisation. BPM has the potential to improve the availability and quality of captured knowledge (due to its formal nature), increase reusability, and consequently reduce the costs of knowledge transfer. The role and contribution of BPM in knowledge management will be discussed in more detail in Section 4.2.

4.1 BPM and KM are related issues

While the methods for developing enterprise models have become established during the 1990s (both for business process analysis and design) these methods have concentrated on how such models can support analysis and design teams, and the question of how these models can be used for effective and efficient sharing of information among other stakeholders (such as line managers and engineering practitioners) has been given less attention.

If enterprise models, such as business process models, embody process knowledge then it must be better understood to what extent and how existing process knowledge can be externalised as formal models, and under what conditions these models may be effectively communicated among stakeholders. Such analysis may reveal why the same model that is perfectly suitable for a business process analyst or designer may not be appropriate for end users in management and engineering. Thus the authors developed a theoretical framework which can give an account of how enterprise models capture and allow the sharing of the knowledge of processes – whether they are possessed by individuals or groups of individuals in the company. The framework also helps avoid the raising of false expectations regarding the effects of business modelling efforts.

4.2 The knowledge life-cycle model

Figure 3 introduces a simple model of knowledge life-cycle, extending (detailing) the models proposed by Nonaka and Takeuchi (1995), and Zack and Serino (1998). Our extension is based on Bernus *et al* (1996), which treat enterprise models as objects for semantic interpretation by participants in a conversation, and establishes the criteria for uniform (common) understanding. Understanding is of course most important in knowledge sharing. After all, if a model of company knowledge that can only be interpreted correctly by the person who produced it, is of limited use for anyone else. Moreover, misinterpretation may not always be apparent, thus through the lack of shared interpretation of enterprise models (and lack of guarantees to this effect) may cause damage. This model (Figure 3) represents relations between different types of knowledge, and will be used as a theoretical framework.

In order for employees to be able to execute production, service or decisional processes they must possess some *'working knowledge'* (e.g. about process functionality, required process inputs and delivered outputs, organisation, management, etc.). Working knowledge is constantly developed and updated through receiving information from the internal environment (based on the knowledge *selection* process) and from the external environment (thought the process of knowledge *acquisition*).

Working knowledge (from the perspective of the knowledge holder) is usually tacit. Knowledge holders don't need to use the possessed knowledge in its explicit, formalised form to support their actions. They simply understand and know what they are doing and how they have to carry out their tasks – having to re-sort to the use of explicit formal knowledge would usually slow down the action.

According to the suitability for formalisation such working knowledge can be divided into two broad groups: *formalisable* and *non-formalisable* knowledge. Such division of knowledge into two broad categories seems to closely correspond to how much the process can be structured, i.e. to be decomposed into a set of interrelated lower level constituent processes. These characteristics can be observed when considering knowledge about different typical business process types.

The formalisation and structural description of innovative and creative processes, such as some management, engineering and design processes (or in general the group of *ad-hoc* processes), is a difficult task, due to the fact that the set of constituent processes is not predefined, nor is the exact nature of their combination well understood by those who have the knowledge. Consequently, knowledge about this type of processes could be considered tacit knowledge (because they are not formalisable unaware processes), i.e. not suitable for formalisation/structuring.

In contrast to the characteristics of the group of *ad-hoc* processes the group of ill-structured and structured (repetitive or algorithmic) processes can be formalised and structured at least to a degree; consequently the knowledge about these processes may become explicit formal knowledge. Examples of such processes are management, engineering and design on the level of co-ordination between activities as performed by separately acting-individuals or groups, and repetitive business and manufacturing activities.

The formalisable part of knowledge (knowledge about structured and ill-structured processes) is extremely important and valuable for knowledge management, because this may be distributed and thus shared with relative ease. Namely, the process of transformation of the formalisable part of tacit knowledge into formal knowledge (the formal part of explicit/aware knowledge) represents one of the crucial processes in knowledge management. The authors believe that the cost of knowledge management (measured by the level of reuse and return of investment to the enterprise) in case of formal explicit knowledge would be lower than in case of tacit (unaware) – or even in case of unstructured explicit – knowledge, simply because the sharing of the latter is a slow and involved process.

To be able to perform the aforementioned formalisation process we need additional capabilities known as *culturally shared* or *situation* knowledge (e.g. knowledge shared by the community that is expected to uniformly interpret the formal models of the target processes). Culturally shared knowledge plays an essential role in the understanding of the process or entity in question and in its formalisation and structuring. E.g. the definition of an accounting process can only be done by an individual who understands accounting itself, but this formalisation will be interpreted by other individuals who must have an assumed prior culturally shared and situational knowledge that is not part of the formal representation (Bernus *et al*, 1996).

As mentioned, one of key objectives of KM is the *externalisation* of participants' knowledge. Regarding the type of knowledge (tacit and explicit) different tools and approaches in knowledge capturing may be used:

- Tacit knowledge (whether formalisable or not) can be transferred through live *in situ* demonstrations, face-to-face storytelling, or *captured informal presentations* (e.g. multimedia records, personal accounts of experience, or demonstrations). Note that tacit formalisable knowledge may be *discovered* through a research process and thus made explicit. Subsequently such knowledge may be captured as described in the bullet point below.
- Explicit knowledge can be captured and presented in *external presentations* (through the process of knowledge capturing also known as knowledge codification). An external presentation may be *formal* or *not formal*. A textual description, like in quality procedure documents (ISO9000) is not formal, while different enterprise models (e.g. functional business process models) are examples of formal external representations of knowledge (knowledge externalisations).

Formal and informal external representations are called *knowledge artefacts*. The advantage of using formal models for process description is the quality of the captured knowledge.

To actually formalise knowledge, *formalisation skills* are needed (in this case business process modelling skills).

The above process of knowledge externalisation has to be complemented by a matching process of knowledge *internalisation* that is necessary for the use of available knowledge resources.

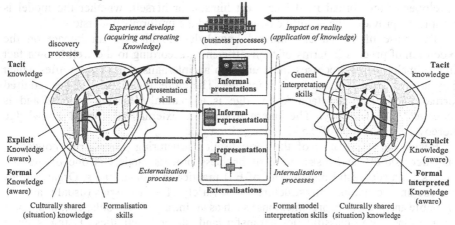

Figure 3: The knowledge life-cycle model

According to the type and form of externalised knowledge, various internalisation processes (and corresponding skills) are necessary. In general, the less formal the presentation / representation, the more prior assumed *situation-specific* knowledge is necessary for correct interpretation. Conversely, more formal representations allow correct interpretation through the use of more generic knowledge and require less situation-specific knowledge. Thus formalisation helps enlarge the community that can share the given knowledge resource.

An informal external presentation of knowledge accompanied with its interpretation (e.g. interpretation of the presented story) can directly build working (tacit) knowledge, however the use of these presentations is only possible in limited situations, and it is difficult to verify that correct interpretation took place as well as the degree of completeness of such knowledge transfer. However, the verification of correct interpretation and completeness is only possible through direct investigation of the understanding of the individuals who internalised this type of knowledge. This is a serious limitation for knowledge sharing through informal means.

A *formal external presentation*, such as a business process model developed in the IDEF0 (ICAM DEFinition) modelling languages (Menzel and Mayer, 1998), must be first interpreted to be of use. To interpret the content, i.e. the information captured in this model, knowledge-processing skills (abilities) are needed. Formal model interpretation skills are generic and not situation dependent, therefore even culturally distant groups of people can share them. Still, such formal representation must be further interpreted by reference to culturally shared, prior assumed knowledge so that the content of the formal knowledge (information captured in the business process model) can be understood and interpreted in the intended way, and thus integrated into working knowledge (to improve competencies). However, to test for correct interpretability it is possible to test whether the primitive concepts in the model (i.e. those not further explained/decomposed) are commonly understood. If this is the case then the formal nature of the model guarantees uniform interpretability. Completeness can be tested without the direct investigation of the understandings of those individuals who internalise this formal knowledge (i.e. the developer of the formal model can test himself or herself, whether the model is complete – provided the primitive concepts used are uniformly understood [32]).

The reuse of formal externalised knowledge could have an impact on the execution of process in terms of their efficiency, according to the well known fact that formally learnt processes must undergo an internalisation process after which they are not used in a step-by-step manner. Therefore, the transfer of the acquired formal knowledge into tacit knowledge is a 'natural' learning process and is necessary for efficiency. The internalisation of externalised formal knowledge thereby closes the loop of the knowledge life-cycle.

Beside the importance of the formalisation/structuring process of knowledge, easy accessibility and distribution of business process models is one of the key factors for a successful deployment of EM practice in organisations. Organisations can use an information infrastructure and a variety of technologies (usually already available and present in organisations) such as an Intranet, web tools, etc., to support storage, indexing, classification, transfer and sharing activities. Using such a distribution mechanism process models can be made available to all stakeholders, and their access can be made platform (software and hardware) independent.

5. CONCLUSION

The great interest in Knowledge Management, as one of the hottest research topics of the past decade, is being conditioned by several driving forces: a) recognition of how difficult it is to deal with complexity in the business environment, b) interest in core competencies, their communication, leverage and possible transfer, c) issues concerning the dissemination of company knowledge in world-wide distributed companies, d) rapid development and adoption of ICT, and e) company awareness of issues concerning individual's knowledge and its externalisation and formalisation.

Companies have already adopted a number of different initiatives, which could become useful components for KM implementation. BPM represents one of these initiatives and a key KM component. BPM as an important tool for KM allows the transformation of informal knowledge into formal knowledge and facilitates its externalisation and sharing.

Beside supporting the knowledge awareness and formalisation process, BPM has the potential to establish the criteria for uniform understanding and improve the availability and quality of captured knowledge (due to its formal nature), increase reusability, and consequently reduce the costs of knowledge transfer.

---

[32] This test is commonly ignored by developers of formal models, probably because they assume that primitive concepts are all known through the users' formal education.

The article developed a further differentiation between various types of knowledge and processes and their mutual relationships (relative to existing knowledge categorisations available in the literature). The proposed knowledge categorisation and definition of key knowledge processes represents the authors' attempt and contribution as a basis for more explicit definitions of key notions in the KM domain. However, further research should be done to create a unified and widely accepted Knowledge Management ontology.

Because business process models embody process knowledge, a better understanding of the extent and effective communication of business process models must be achieved. Therefore, by use of the presented theoretical framework this article gave an account of how enterprise models capture and allow the sharing of the knowledge encapsulated in processes. The framework also:

- helps to avoid the raising of unrealistic expectations regarding the effects of business modelling efforts
- presents major knowledge categories, stages in knowledge transformation and activities in this process
- defines the correlation between the formalisable and non-formalisable knowledge categories and process types and
- emphasises the importance of the transformation process on the formalisable part of the knowledge, into its formal presentation as one of the crucial processes in knowledge management.

REFERENCES

Alavi, M., Marwick, P. (1997) One Giant Brain. Boston (MA) : Harvard Business School. Case 9-397-108

Baker M., Baker, M., Thorne, J., Dutnell, M. (1997) Leveraging Human Capital. Journal of Knowledge Management. MCB University Press. 01:1 pp63-74

Beijerese, R.P. (1999) Questions in knowledge management: defining and conceptualising a phenomenon. Journal of Knowledge Management. MCB University Press. 03:2 pp94-110

Bennet, D., Bennet, A. (2002) The Rise of the Knowledge Organisations. in Holsapple, C.W. (Eds.) Handbook on Knowledge Management 1. Berlin : Springer-Verlag. pp5-20

Bender, S., Fish, A. (2000) The transfer of knowledge and the retention of expertise: the continuing need for global assignments. Journal of Knowledge Management. MCB University Pres. 04:2 pp125-137

Bernus P., Nemes, L., Moriss, B. (1996) The Meaning of an Enterprise Model. in Bernus, P., Nemes, L. (Eds.) Modelling and Methodologies for Enterprise Integration. London : Chapman and Hall. pp183-200

Chen, D., Doumeingts, G. (1996) The GRAI-GIM reference model, architecture and methodology. in Bernus, P., Nemes, L. and Williams, T.J. (Eds.) Architectures for Enterprise Integration. London : Chapman & Hall. pp102-126

Conner, K., Prahalad, C.K. (1996) A resource-based theory of the firm: Knowledge versus opportunism. Organization Science. Vol. 7 pp477-501

Davenport, T.H. (1993) Process innovation: reengineering work through information technology. Boston (MA) : Harward Business School Press

Davenport, T. H., Prusak, L. (1998) Working Knowledge: How Organizations Manage What They Know. Boston (MA) : Harvard Business School Press. pp16

Holsapple, C.W., Joshi, K.D. (2002) A Knowledge Management Ontology. in Holsapple, C.W. (Eds.) Handbook on Knowledge Management 1, Berlin : Springer-Verlag. pp89-128

Holsapple, C.W., Whinston., A.B. (1987) "Knowledge-based Organizations." Information Society. (2) pp77-89

ISO/TC 176/SC2 (2000) ISO9004:2000 Quality management systems – guidelines for performance improvements

Junnarkar, B., Brown, C.V. (1997) Re-assessing the Enabling Role of Information Technology in KM. Journal of Knowledge Management. MCB University Press. 01:2 pp142-148

Menzel, C., Mayer, R.J. (1998) The IDEF family of Languages, in: Bernus, P., Nemes, L. and Williams, T.J. (Eds.) Architectures for Enterprise Integration. London : Chapman & Hall. pp102-126

Nonaka, I., Takeuchi, H. (1995) The Knowledge – Creating Company: How Japanese Companies Create the Dynamics of Innovation. New York : Oxford University Press

Oxford University Press (1999) The Oxford English dictionary. Version 2.0

Polanyi, M. (1958) Personal Knowledge. University of Chicago Press

Polanyi, M. (1966) Tacit Dimension. New York : Doubleday

Rouggles, R. (1998) The State of the Notion: Knowledge Management in Practice. California Management Review. 40(3) pp80-89

Schultze, U. (2002) On Knowledge Work. in: Holsapple, C.W. (Eds.) Handbook on Knowledge Management 1, Berlin : Springer-Verlag. pp43-58

Skyrme, D., Amidon, D. (1997) The Knowledge Agenda. Journal of Knowledge Management, MCB University Press. 01:1 pp27-37

Spender, J.C. (2002) Knowledge Fields: Some Post-9/11 Thoughts about the Knowledge-Based Theory of the Firm. in: Holsapple, C.W. (Eds.) Handbook on Knowledge Management 1. Berlin: Springer-Verlag. pp59-72

Teece, D.J. (2002) Knowledge and Competence as Strategic Assets. in: Holsapple, C.W. (Eds.) Handbook on Knowledge Management 1. Berlin : Springer-Verlag. pp129–152

Vernadat, F. (1996) Enterprise Modelling and Integration – Principles and Applications. Chapman & Hall

Vernadat, F. (1998) The CIMOSA Languages. in: Bernus, P., Mertins, K. and Schmidt G. (Eds.) Handbook on Architectures of Information Systems. Berlin : Springer – Verlag. pp243-264

Zack, M.H., Serino, M. (1998) Knowledge Management and Collaboration Technologies. Lotus Development Corporation.

6. ATHENA Integrated Project and the Mapping to International Standard ISO 15704

[1]David Chen, [2]Thomas Knothe and [3]Martin Zelm[33]
[1]LAP/GRAI, University Bordeaux 1, France, Email: chen@lap.u-bordeaux1.fr
[2]FhG-IPK, Berlin, Germany, Email: Thomas.knothe@ipk.fhg.de
[3]CIMOSA Association, Germany, Email: martin.zelm@cimosa.de

This paper aims at presenting an overview of a European Integrated Project ATHENA to develop interoperability of enterprise applications and software. The first part of the paper tentatively maps the expected ATHENA solution components to ISO 15704 which is an important standard in the area of enterprise integration. This mapping allows categorising expected ATHENA research results according to ISO 15704 and evaluating the consistency and completeness of ATHENA solutions with respect to the ISO 15704 framework. The second part of the paper focuses on one solution component: enterprise modelling language. Possible use of UEML v1.0 in ATHENA A1 project and related work to develop UEML 2.0 in INTEROP NoE will be discussed. Conclusions are given at the end of the paper.

1. INTRODUCTION

ATHENA (Advanced Technologies for Interoperability of Heterogeneous Enterprise Networks and their Applications) aims at a holistic approach to develop interoperability of enterprise applications and software (Athena, 2004). It puts emphasis on the integrated research in three relevant domains to tackle interoperability problems: Enterprise Modelling (EM), Architecture and Platform (A&P), and Ontologies (ONTO). ATHENA is actually a research program which consists of a set of projects, providing an overall interoperability solution in terms of prototypes, specifications, guidelines and best practices. One strategic orientation of ATHENA is to actively interact with standardisation bodies not only to use available standards whenever possible, but also to contribute further standard developments. As a starting point and at a high level abstraction, one relevant standard identified is the ISO 15704 (Requirements for Generalised Enterprise Architectures and Methodologies). This standard defines the generic concepts and components to use for enterprise integration and engineering projects. The first part of the paper tentatively evaluates, on the one hand the appropriateness of ISO 15704 to ATHENA approach; and on the other hand the consistency and the completeness of expected ATHENA solutions with respect to ISO 15704 framework. The mapping is developed on the basis of the analysis and comparison between ATHENA and ISO 15704. The second part of the paper is concerned with one ATHENA result

---

[33] Dr Martin Zelm is member of INTEROP Network of Excellence (NoE).

component: the development of a language for modelling collaborative enterprises. The possible use of UEML will be discussed and on-going work outlined.

2. ATHENA RESEARCH ACTIVITIES

ATHENA Integrated Project consists of three action lines in which the activities take place. In Action Line A, the R&D activities are carried out. Action Line B takes care of the community building. Action Line C hosts all management activities. Under Action line A, six research projects were defined and launched.

- *Enterprise Modelling in the Context of Collaborative Enterprises* (A1) develops methodologies for management and modelling of situated processes, flexible resource allocation and assignment for work management and execution monitoring. This project will enable scalable EM methodologies and infrastructures, repository services and portal server services.
- *Cross-organisational Business Processes* (A2) deals with modelling techniques to represent business processes of different organisations on a level that considers the privacy requirements of the involved partners. Such models, enriched with ontologies, will have two perspectives: an enterprise modelling aspect that assigns a process to its context in the enterprise, and a formal aspect to perform computational transformations to allow re-use of a process in a cross-organisational environment.
- *Knowledge Support and Semantic Mediation Solutions* (A3) aims at developing methods and tools for enterprise knowledge management, to support enterprise and application software interoperability. Focus is to use formal knowledge, organised in domain ontologies, to annotate the business processes and the software components in order to reconcile the possible mismatches in unanticipated cooperation activities.
- *Interoperability Framework and Services for Networked Enterprises* (A4) is concerned with the design and implementation of the infrastructure supporting interoperability adopting the Integrated Paradigm (i.e. where there is a standard format for all constituent sub-systems). The resulting toolset will be a set of software and engines that prepare any enterprise in the adoption and exploitation of interoperability support infrastructures.
- *Planned and Customisable Service-Oriented Architectures* (A5) will develop the understanding, tools and infrastructures required for service-oriented architectures. Although the project will consider existing infrastructures, an emphasis will be on the development of an environment for easier application development that natively provides better customisation.
- *Model-driven and Adaptive Interoperability Architectures* (A6) develops innovative solutions for the problem of sustaining interoperability through change and evolution, by providing dynamic and adaptive interoperability architecture approaches. It aims to advance the SoA in this field by applying the principles of model-driven, platform independent architecture specifications, and dynamic and autonomous federated architecture approaches, and the usage of agent technologies.

The research results will be structured in the ATHENA Interoperability Framework (AIF). The framework has three parts: (1) **Conceptual Integration**: definition of the

Interoperability Reference Architecture and associated Interoperability Methodology; (2) **Applicative Integration**: definition of Best Practices, Guidelines and Handbooks; and (3) **Technical Integration:** definition of an Interoperability Support Infrastructure and tools, and the Technical Architecture.

3. MAPPING ATHENA SOLUTION COMPONENTS TO ISO 15704 FRAMEWORK

This section presents the GERAM (Generalised Enterprise Reference Architecture and Methodologies) framework defined in ISO 15704 and the mapping of ATHENA expected solution components to GERAM. Enterprise reference architectures and methodologies shall be capable of assisting and structuring the description, development, operation, and organisation of any conceivable enterprise entity, system, organisation, product, process, and their supporting technology (ISO 15704, 2000).

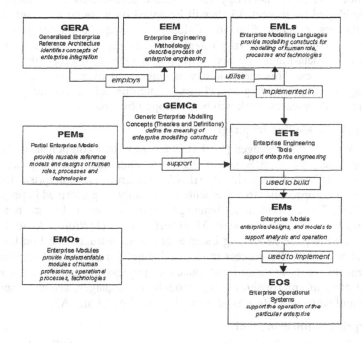

Figure 1. GERAM framework and its components (ISO 15704)

3.1 Enterprise Reference Architecture

(Generalised) Enterprise Reference Architecture (GERA) defines the enterprise related generic concepts recommended for use in enterprise engineering and integration projects. In ATHENA, the reference architecture aims at providing an appropriate categorisation of interoperability concepts in relations to developed technologies and applications. This research is carried out in project A4 (Interoperability Framework and Services for Networked Enterprises). The reference architecture will also provide the conceptual integration of research results of

ATHENA and is embedded into the AIF. External standards and knowledge will also find their place in the reference architecture based on an analysis of their appropriateness for resolving interoperability issues (Athena, 2004).

3.2 Enterprise Engineering Methodology

In GERAM, Enterprise Engineering Methodologies (EEM) describes the processes of enterprise engineering. It provides methods of progression for every type of life-cycle activity. The ATHENA interoperability methodology is associated to the reference architecture and is also developed by project A4. This interoperability methodology is a set of methodologies with the following components: (1) A methodology for gathering, structuring and representing requirements, elaborated and used by Activity B4 (Dynamic Requirements Definition); (2) A methodology for collaborative business process modelling developed by project A1 and will make use of UEML 1.0 for process model exchange; (3) A methodology specified by project A2 to model cross-organisational business processes and its implementation for execution; (4) A methodology for semantic annotations to business process models, developed by project A3. It will be based on a sound mathematical basis (such as Process Algebra, Situation Calculus or Graph Grammars) to make it independent of the specific user-oriented notations; (5) A methodology for specification of meta-modelling and to represent mature, interoperable and high-quality web services. This methodology is elaborated by project A5; and (6) A methodology for implementing model-driven interoperable agent and peer-to-peer architectures. This research work will be developed by A6.

3.3 Enterprise modelling language

Enterprise Modelling Languages (EMLs) is an important component of GERAM. To develop enterprise models potentially more than one modelling language is needed (ISO 15704, 2000). In ATHENA, research on enterprise modelling languages vs. interoperability requirements will mainly be performed in projects A1, A2 and A3. These modelling languages will be used by interoperability methodology to build various models. More particularly: (1) Project A1 will develop Collaborative Enterprise Modelling Languages and Constructs based on UEML 1.0 meta model and others for process model exchange; (2) Project A2 focuses on the development of cross-organisational business process modelling language; (3) Project A3 aims at enterprise ontology modelling languages, semantic annotation techniques and languages. Results of A3 will be used in A1 and A2.

3.4 Enterprise engineering tool

Enterprise Engineering Tools (EETs) deploy enterprise-modelling languages in support of enterprise engineering methodologies, and specifically support the creation, use, and management of enterprise models (ISO 15704, 2000). The development of modelling tools in ATHENA is mainly concerned with projects A1, A2 and A3 as well as A6. Project A1 develops customisable tools for enabling the rapid adoption of collaborative business models, especially for use in SMEs. Project A2 develops (Cross-organisational Business Process) modelling tool to support the cross-organisational business process modelling language, methodology and its enactment. Project A3 will research an ontology-based semantic annotation and reconciliation tool to support the language and methodology developed for the same purpose to capture domain knowledge. It consists of tools supporting languages for Semantic Annotation of: (1) Business Processes, and (2) e-Services. Project A6 will use results of project A3. A semantic UML mapping tool – based on UMT (UML Model Transformation) open source toolkit will be developed. This tool is used to describe, integrate and relate platform independent service and information models, with a corresponding execution support on the platform specific level for UML system models.

3.5 Enterprise modelling concepts

(Generic) Enterprise Modelling Concepts (GEMCs) are the most generically used concepts and definitions of enterprise engineering and integration. The three forms of concept definition are, in increasing order of formality (ISO 15704, 2000): (1) glossaries, (2) meta-models, and (3) ontological theories. In ATHENA, generic enterprise modelling concepts are mainly developed in A1 and A2 projects in collaboration with A3. Besides of existing concepts identified in some standards (ISO 15704, EN/ISO 19439, EN/ISO 19440, etc.), concepts related to modelling interoperability requirements and solutions will be developed.

3.6 Partial enterprise model

Partial Enterprise Models (PEMs) (reusable reference models) are models which capture concepts common to many enterprises. The use of PEMs in enterprise modelling will increase modelling process efficiency (ISO 15704, 2000). One of the key results of ATHENA is to define a technologically neutral reference model that provides a stable, generic foundation for specific technical innovations. It will provide Guidelines and Best Practices, incorporating results from Technology Testing and implementation of this model. In particular the elaboration of this technologically neutral reference model will be based on semantic mediation solutions and provides components of interoperability infrastructures.

3.7 Enterprise models

Enterprise Models (EMs) are expressed in enterprise-modelling languages and are maintained (created, analysed, stored, distributed) using enterprise engineering tools (ISO 15704, 2000). In other words, enterprise models are models of particular enterprises and maybe created from enterprise reference models by instantiation or particularisation. In ATHENA, generic interoperability solutions proposed by A projects (A1-A6) will be moved to B5 (industrial test-sites) for testing and validation. To do this, enterprise models will be created to represent various industrial scenarios. For examples enterprise models representing collaborative enterprise interoperations (project A1) and cross-organisational business process interoperations (project A2) etc.

3.8 Enterprise module

Enterprise Modules (EMOs) are physical entities (systems, subsystems, software, hardware, and available human resources/professions) that can be utilised as common resources in enterprise engineering and integration. In general EMOs are implementations of partial models identified in the field as the basis of commonly required products for which there is a market. One set of enterprise modules of distinguished importance is the Integrating Infrastructure that implements the required Integrating IT Services (ISO 15704, 2000). In ATHENA, the interoperability infrastructure is a key result developed by A4 in collaboration with some other A projects (for examples, Model-driven and Adaptable Interoperability Infrastructure by A6, and Intelligent Infrastructure to implement core Enterprise Modelling languages and meta-model templates by A1).

3.9 Enterprise operational system

Enterprise Operational Systems (EOS) support the operation of a particular enterprise. They are all the hardware and software needed to fulfil the enterprise objective and goals (ISO 15704, 2000). In ATHENA, four operational systems representing four scenarios will be implemented to validate project results: (1) Supply Chain Management in Aerospace industry (EADS), (2) e-Procurement in Furniture SMEs (AIDIMA), (3) Collaborative Product Development in Automotive (FIAT), and (4) Product Portfolio Management in Telecommunications (INTRACOM). For example, Project A1 will implement an operational system in INTRACOM to experience product portfolio management interoperability.

3.10 Summary of mapping

Figure 2 shows the mapping of ISO 15704 GERAM components to the ATHENA Interoperability Framework (AIF). Reference architecture, methodology, modelling languages and concepts as well as reference models are all conceptual elements. They are used to build particular enterprise models of studied company. Modelling tools are technology component which support the model construction. The 'particular' enterprise model(s) is conceptual model(s) and is applicative in a particular domain. Enterprise models are then implemented in operational systems with enterprise modules (infrastructure for example) to support operational systems that perform daily enterprise operations. Both enterprise modules and enterprise operational systems are concrete technical (technological) elements.

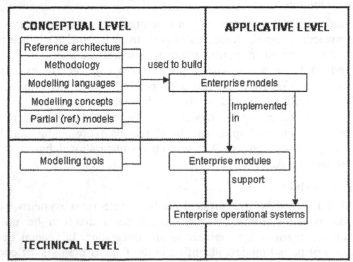

Figure 2. Mapping ISO 15704 framework to ATHENA Interoperability Framework

The table below summarizes the mapping and intuitively gives an evaluation on the degree of conformance. The '+++' means a perfect mapping and '+' indicates that the mapping is poor. '++' is in between.

Table 1: Summary of mapping

ISO 15704 (GERAM Framework)	ATHENA Research solutions	
Enterprise reference architecture identifies concepts of enterprise integrations	ATHENA reference architecture (A4), but also IT oriented architectures (A5, A6)	++
Engineering methodology describes process of enterprise engineering	ATHENA methodologies (A4 but also A1, A2, A3, A5, A6, B4)	+++
Enterprise modelling languages provide modelling constructs	ATHENA enterprise modelling languages (A1, A2 and A3 (semantic))	+++
Enterprise modelling tools support enterprise modelling and engineering	ATHENA enterprise modelling tools (A1, A2, A3), also A6.	+++
Enterprise modelling concepts define meanings of enterprise modelling constructs	ATHENA enterprise modelling concepts describing interoperability requirements/solutions (A1, A2, A3)	+++
Partial enterprise models provide reusable reference models for designing enterprise models	ATHENA technologically neutral reference model as generic foundation for specific technical innovations (A4)	+
Enterprise models are designed for a particular enterprise	ATHENA enterprise models for testing solutions (mainly A1, A2 and B5)	+++
Enterprise modules are implemented common enterprise system's components	ATHENA interoperability infrastructure (A4) but also Model-Driven adaptable infrastructure (A6)	++
Enterprise operational systems supports operations of a particular enterprise	ATHENA operational systems implemented by B5 in industrial sites (testing)	++

It should be noted that ATHENA can provide more value to the ISO 15704 Standard via the thorough and consequent using of the GERAM concepts and terminology – which might also lead to discovering open issues – as well as via a broad dissemination of the pilot and test bed results. The further development of the ATHENA Interoperability framework aiming at categorising and integrating ATHENA solution components (project A4) will also provide valuable inputs for future ISO 15704 revision.

4. ENTERPRISE MODELLING LANGUAGE (EML)

Among various research components, Enterprise Modelling Language (EML) has a special position because of increasing attention to model-driven or model-based architecture and application developments.

4.1 Initial result – UEML v1.0

The concept of UEML was born in 1997 in the frame of ICEIMT (Torino conference) organised in cooperation with NIST. UEML thematic network project

(UEML, 2001) was the first concrete action to develop UEML involving key Research Centres and some European tool providers in the domain of Enterprise Modelling. The aim of this first activity was to: (1) form an interest group of important Enterprise Modelling Stakeholder; (2) identify requirements on UEML and Enterprise Modelling in general; (3) define the first version of the UEML 1.0 meta model; (4) elaborate an UEML exchange format as the first prototype in order to analyse the feasibility of the UEML concepts.

The UEML 1.0 meta model was derived from the analysis of commonalities between the three involved Enterprise Modelling languages: GRAI, EEML and IEM. So for instance the GRAI GRID for decision support is not covered by the UEML 1.0. An initial set of UEML constructs (Berio, 2003) were identified as shown in Figure 3. The exchange of process models via the implemented XML format was mostly successful. However the exact transformation of the graphical data from one tool to the other was difficult to achieve.

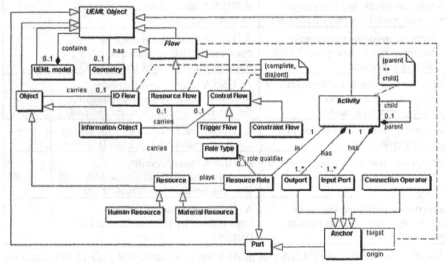

Figure 3. UEML constructs – UEML v1.0 meta model

4.2 Use of UEML v1.0 in ATHENA A1 project

The ATHENA A1 project (Knothe, 2004) will use experiences and results of the UEML Thematic Network Project in order to elaborate among others the Modelling Platform for Collaborative Enterprises (MPCE). The project is led by FhG-IPK of Berlin. Today there are three main points for adapting the approach to achieve enterprise modelling interoperability (Mertins *et al.*, 2004) (also see Figure 4):

- *Select and adapt a common Meta Meta Model.* This approach will ensure easier mapping by common basic concepts without restricting single Enterprise Modelling languages. Possible useful concepts are MOF (Meta Object Facility), RDF (Resource Description Framework) or OWL (Web Ontology Language). Whereas MOF provides a rigid framework for the meta model extension, RDF is a language for representing information about resources in the World Wide Web. RDF is used for identifying elements by using Web identifiers (called *Uniform*

Resource Identifiers, or *URIs*), and describing resources in terms of simple properties and property values. So the RDF language can be influence the principles implemented in the repository management system for finding modelling elements inside the repository. So the complexity of the interface to the modelling tools could be reduced. OWL is a W3C specification to define domain ontology's according to a formal description language. The advantage to RDF is the capability to define expressions similar to first order logic. For the extensibility of the POP* Repository in order to define domain oriented reference structures for easier model exchange this language could be a candidate for further analysis.

- *Define a wide range repository structure for storing enterprise models*. Common and non common modelling elements could be stored. So linked enterprise modelling tools must not cover the complete model. It should be possible to change only some dimensions of an enterprise model. The repository services have to ensure consistency of the enterprise model inside the repository structure by using reflective views.
- *Analyse the existing enterprise modelling standards to implement the repository structure*. E.g., EN/ISO 19440 or 19439 should influence the definition work. On the other hand the currently new emergent methodologies like BPDM (Business Process Definition Model) from OMG will be taken into account. The link of the repository to BPDM can enable direct links to the execution oriented levels. The objective here is to support execution of models parts, stored in the common repository. Especially the existing UML models will be useful for analysis.

A preliminary set of constructs focusing on process related concepts has been identified, based on the inputs from EN/ISO 19440, the OMG BPDM and UEML. These constructs will be refined and further extended to cover other concepts relating to interoperability of collaborative enterprises.

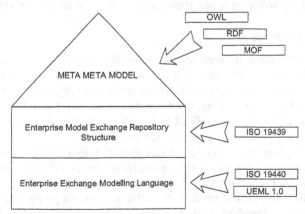

Figure 4: Athena A1 approach and related existing work that will influence the Exchange System development

The use of UEML in project A1 aims at tackling the interoperability problem between enterprise models and modelling tools (at higher abstraction level) providing a mapping mechanism. It also aims at vertical interoperability to allow generating workplaces from high abstraction enterprise models. Finally solutions provided by A1 should allow interoperability between workplaces so generated at run time level.

4.3 Develop UEML v2.0 in INTEROP

Besides of the use of UEML v1.0 in ATHENA project A1, within INTEROP Network of Excellence (Interop, 2003), UEML v1.0 is being further developed by WP5 (Common Enterprise Modelling Framework) led jointly by University of Torino and FhG-IPK of Berlin. Main objective of UEML v2.0 is to define constructs for modelling distributed enterprises for interoperability on the one hand, and on the other hand evaluate how the UEML can be used to support synchronisation of different distributed enterprise models. UEML v2.0 will be released at the month 18 (May 2005) and UEML v3.0 at the month 36 (November 2006). Furthermore, a strategy for 'UEML' extension and its assessment will be developed. Two main open issues/questions raised are (Berio, 2004): (1) Should UEML be an ontology for evaluating Enterprise Modelling Languages? (2) If UEML is not an ontology, which ontology should be used? Concerning the use of ontology, the State-of-the-Art indicated that existing ontology solutions which exhibit a high semantic adequacy such as MIT Process Handbook or Toronto Virtual Enterprise (TOVE) are too complicated for a practical and extensive industrial use. Current ontology languages are fairly weak in representing enterprise and business concepts (Ideas, 2003) As INTEROP WP 5 will only define some possible strategies new projects are needed for further development. Here the experiences of ATHENA A1 can be used in order to develop concepts that are suitable for the industry..

Another research activity relating to UEML v2.0 is currently performed by WP7 (led by GRAISOFT) in INTEROP. It aims to generate Customised Enterprise Application from enterprise models. The use of UEML v2.0 is expected to allow not only the mapping between enterprise models, but also providing transformation mechanism linking software application components to enterprise model content.

5. CONCLUSION

This paper has presented an overview on ATHENA Project and tentatively mapped ATHENA to ISO 15704. The mapping developed is rather straightforward using the most salient characteristics known today for each category, and is expected to be further refined. Globally speaking, the mapping is successful and ATHENA solution components fit well within ISO GERAM framework. However the mapping also encountered some terminology problem. On the one hand, terms used in ATHENA are not fully compliant with ISO 15704 (for example, tool in ATHENA has a broad meaning and not only refer to enterprise modelling tool). On the other hand, some terms used in ISO 15704 may lead to some misunderstanding (for example enterprise module and partial model). Nevertheless this mapping contributes to the use of ISO 15704 standard and a better categorisation of ATHENA project solutions.

Acknowledgments

This paper is partly funded by the E.C. through the ATHENA IP. It does not represent the view of the E.C., and authors are responsible for the paper's content.

ATHENA IP is funded by the European Commission, under the 6th Framework R&D Programme, contract n° 507849. The authors thank and acknowledge the members of the ATHENA consortium: SAP (D), AIDIMA (E), Computas (NO), CR Fiat (I), DFKI (D), EADS (F), ESI (E), Formula (I), IPK (D), Graisoft (F), IC Focus (UK), Intracom (EL), LEKS (I), SINTEF (N), TXT (I), Univ. Bordeaux 1 (F), UNINOVA (POR), IBM (UK), SIEMENS (D).

REFERENCES

ATHENA (2004), Advanced Technologies for Interoperability of Heterogeneous Enterprise Networks and their Applications, FP6-2002-IST-1, Integrated Project Description of Work.

Berio, G. *et al.* (2003), UEML Deliverable D 3.1 Requirements analysis: initial core constructs and architecture, 2003.

Berio, G. (2004), Report on the received contributions from INTEROP WP5 Partners, Version 1.0, 10 April 2004.

Karlsen, D.; Lillehagen, F. (2004): Implementing the AKM technology.

Knothe, T., *et al.* (2004), ATHENA A1 project description, ATHENA project internal document, 2004.

IDEAS (2003), IDEAS Project Deliverables (WP1-WP7), Public Reports, www.ideas-road map.net

INTEROP (2003), Interoperability Research for Networked Enterprises Applications and Software, Network of Excellence, Proposal Part B, April 23, 2003.

ISO 15704 (2000), Industrial automation systems - Requirements for enterprise-reference architectures and methodologies, ISO 15704:2000(E).

Mertins, K.; Knothe, T.; Zelm, M. (2004), User oriented Enterprise Modelling for Interoperability with UEML, the EMMSAD'04, RIGA, Latvia, June, 2004

OMG (2004), Meta Object Facility (MOF) Specification, http://www.omg.org/docs/formal/02-04-03.pdf

UEML (2001), UEML Thematic Network - Contract N°: IST – 2001 – 34229, Description of Work.

W3C (2004), Resource Description Framework (RDF): Concepts and Abstract Syntax http://www.w3.org/TR/rdf-concepts/

7. Architectural Principles for Enterprise Frameworks: Guidance for Interoperability

Richard A. Martin[1], Edward L. Robertson[2], John A. Springer[3]

1 Tinwisle Corporation Email: tinwisle@bloomington.in.us
2 Computer Science Department, Indiana University Email: rbstn@cs.indiana.edu
3 Computer Science Department, Indiana University Email: jospring@cs.indiana.edu

This paper presents a number of principles related to the construction and use of enterprise architecture frameworks. These principles are intended to guide the development of a formal foundation for frameworks but also serve as guidance for efforts to enable the interoperability of enterprise models and model components. The principles are drawn from analyses of a number of existing frameworks and from observation of and participation in framework development.

1. INTRODUCTION

An *enterprise architecture framework* is a means to understand an enterprise or class of enterprises by organizing and presenting artifacts that conceptualize and describe the enterprise. An *enterprise*[34] is a collective activity in a particular domain, with actors sharing a common purpose; an enterprise can be a business, a collection of businesses with a common market, a government agency, etc. *Architecture* is a metaphor to the realm of office towers and bridges, intended to capture the use-oriented, as opposed to construction-oriented, aspects of the design of those structures. A *framework* is a structured container for holding and interconnecting things[35] – in the remainder of this document those things are *artifacts* that comprise the enterprise architecture. In framework contexts, artifacts are almost always models of some kind, which we sometimes call "components" to indicate that they are pieces of the entire framework. These artifacts are conceptual, logical, and physical representations at all levels of the enterprise and range from simple lists through elaborate data models, tools supporting methodologies, and operating procedures. In the following, "framework" will always be shorthand for "enterprise architecture framework".

Frameworks have been widely used. The Information Technology Management Reform Act of 1997 led to the U.S. Government's Federal Enterprise Architecture

---

[34] The word "organization" is a common synonym for enterprise, but we must often use "organization" to denote the way things are organized and thus restrict it to that use.

[35] As another metaphor, think of a framework for electronic components which both holds circuit boards and provides for wiring between those boards.

Framework (FEAF), which "describes an approach, including models and definitions, for developing and documenting architecture descriptions" (U.S. GAO, 2003). It is being deployed in all non-military agencies of the U.S. Government. The annual ZIFA Forums (ZIFA, 2004) have included nearly 100 case studies highlighting the benefits of frameworks. Bernus *et al.* (Bernus, 2003) give several thorough case studies (along with an extensive discussion of enterprise architecture issues). Whether the frameworks address manufacturing operations, process control, information systems, or government bureaucracy, the artifacts produced to describe the enterprise comprise a valuable asset requiring its own distinct management. Managing and gaining full value from that asset is the reason enterprise architecture frameworks are conceived, built, and used.

Professional practice has taught us about the fragility of isolated application silos on islands of automation and about the difficulty in achieving interoperability under such circumstances. While these are typically called "data silos," the significant problem is that they are in fact model silos. That is, the mismatch of underlying models is the greatest impediment to integration and interoperability.

In spite of their wide use and importance, frameworks have all been defined only descriptively. This means that it is currently impossible to formally relate different frameworks, to say nothing of implementing tools that properly support these frameworks.36

This work is about frameworks in general and not about any one particular framework. Although our original motivation was the Zachman Framework for Enterprise Architecture (Zachman, 1987, ZIFA, 2004), we examined and incorporated several other frameworks, which are itemized in Section 2. Moreover, this work is about structure and not about contents. Thus "framework" by itself indicates a collection of descriptions and principles for organizing framework contents while "framework instance" indicates the use of a framework describing one particular enterprise.

The primary goal of this paper is to identify the guidance for interoperability that the principles elicit. Such guidance follows from the understanding of frameworks and framework formalization that led us to the use of frameworks to support organization and interaction of the many models associated with an enterprise. This work continues our effort to formalize the ways in which these particular frameworks manifest the architecture of an enterprise (Martin, 1999), with an eye toward (i) connecting a framework instance's contents, (ii) manipulating those contents and connections, and hence (iii) relating different frameworks and recasting instances from one framework standard to another. While our primary motivation for developing these principles is to use them to guide our formalization activities, we believe that many are directly useful in the development of individual frameworks and for enabling interoperability among framework instances.

Section 2 begins this paper with a discussion of the origin and (to the extent possible) validation of the principles. Section 3 introduces a few principles that are

---

36 There are software packages that purport to implement various frameworks, but these packages only implement the "holding" aspect of frameworks. That is, they are tools for editing and managing representations which populate a framework instance, without respect to the semantics that the framework provides.

general in nature, applicable to any modelling and analysis endeavour,37 while Section 4 discusses principles especially pertinent to frameworks. We then conclude this document by considering how these principles guide the formalization of frameworks and efforts to enable interoperability.

2. ORIGINS OF THE PRINCIPLES

The principles described below come from (i) evaluation and comparison of different frameworks, (ii) observation of the process of defining frameworks, and (iii) participation in this same process.

Principles are largely based on analysis of the framework architectures: Zachman (Zachman, 1999), an ISO draft standard titled Enterprise Integration - Framework for Enterprise Modelling (ISO 19439, 2004), ISO Standard 15288 Information Technology – Life Cycle Management - System Life Cycle Processes (ISO/IEC 15288, 2002), and the U.S. Department of Defense C4ISR Architecture Framework (US DoD, 1997), an analysis which we reported in (Martin, 2002, Martin, 2003).38

Principles are also based on professional observation and participation -- often experience of the difficulties which arise when these principles are not followed. Meeting minutes from ISO efforts illustrate such difficulties, as in the statement "Something is not very clear the distinction between the interoperability of process models and the interoperability of processes" (WG1, 2003), which reflects principle 3.4 about meta-levels. Our own professional experience includes constructing and analyzing models in an enterprise context, teaching modeling, and participating in the development of international standards for enterprise architectures.39

We do not claim to have originated all these principles. Several are simply our statements of well-established suggestions (e.g. 3.6, "Do not hide architecture in methodology", which is a rephrasing of the data independence principle (Date, 1981)). Principles reflecting some of the same concerns as ours have been identified elsewhere (Greenspan, 1994, ISO TR 9007, 1987, Totland, 1997), although these other principles are largely directed at ensuring the fidelity of the modelling process.

Occasionally specific facts are given in evidence. Only a few principles can be supported so concisely. One such principle 4.6, that states the independence of three commonly correlated scales, is supported by examples high in one scale but low in another. Unfortunately, principles that describe general behaviour do not admit such concise support. This is very loosely similar to the difference between existential and universal propositions, in that one instance proves the former.

Perhaps the most insightful principle is principle 4.4, which recognizes that analytical partitioning uses both grids and trees. We first observed this duality in the context of adding detail within a Zachman framework (Inmon, 1997), necessitating the use of recursion within a frame. This principle has been validated by its use in comparing frameworks (Martin, 2003) and its value in the development of

---

[37] We are still using "framework" as shorthand for "enterprise architecture framework", but it would be a valuable exercise to see which of these principles hold for other classes of frameworks.

[38] Space limitations make it impossible to repeat that analysis here.

[39] Richard Martin is convener of TC 184/SC 5/WG 1, "Modeling and architecture", of the International Standards Organization.

international standards (Bernus, 1996), particularly ISO 15704:2000 Industrial Automation Systems Requirements for Enterprise Reference Architecture and Methodology(ISO 15704, 2000).

Many principles focus on highlighting and refining distinctions (such as principle 3.5, which distinguishes dependency and temporal order). They arise from observation of the ways in which people model, and the successes and the difficulties encountered therein.

Principles may be descriptive, describing the way that model artifacts are constructed and organized, or prescriptive, recommending how they should be. However, prescriptive principles all began as observations of the form "People have trouble with ...". Prescriptive principles of course guide practice; but they also guide the formalization effort, indicating what should be facilitated or discouraged.

3. GENERAL PRINCIPLES OF MODELLING

Modeling as we mean it is a conceptual exercise, only analogously related to physical modeling as in, say, model railroads.[40] Conceptual modeling does yield representations in a particular medium, not necessarily a medium with physical manifestations, but these are representations of the modeled concepts. Thus principles apply to both concepts and representations.

Each of the following principles begins with a short phrase (indicated in that manner) which identifies and hopefully summarizes the principle. More extensive discussion of the respective principles, including evidence for them, is given in our EMMSAD04 paper or technical report (Martin, 2004a, Martin, 2004b). Much of this paper originates in those works as well.

3.1 Communication is a goal of modeling

Models (including frameworks) are formal artifacts but they are developed and used by people. Therefore any modeling formalism must be robust and tractable in interaction with non-formal components – people. This principle is discussed at great length in (Totland, 1997) and related psychological factors are discussed in (Siau, 1999).

3.2 Complexity tradeoff

There is typically a tradeoff between complexity in the modeling medium and complexity in model instances constructed using that medium. Modeling mechanisms therefore should be defined with an attempt to find a "sweet spot" where these complexities are in balance.

3.3 Naming matters

Naming, i.e. the assignment of a string[41] to a concept or artifact, serves as the bridge between formal artifacts and human interpretation. That is, there are two sides to naming: "external" (relating to the real world) and "internal" (relating to the

---

[40] We draw this distinction because, for most people, the first connotation of "make a model" is to construct a model railroad or something similar. Model railroads diminish function but primarily reduce physical scale; indeed, the first descriptor applied to a model railroad is its "gauge", or physical scale.

[41] We do not use "label" because we want to restrict that term to a specific use.

mechanism and models of a framework). Said another way, internal naming involves formal meaning while external naming involves human understanding of that meaning.

Both sides of this principle impact interoperability. Internally, interoperable components must interpret names consistently across the interaction, hence the emerging emphasis on formal ontological methods to resolve semantic consistency. Externally, human mediated interoperability depends upon the correct assignment of actions to messages received and the creation of messages that convey the intended semantics to the receiver. Whereas the internal context should be well defined, the external context is often ambiguous.

3.4 Use "meta" with great care, because the term is seriously overloaded

This particularly applies when discussing meta-levels. This is particularly true because "meta" is a relative term, not an absolute.

One obvious example of the relativeness of "meta" is observable in the realm of ER modeling. There, the meta-model level decomposes all models into Entities and Relationships; the model level may decompose a particular model for corporations into Department, Employee, Project (instances of Entity), Works For (instance of Relationship), etc.; the model population level (for a fixed corporation) into Sales, Human Resources, Accounting, etc. (instances of Department). Thus the model level is meta with respect to the model population and ER notation is the meta-meta level for the model population. Notice that "instance" is also a relative term, in that it does not indicate an absolute level but only the level below X when used in the phrase "instance of X". Also, "meta" is roughly the inverse of "instance of", in that the meta of an instance of X is in fact X . However, since our interest focuses on models and meta-models, henceforth "instance" shall denote artifacts at the model level; that is, Department, Employee, Works for, etc in the above example.

3.5 Dependency is not chronology

That is, just because B depends upon A, it is not necessary that B follows A in time. While much of the evidence for this principle comes out of difficulties arising when it is not followed, ISO 14258 Industrial automation systems – Concepts and rules for enterprise models, makes this distinction explicit (ISO 14258, 1998).

3.6 Do not hide architecture in methodology

It is wrong to bury characterizations of things in methods that are used to construct them. This is not to claim that methods do not constrain results (to claim so would be most foolish) but rather to observe that such constraints must be made explicit and external to the construction process. In particular, the architectural form should survive changes in method and technology. Thus the link between architectural form and interoperability is very strong. Robust interoperability should also survive changes in method and technology.

4. PRINCIPLES SPECIFIC TO FRAMEWORKS

4.1 Frameworks organize artifacts

A framework is a means to facilitate understanding of enterprises and to communicate that understanding, principally by organizing and connecting artifacts used to represent a particular enterprise. Frameworks help us to take very richly textured descriptive and prescriptive artifacts and arrange them for practical understanding. Frameworks help to simplify complex artifact collections that are composed of many inter-related components. The organizational mechanism of a framework is primarily a collection of dimensions along which the artifacts are placed and hence classified. It is in the number and different natures of these dimensions that frameworks vary. Many further principles relate to the characterization of these dimensions.

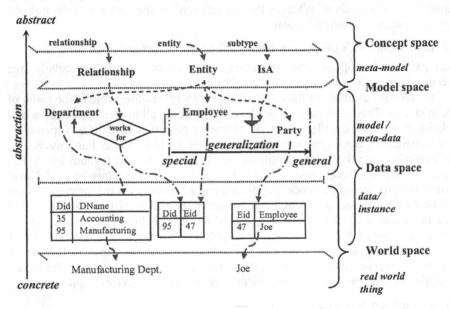

Figure 1 Relative meta-

4.2 Distinguish structure from connectivity

Structure and connectivity are distinct aspects of frameworks42 and a framework formalization (or standard) should distinguish them. The clarity of this distinction directly impacts the quality of a framework; unfortunately many frameworks do not achieve their intended impacts because they do not exhibit this distinction with sufficient clarity. Furthermore, useful reorganizations of a framework, one of many viewing mechanisms, can be tractably expressed when phrased in structural terms, whereas desired views involving connections may be difficult to specify and expensive to compute.

4.3 Separate policy from mechanism

That is, policy should be found in framework contents and not framework structure.

---

[42] We find it helpful to visualize a computer room where frames both hold devices (servers, disk drives, communications interfaces, etc.) and provide channels for wiring these devices together. A second metaphor is between bone (structure) and muscle (connection); this emphasizes that operation largely occurs through the connections.

4.4 Two aspects of organization

There are two general ways in which items within a framework are (typically) arranged: (i) in an ordinant structure (that is, a table, grid, or matrix) or (ii) in a decompositional structure (that is, a tree). We call either of these dimensions of the arrangement. Dimensions of either kind are discrete43 and ordinant dimensions typically have only a few coordinate positions. The coordinate positions of an ordinant dimension may be ordered (e.g. rank) or unordered (e.g. gender), while a decompositional dimension is always ordered only by its containment relation.

An important step in organizing artifacts is to identify and characterize (as ordinant or decompositional) the dimensions that define the structure. The definition of an ordinant dimension is the identification of its coordinates and, where relevant, the order of those coordinates. Recall that dimensions only describe the placement of items (in a real or conceptual space) and not the interconnection of these items, which is typically much richer and more complex.

Given this distinction in structural arrangement and the two principles that follow, it seems critical that structural alignment be essential for interoperability. Context is a structural characteristic of frameworks and the semantic interpretation of content is highly dependent upon context.

4.5 Decomposition may occur at many meta-levels

That is, it is natural and expected that there be meta-level and model-level decompositions (from whatever perspective "meta" is considered). For example, saying that the <conceptual; what> cell of a Zachman frame contains Entities and Relationships is a meta-level decomposition of that cell, while saying that Employee and Department are Entities is a model-level decomposition.

4.6 Three aspects of scale

There are (at least) three distinct dimensions that reflect conceptual (as opposed to physical) scale: (i) abstractness, ranging from abstract to concrete, (ii) scope, from general (generic) to special (specific), and (iii) refinement, from coarse to fine. Using the terminology of principle 4.4, abstractness, and scope are ordinant-ordered and refinement is decompositional.[44]

Because it is common to have co-occurrence of the origin or extreme endpoints in all three dimensions (as a module that is concrete, specific, and finely refined), these three dimensions are often confused. Understanding (and distinguishing) conceptual scales is essential because they govern the ways in which framework dimensions are conceived, ordered, populated, and constrained.

---

[43] This statement necessarily holds for decompositional dimensions but is sometimes relevant to distinguish meta-coordinates from instance coordinates where ordinant dimensions are involved.

[44] In fact, refinement is often the canonical hierarchy.

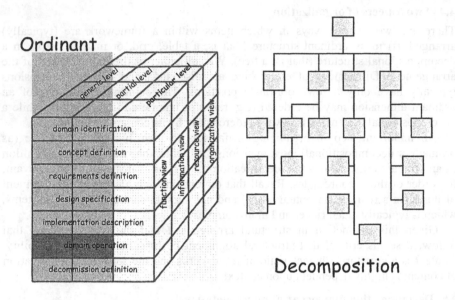

Figure 2 Two aspects of organization

4.7 One dimension manifests purpose within a framework

One, and typically only one, of a framework's ordinant-ordered dimensions reflects the purposive nature expressed within a framework. Note that such a "purposive dimension" does not represent the purpose of the framework but instead represents the fact that artifacts derive their purpose from artifacts earlier in the dimension's order (most often through elaboration). Derived dimensions, produced through views (see principle 4.11 below), may also exhibit a purposive order; the C4ISR's "Force Integration" dimension, derived from a command-structure hierarchy, exhibits the purpose inherent in any chain of command.

The ordering of a purposive dimension often manifests itself as causality, dependency, or chronology. However, it is not merely a time dimension, even though purpose in a framework often leads to temporal ordering in the operations of the enterprise. This indeed follows from general principle 3.5.

4.8 Refinement is recursive

The decompositional scale dimension, refinement, is fundamentally different in that it works (or at least works best) through decomposition and successive refinement. Thus frameworks should be recursive in their application. Unfortunately, practice often foreshortens the recursion, forcing a fixed (albeit hierarchical) or flattened structure.

Recursion also has an impact on contextual alignment for interoperability. Erroneous assumption of recursive level during interactions is as destructive to automation outcomes as it is to human mediated activities.

4.9 All context is relevant

It seems necessary, as one moves through a framework along its purposive dimension, from row to row in a Zachman framework for example, that the entire framework structure at one row is *potentially* relevant when describing a component at the next. This is not to claim that an entire row is in fact materially relevant for each component in the next; it is merely recognition that all of the models from prior coordinates can be useful in understanding and constructing the next. Moreover, it is sometimes as important to know which concerns are not needed as it is to know which are. Perhaps this principle just reflects the fact that frameworks, and the enterprise domains with which they are concerned, are not suitable for minimally descriptive artifacts.

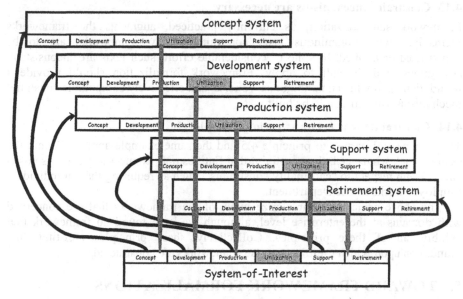

Figure 3 Recursion in ISO 15288 cf.

4.10 Connections can be of arbitrary arity

Connections between framework artifacts can be of arbitrary arity, although binary ones are most common. However, it is sufficient to provide for the construction of arbitrary connections using binary ones. For example, a Relationship in an ER model may be constructed to have any degree, but the basic connections are always between a single Entity and a single Relationship.

4.11 Views are important in standards and methodologies

A framework formalism should provide a general mechanism for defining views. Views are used in enterprise modeling because the complexity of an enterprise makes it impossible for a single descriptive representation to be humanly comprehensible in its entirety. The notion of view is inherent in any large, complex structure observed and managed by many individuals who neither can nor should attempt to analyze, design, or implement the entire structure.

The view mechanism should be general and dynamic. It must be general because there is little commonality of particular views across frameworks. It must be dynamic both because new views arise as standards are extended and because ad hoc views are requested. Just as view mechanisms provide the content for integration, so too will systems and components become interoperable through view mechanisms. To act on behalf of another seems to require some mechanism for perception that goes beyond simple enumeration of content.

4.12 Construction through views

Views are not merely used for viewing; they are often used for constructing and populating frameworks.

4.13 Constraint mechanisms are necessary

Framework standardization, as currently practiced, augments the frameworks themselves with voluminous texts constraining how frameworks are to be constructed or applied. In spite of considerable effort, such texts are inconsistent, ambiguous, and difficult to apply. Framework formalization should provide a foundation upon which unambiguous, concise, and effectively computable constraint mechanisms can and should be built.

4.14 Constraints may occur at various meta-levels

This is a natural partner to principle 4.5 and the same example applies. Within the <conceptual; what> cell, the constraint that Entities only connect to Relationships is "meta" with respect to "cardinality constraints", such as requiring that an individual Employee works in one Department.

The above principles characterize many of the frameworks that are concerned with domains at the enterprise level, although we have found no framework that exhibits all of these principles. Collectively, these principles constitute the foundation upon which useful enterprise frameworks are constructed.

5. TOWARD FRAMEWORK FORMALIZATIONS

While the previous sections discussed principles obtained from observation and analysis of existing frameworks, this section outlines how these principles guide formalizing enterprise frameworks. Although the individual framework instance is of course the formalized artifact, the following discussion is directed toward "architectural" standards that prescribe how a collection of frameworks is to be formalized.

There are four major aspects of a formalism that follow from the above principles. We itemize these four and justify why they should be treated distinctly. The long version of this paper then delves more deeply into these four aspects (Martin, 2004b).

- **structure**: the way that components and sub-components of an enterprise are placed within a framework. Principles 4.4 – 4.8 guide the elaboration of this aspect.
- **connections**: the manner in which components and sub-components of an enterprise are interconnected within a framework. It is through these connections that the operations of an enterprise are manifest.

- **views**: formal mechanisms for restructuring a framework to emphasize features from a particular conceptual or operational perspective.
- **constraints**: formal mechanisms by which the conformance of a particular instance to a standard or architecture may be evaluated.

The deliberate separation of structure and connections is a direct consequence of principle 4.2. A framework is thus a structure for holding artifacts and a mechanism for connecting them.

The needs for views and constraints are enunciated in principles 4.11 and 4.13 respectively. While it is necessary to draw distinctions between structure and connections, it is advantageous to do the opposite, drawing parallels between views and constraints. In particular, the ability to define views immediately enables constraints definable in terms of views, as in "view A is a subset of view B".

A formalism for framework structures provides the foundation upon which formalizable, and therefore precise and coherent, view mechanisms can be built; and, conversely, view mechanisms provide the formalism through which one single overarching structure is coherently and consistently created by these many individuals.

6. CONCLUSION

We have identified twenty principles about the ways in which enterprise frameworks are or should be constructed and used, but this is only one step on a longer path. These principles will guide the formalization of frameworks, as discussed in section 5, but we are early in the work of that formalization. It is evident that the structure of a framework is carried by a tree whose nodes have a tabular, dimensional form, but many details governing the expression of structure and the interaction of this expression with connections, views, and constraints are yet unknown. Because existing frameworks do not treat connections in a disciplined manner, there is less guidance concerning connections from existing practice.

Interoperability, that is the automatic operation of agents from one enterprise in the context of a second, can be facilitated through enterprise framework principles in two ways. The first, and by far the preferable, way is to enable the automated agent to "understand" the second enterprise's context. Unfortunately, the mere presence of frameworks does not guarantee this. The second, and always available, way is that the frameworks facilitate true human understand even if such understanding is not immediately automatable. That is, a variety of implications of the above principles, such as having model artifacts specifically identified through frameworks, knowing the dimensional structure of frameworks, and having constraints articulated, facilitates human specification of the integration that is a precursor to interoperability.

Because these are principles, we expect situation specific exceptions. Models of every kind are most often incomplete and imprecise representations expressed using available tools and media. To the extent that these principles guide a better understanding of the structure, connections, views and constraints embodied in a modern enterprise, they can add precision and completeness to the expression of that enterprise. And finally, it is important that the formalization attempts to reach "sweet spots", as discussed in principle 3.2.

In as much as the principles enunciated herein are the core of a "requirement specification" for analysis and formalization of enterprise frameworks, we welcome all suggestions and comments.

REFERENCES

Bernus P, Nemes L, Williams T J, editors (1996) Architectures for Enterprise Integration, Chapman and Hall, London

Bernus P., Nemes L., & Schmidt G., editors (2003) Handbook on Enterprise Architecture. Springer Verlag, Berlin

Date, C. J. (1981) *An Introduction to Database Systems.* Addison-Wesley

Greenspan S. J., Mylopoulos J., & Borgida A. On formal requirements modeling languages: RML revisited. In *International Conference on Software Engineering*, pages 135--147, 1994.

Inmon W., Zachman J., & Geiger J. (1997) Data Stores, Data Warehousing, and the Zachman Framework. McGraw-Hill

ISO 14258 (1998) Industrial Automation Systems – Concepts and rules for enterprise models, International Oragization for Standards, Geneva.

ISO 15704 (2000) Industrial Automation Systems – Requirements for Enterprise Reference Architecture and Methodology, International Organization for Standardization, Geneva

ISO 19439 (2004) FDIS Enterprise Integration – Framework for Enterprise Modelling, International Organization for Standardization, Geneva.

ISO/IEC 15288 (2002) Information Technology - Life Cycle Management - System Life Cycle Processes. International Organization for Standardization and International Electrotechnical Commission, Geneva

ISO TR 9007 (1987) Concepts and Terminology for the Conceptual Schema. International Organization for Standardization, Geneva, No long available through www.iso.ch.

Martin R. & Robertson E. (1999) Formalization of multi-level Zachman frameworks. Technical Report 522, Computer Science Dept., Indiana Univ., 1999. www.cs.indiana.edu/ftp/techreports/TR522.html.

Martin R. & Robertson E. (2002) Frameworks: Comparison and correspondence for three archetypes. In ZIFA 2002 Enterprise Architecture Forum, 2002.

Martin R. & Robertson E. (2003) A comparison of frameworks for enterprise architecture modeling. In ER2003 22nd Intl. Conf. on Conceptual Modeling, pages 562--564.

Martin R. A., Robertson E. L., & Springer, J. A. (2004a) Architectural principles for enterprise frameworks. In *CAiSE Workshops Knowledge and Model Driven Information Systems Engineering for Networked Organizations*, Janis Grundspenkis & Marite Kirikova (Eds.), Riga Technical University, Lativa

Martin R. A., Robertson E. L., & Springer J. A. (2004b) Architectural principles for enterprise frameworks. Technical report, Computer Science Dept., Indiana Univ., www.cs.indiana.edu/ftp/techreports/TR594.html.

Siau K. (1999) Information modeling and method engineering: A psychological perspective. *J. of Database Systems*, 10(4):44--50.

Totland T. (1997) *Enterprise Modeling as a Means to Support Human Sense-making and Communication in Organizations*. PhD thesis, Norwegian University of Science and Technology, Department of Computer and Information Science.

U.S. Department of Defense - Architecture Working Group (1997) Command, Control, Communications, Intelligence, Surveillance, and Reconnaissance (C4ISR) Architecture Framework, Version 2.0

U.S. General Accounting Office (2003) GAO-03-584g Information Technology: A framework for assessing and improving enterprise architecture management. Washington, D.C.

WG1 (2003) Meeting Minutes St. Denis. International Organization for Standardization TC 184, SC 5, WG1, available at forums.nema.org/~iso tc184 sc5 wg1.

Zachman J. A. (1987) A framework for information systems architecture. IBM Systems Journal, 26(3), 1987.

ZIFA (2004) The Zachman Framework. Zachman Institute for Framework Advancement. Various pages at www.zifa.com; the "Quickstart" is particularly relevant.

8. UEML: a Further Step

Giuseppe Berio

Dipartimento di Informatica Università di Torino berio@di.unito.it

This paper presents a further step towards a UEML (Unified Enterprise Modelling Language) starting from the result of the UEML project, funded by the European Commission under the IST-Vth Framework Programme of Research. Specifically, the paper provides the basic theories and thinking underlying the project work as well as current improvements based on a data-integration perspective.

1. INTRODUCTION

Many problems raising in *Enterprise Integration* (EI) and *Enterprise Engineering* (EE) are certainly due to the fact that there are many *Enterprise Modelling Languages* (EMLs) and *Enterprise Modelling Tools* (EMTs), spanning from industrial analysis, management analysis, strategic planning, human management, budget and so on. Probably, the reasons are: because there is no one language covering all the aspects required for modelling and on the other hand, each EMT with the specific language is able to perform specific analysis or have more or less direct link to *enterprise software tools (i.e. enterprise software applications)*. The major need facing to this situation is probably an environment in which both *enterprise models* and *enterprise tools* can be integrated. In this way, integrated models have more chance to be used and maintained in a consistent ways. However, while *integration of tools* is usually perceived as really useful in practice and feasible, *integration of models*[45] is the major challenge because focusing on the integration of the *content of models*.

The idea of a UEML (*Unified Enterprise Modelling Language*) for improving the situation described above was born in the context of *ICEIMT initiatives* (Petit *et al.*, 1997), further advocated and explored in recent papers, e.g. (Vernadat, 2002). However, the *first project* on such a UEML started in 2002, funded by the *European Commission* under the *IST-Vth Framework Programme of Research*.

This paper reformulates the approach undertaken in the UEML project in term of *data-integration* (Calvanese, *et al.*, 2002, 2003) In fact, results of the UEML project are really close to some data-integration approaches. Then, the paper presents a possible further step towards the introduction of *Enterprise Reference Architectures* (ERAs) (ISO 1998, IFAC-IFIP Task Force, 1999). The interest of ERAs is to represent a coherent set of distinct modelling purposes for the same language (such

---

[45] In this paper integration of models comprises the *exchange of models* between distinct EMTs.

as a UEML). With ERAs, it is therefore possible to differentiate between the language and its various purposes.

The paper is organised as follow. Section 2 describes problems to be approached by a UEML. Section 3 summarises the foundations of a UEML as defined in the UEML project: the section especially describes the links between modelling languages and databases. Section 4 provides an overview about data-integration approaches; it also describes some simple examples. Section 5 describes how the approach undertaken in the UEML project can be reformulated in term of data-integration. Section 6 describes how data-integration approaches allow to easily introduce ERAs with a UEML. Finally, section 7 summarises the contributions of this paper.

2. UEML: PROBLEMS

The state of the art (Petit *et al.*, 2002b) issued from the UEML project reveals that distinct tools and languages are required because they can be used for achieving, probably in the best way, specific objectives (i.e. UEML should not substitute existing languages: though, UEML should be focused on model integration);

- most of the languages for enterprise modelling are not formalised in their semantics (i.e. the semantics is not "mathematically" described) but they are in their syntax i.e. it is known what is a model and what is not a model but it is less known if the model is meaningful or not;
- this lack of semantics is managed by a correct understanding and usage of modelling methodologies (methods) which allows to make right models;
- some semantics is added to models under specific purposes (for instance, simulation of models) but this semantics is not explicitly stated (e.g. it is part of simulation tools);
- it is very difficult and probably impossible to provide one formal semantics which is good for every purpose;
- enterprise models are intended for a broad usage even by humans; models can be used for teaching how the work should be, what an enterprise is, how it evolves, why it evolves, how it can be improved and so on.

As a consequence, the first problem to be approached by a UEML is that *models made in languages with no semantics* or an *informal semantics* or *hidden semantics*, are *more or less free of interpretation*. The second problem is that, having some *formal semantics* for languages and models is not enough. In fact, two distinct models with the same *mathematical semantics* (i.e. formal) may be partially equivalent in term of the *real world phenomena they represent*. A practical test for understanding this problem is a *process with just three activities*: mathematically, it may mean that there are three *activities in sequence*. However, two employees looking to this process may interpret it in distinct ways, just because the *names of the activities* suggest that the right interpretation *is not a sequence* but it is a *sequence of requests* that may be not fulfilled. There, w*here is the semantics* (Ushlod, 2003)? Part of the semantics is likely to be in the names of the activities which can only be interpreted by the employees (because of their knowledge). Therefore, given an enterprise model, its *context of interpretation*, mainly distinguished in *machines* and *humans*, still remains an important aspect.

Researches about *ontologies* (Gruber, 1993, Guarino (Ed.), 1998) try to take into account the context of interpretation of a model both for humans and machines. In fact, this is part of the following definition: *"an ontology is a specification of a conceptualisation shared by a community"*. Much of the *power of ontologies* is in the fact every ontology specification should be "shared by a community". However, what does "sharing by a community" mean? What is a "community" and how is it organised? Interesting examples come from the *laws genesis*. You have who defines the laws, you have structures able to interpret the laws in the *context, at different levels*, in a *continuous cycle*: however, contradictions are possible and acceptable. Therefore, generally speaking, *full sharing* by a *community of humans* seems to be rather difficult to be achieved. However, we may constrain as much as possible the conceptual domains and try to achieve full sharing for a very limited number of concepts. Nevertheless, it is not clear if the complexity is in the number of concepts or is in the inherent complexity of the concepts (Corrrea da Silva, *et al.*).

Table 1: Analogies between modelling languages and databases

Database Glossary	**Modelling Language Glossary**	**Samples of Model, Model artefacts and Meta-model**
Data (instances)	Model artefacts	"Mount", "Employee", "Robin"
Database (a set of related instances)	Model (a set of related model artefacts) Models (a set of models)	{"Mount", "Dismount", "Mechanical Part", "Employee", "Robin", "Edgar"}
Schema	Meta-model (representing the way for providing an *abstract syntax of a language*)	Activity, ObjectClass, Role, Object
Integrated schema	Integrated meta-model	Activity, ObjectClass, Role, Object

The UEML project team has performed a study concerning possible solutions to the problems mentioned above. On one hand, this study especially recognises that the interpretation of an *enterprise model* is provided by its *context of interpretation* (machines or humans). Whenever there are various contexts of interpretation and various models, an *integrated enterprise model* (better defined in the remainder) is a way to guarantee a *consistent usage of distinct models within distinct contexts*. To make integrated enterprise models, a UEML is a prerequisite. On the other hand, the study recognises the fact that *extensive formalisations of semantics* of a UEML is probably not a key point towards effective solutions to model integration. Some of the reasons are technical (Berio, Petit, 2003). In fact, if it is needed to *formally check a property*, the techniques for checking this property are really various and based on distinct formalisations (because *incompleteness* of some formalisations or *complexity* of the checking technique). Thus, while formalisations are needed for proving if an integrated model makes sense (i.e. no contradiction occurs), satisfies some properties, and they might also be useful for driving the model integration, they do not allow to infer how model integration should be performed and what an integrated model should be.

3. UEML: FOUNDATIONS

Table 1 states some analogies between databases and modelling languages. These analogies suggest that within the *database area*, *database integration*, *database architectures* and *schema integration techniques*, have already approached the two problems that should also be approached by a UEML.

These analogies can be stated because a syntax of a EML has not meaning per se: its meaning is given elsewhere (i.e. by the context of interpretation). As data, a syntax of a EML is interpreted by users or software components in various ways: some of these ways are correct, some ones are not; some ones are mutually consistent, some ones are contradictory. The main difference between databases and languages is probably that databases have a narrow scope than languages.

4. DATA-INTEGRATION

The objective of this section is not to present data-integration *per se*: though, this section describes some important aspects which are required for understanding the contributions of this paper to the UEML development.

Data-integration (Calvanese, *et al.*, 2002, 2003) provides the theoretical base for *database integration. Database integration is much more than the well known classical schema integration* because it directly works on data: in fact, schema-integration and data-integration may refer to *distinct phases* of the *lifecycle*, respectively *design* and *implementation*. Within data-integration, a *query result* is based on *data currently stored in several databases.* Data-integration is very relevant whenever these databases are *highly autonomous* both at schema and data levels. There are *four approaches* to data-integration:

- GAV (global as view),
- LAV (local as view),
- GLAV (generalisation of GAV and LAV),
- P2P (peer to peer).

Apart the P2P approach, the other ones need an explicit *integrated schema* (i.e. schema integration is a prerequisite to data-integration).The LAV and GLAV approaches explicitly acknowledge that it is not possible *to integrate distinct databases* but these distinct databases (qualified as local) can be understood as *views* on a *global database* (and also *vice-versa* in the GLAV approach). In other words, they do not provide effective ways for directly making *one integrated database*, i.e. a database which represents some *equivalence of data* stored in the various databases (as we will see in the remainder). Therefore, Table 1 can be completed by the following analogy:

Database Glossary	Modelling Language Glossary
Integrated database	Integrated model

being an *integrated model* defined as a model where two *distinct model artefacts* should never be *equivalent* (in term of the real world phenomena they represent).

Because of their relevance to the UEML, two important aspects concerning data-integration need to be carefully analysed:

- differences with schema integration,
- mappings.

Both aspects are discussed using some simple examples which are also useful in other sections of the paper.

Let suppose to have *three relational schema* (two qualified as local and one qualified as integrated) containing respectively three tables (T1, T2, T3), and *three databases* (two qualified as local and one qualified as global) (Figure 1). In the integrated schema, we have deliberately added the column *Database* to clearly show differences with schema integration.

Following the *LAV approach*, it is possible to represent *mappings* between one database and another one, by using any available *query language* and schema information. In the example of Figure 1 below, *mappings* may be as follows:

Select(T2(OrderN, Supplier Name) where database=1) = T1(OrderCode, Supplier);

Select(T2(OrderN, Supplier Name) where database=2) = T3(Order, Supplier ID).

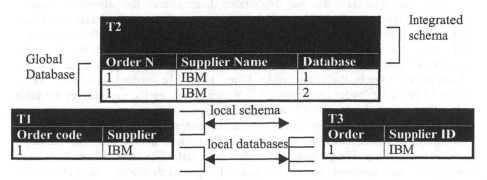

Figure 1: Example of data-integration

In general, the LAV approach allows:

- given one global database, *to derive* the local databases; i.e. mappings can be used as kind of *export mechanism*;
- given the local databases *to partially derive* a global database; i.e. mappings can be used as kind of *import mechanism*.

In the example of Figure 1, for instance, the export mechanism allows to derive data in T1 by applying the related query (because of the equivalence "="). The import mechanism is more interesting. In fact, the equivalence "=" in the mappings makes possible to fully derive the global database from the local ones: however, from a strictly theoretical point of view, this derived global database is only one of the possible global databases (for instance, in the global database we may freely add data unrelated with the two local databases i.e. with *database<>1,2*). In the general case of LAV (and GLAV), the '=' can be a *set inclusion* '⊇'.

In the *GLAV approach* both import and export mechanisms are partial. To illustrate this point, the previous example is further extended to the GLAV approach by introducing the couple of mappings below:

Select(T2(OrderN, Supplier Name) where database=1) =
                                        SomeQuery1(T1(OrderCode, Supplier))

Select(T2(OrderN, Supplier Name) where database=2) =
                                        SomeQuery2(/T3(Order, Supplier ID)).

where *SomeQuery1* and *SomeQuery2* indicate some queries involving available tables and columns (for instance, *SomeQuery1* may be *Select(T1(OrderCode, Supplier) where Supplier=IBM))*).

With these set of mappings, on one hand, we may freely add data to the global database (as before) and, on the other hand, we can add data to the local databases which do not satisfy both SomeQuery1 and SomeQuery2 respectively (for instance, orders of any supplier which is not IBM).

Now, the field Database has deliberately been added: this allows to manage situations in which we do not know how much data in the local databases are related (e.g. referring to the example in Figure 1, if the two orders represent the same order) but we are able to integrate, at some extent, the schema. In this sense, the integrated schema in Figure 1 might be "correct" but while "IBM is a supplier", the meaning of orders might not be the same: "order 1" in T1 could be "order to supplier", "order 1" in T3 could be "order from supplier". Which would be the meaning of the table T2 in the integrated schema? T2 would be a *generic relationship* between *order* and *supplier* with specific interpretation in specific contexts (i.e. local databases): thus, the column Database takes into account these contexts of interpretation for the same generic relationship. The previous examples show that while local schema have been integrated, local databases remain distinct in the global database: if it is known the Database 1 only contains "orders to supplier" and Database 2 only contains "orders from supplier", the global database is also an integrated database.

So far, the global database depicted in Figure 1 still distinguishes between the two orders stored respectively in T1 and T3. It is possible that after some extensive analysis, it is decided (by contexts of interpretation) that there is a final *integrated database* in which these two orders have been compared and their equivalence has finally been stated.

The *equivalence of the two orders* can be represented within the *GLAV approach* by a couple of mappings involving an *integrated database* (numbered as 3 on Figure 2) in which equivalent data are never replicated:

Select(T2(OrderN, Supplier Name) where database=3) =
                                        SomeQuery1(T1(OrderCode, Supplier));

Select(T2(OrderN, Supplier Name) where database=3) =
                                        SomeQuery2(/T3(Order, Supplier ID)).

Therefore, the table T2 in the integrated schema is representing the *same relationship* which exists in the two local schema.

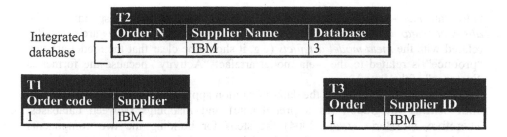

Figure 2: Example of integrated database

As it can be noted, the same *integrated schema* (i.e. T2) is able to "host" both (the interesting data of) the local databases and *integrated databases*. Therefore, the following GLAV mappings *without the Database column* generalises the situations depicted in Figures1 and 2:

Select(T2(Order N, Supplier Name)) ⊇ SomeQuery1(T1(OrderCode, Supplier));
Select(T2(Order N, Supplier Name)) ⊇ SomeQuery2(T3(Order, Supplier ID)).

The couple of mappings above which does not differentiate between situations depicted in Figures 1 and 2, can be specialised if some information concerning the contexts of interpretation suggest that some data(bases) should never be integrated (as the case depicted in Figure 1). On the other hand, integrated databases cannot be inferred from the two local databases by using this last couple of mappings: additional external information is eventually required.

5. UEML AND DATA-INTEGRATION

The UEML project states that at least two main components should be developed for a UEML:

- An integrated meta-model called *UEML meta-model* which should be able to accommodate both models to be integrated and results of integration; however, why and how to integrate models is provided elsewhere;
- A *set of mappings* which relates *meta-models of EMLs* (called *originating meta-models*) to the UEML meta-model, and *traces* how meta-models of EMLs contribute to the UEML meta-model.
- The *UEML set of mappings* is characterised by two facts:
- Mappings should be *stable* i.e. they should remain valid across specific situations;
- Mappings should be *standardised*.

In the UEML project, three (originating) EMLs, IEM (Jochem and Mertins, 1999), EEML (External, 2000) and GRAI/Actigrams (Doumeingts, *et al.*, 1992, Doumeingts, *et al.*, 1998), have been selected for making a first version of a UEML named *UEML 1.0*. An *integrated meta-model* has been defined by following some steps (Berio *et al.*, 2003) which are based on database schema integration suggestions (Petit, 2002a). The key point is the usage of a *scenario* (i.e. a set of models, analogous to databases in Table 1, in which it is possible to recognise equivalent model artefacts belonging to the distinct models). Specifically, given two models (part of the scenario) represented in two modelling languages, the first step is to make *meta-models* (as UML class models (OMG, 2002b)) corresponding to the *abstract syntax* of these *modelling languages*. Then, *model artefacts* are explicitly related with the *meta-model artefacts* (e.g. it should be clear that the model artefact "produce" is related to the meta-model artefact "Activity" because the former is "instance" of the latter).

As explained in section 4, the data-integration approaches take into consideration both schema integration (as a prerequisite) and mappings between databases. Therefore, in (Berio *et al.*, 2004) the steps for building the two components mentioned above have been reformulated in term of data-integration by using the *P2P and GLAV approaches*. The resulting rule for building an *integrated meta-model* and *a set of mappings* is as follow:

Given two meta-models artefacts, C and K, belonging to two distinct EMLs respectively, if there exist two predicates Query1 and Query2 (expressed, for instance, in OCL (OMG 2002a) if meta-models are represented in UML) such that *Query1(C)* ⊇ *Query2(K) or Query1(C)* = *Query2(K)* are true according to the *scenario*, then a *meta-model artefact H* is introduced in the *UEML meta-model*;

H satisfies the two *GLAV mappings*

Select(H) ⊇ *Query1(C);*
Select(H) ⊇ *Query2(K).*

The next example (Figure 3 below) is very similar to the previous ones on databases. It is however based on "names" often found in EMLs and the rule provided above (it should be noted that, even possible, we do not add any identifier that may be used to identify distinct model artefacts inside each model).

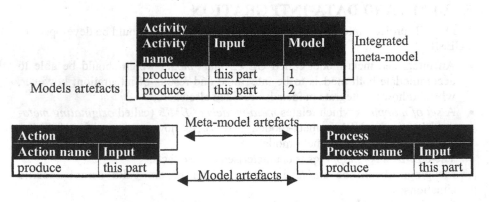

Figure 3: Example of integrated meta-model

A set of GLAV mappings is:

Select(Activity(Activity Name, Input)) ⊇
SomeQuery1(Action(Action Name, Input));
Select(Activity(Activity Name, Input)) ⊇
SomeQuery2(Process(Process Name, Input))

6. UEML AND ENTERPRISE REFERENCE ARCHITECTURES

The column *Model* on Figure 3 is a generic way to maintain the context of interpretation of models as in previous examples with the *Database* column. As explained in section 4, any information about the context of interpretation is important to differentiate data in a set of databases (then to differentiate model artefacts in a set of models). A possible generalisation simply states that *meta-model artefacts* should be related to an *Enterprise Reference Architecture*. In fact, ERAs concepts can be used for representing information about contexts of interpretation. For instance, concepts as *lifecycle (phases), life history, enterprise entity type* and *modelling framework* (with views) within the *GERA nomenclature* (IFAC-IFIP Task Force, 1999), are really useful for distinguishing if a generic meta-model artefact such as *Activity* is referring to processes built in the requirement phase. Other concepts, such as *domains, layers, usage context*, may also be introduced.

A UEML meta-model and an ERA can be related by using some kind of *projection* of meta-model artefacts (e.g. *Activity*) onto the *ERA* (instead of using additional columns as in the examples about databases in section 4). For instance, we can project *Activity* that we are assuming part of the *UEML meta-model*, to phases such as *requirement* or *design*. These projections of *Activity* state that one meta-model artefact (i.e. *Activity*) is used according to the specifically related ERA concepts.

These projections essentially make *copies of the specified UEML meta-models artefact*. This also means that one *meta-model artefact can be projected several times*. The idea underlying copies is to make explicit as much as possible that the same meta-model artefact (e.g. *Activity*) whenever used for distinct purposes (i.e. distinct contexts of interpretation) also requires distinct model artefacts. For instance, *Activity* whenever projected onto *requirement* and *design* phases, makes two copies of *Activity* itself (possibly renamed) which should not share model artefacts but they are the same in term of the original meta-model artefact.

Copies of a meta-model artefact can be used to define *new mappings* which should be consistent with the previously stated (i.e. part of the set of GLAV mappings associated to a UEML). New mappings are useful to represent specific situations and they should represent *explicit decisions* on how to use languages (such as a UEML) inside a given ERA. The *consistency condition* is:

If *Query1(C)* $\supseteq$ *Query2(K)* is in the *UEML set of GLAV mappings* and
Query3(H) $\supseteq$ *Query4(K)* or *Query3(H)* = *Query4(K)* is the *additional set of mappings* involving copies,
Then *H=Projection(C; ...)* (i.e. H should be a copy of C)

The consistency condition is represented by the diagram below (Figure 4).

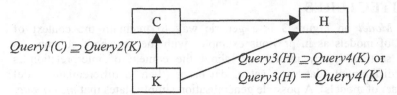

Figure 4: Consistency condition on set of mappings involving copies

referring to the GLAV mappings corresponding to the example shown in Figure 3, if the following projections are defined (represented as arrows and boxes, still named as *Activity* on Figure 5 below):

- **Projection(Activity; Design, Process View)** defining a copy of Activity which is used in the design phase for modelling processes;
- **Projection(Activity; Requirement)** defining a copy of Activity which is used in the requirement phase;
- it is not allowed to map Action onto any copy of Object.

Copies can explicitly be related by using *new relationships*: in fact, the GLAV approach allows to define new relationships in the integrated schema that do not appear in the local databases and schema. These new relationships are really relevant in case of modelling languages because they allow to *situate models* in *specific contexts of interpretation*. For instance, if there are two languages, one for representing (aspects of) *processes* and the other one for representing (aspects of) *services*, a new relationship might be introduced between *process* and *services* with the following meaning: "*process delivers service*". These new relationships might also be due to *methodologies* used with ERAs. This point is shared with the *method engineering discipline* (Brinkkemper, *et al.*, 1999).

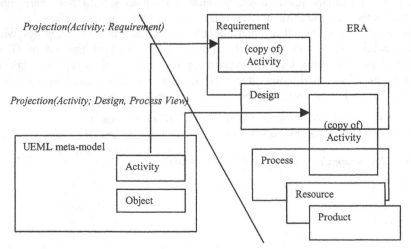

Figure 5: Example of UEML with ERA

7. CONCLUSIONS

This paper discusses how the perspective of *data-integration* can be used for improving the approach undertaken in the UEML project to a UEML development. The data integration perspective allows:

- To make clearer the distinction between *integrated meta-model* (analogous to integrated schema) and *integrated model* (analogous to integrated database) (sections 3 and 4);
- To clarify the role of *mappings* between a UEML and other EMLs as
  - mechanisms to trace how a UEML meta-model has been generated from meta-models of other EMLs (which is often not traced in classical schema integration) (see Section 5);
  - mechanisms to help model integration (section 4);
- To provide a base for taking into account information provided by ERAs about the *contexts of interpretation* of enterprise models (by re-using the mechanisms of mappings); in this way, model integration can become safer because model artefacts are much better differentiated (section 6).

Acknowledgement

The author would like to thank all the UEML IST-2001-34229 core members for their scientific contribution to the work. This work is partially supported by the Commission of the European Communities under the sixth framework programme (INTEROP Network of Excellence, Contract N° 508011, <http://www.interop-noe.org>).

8. REFERENCES

Berio, G. Anaya Fons, V. Ortiz Bas, A. (2004). Supporting Enterprise Integration through (a) UEML. To appear in Proceedings of EMOI Workshop joint with CaiSE04 – Riga – Latvia – June 2004.

Berio, G. Petit, M. (2003). Enterprise Modelling and the UML: (sometimes) a conflict without a case. In Proc. of Concurrent Engineering Conference 03, July 26-30, Madeira Island, Portugal.

Berio, G. *et al.* (2003). D3.1: Definition of UEML – UEML project, IST–2001–34229, www.ueml.org.

Brinkkemper, S. Saeki, M. and Harmsen, F. (1999). Meta-modelling based assebly techniques for situational method engineering, Information Systems, Vol 24, N.3, pp.209-228.

Calvanese, D. De Giacomo, G. Lenzerini, M. (2002). Description logics for information integration. In Computational Logic: From Logic Programming into the Future (In honour of Bob Kowalski), Lecture Notes in Computer Science. SpringerVerlag

Calvanese, D. Damaggio, E. De Giacomo, G. Lenzerini, M. and Rosati, R. (2003). Semantic data-integration in P2P systems. In Proc. of the Int. Workshop on Databases, Information Systems and Peer-to-Peer Computing.

Corrrea da Silva, F. Vasconcelos, W. Robertson, D. Brilhante, V. de Melo, A. Finger, M. Augusti, J.. On the insufficiency of Ontologies. Knowledge Based Systems Journal.

Doumeingts, G. Vallespir, B. and Chen, D. (1998). Decision modelling GRAI grid. In, P. Bernus, K. Mertins, G. Schmidt (Eds.) Handbook on architecture for Information Systems, Springer-Verlag.

Doumeingts, G. Vallespir, B. Zanettin, B. and Chen, D. (1992). GIM, GRAI Integrated Methodology - A methodology for designing CIM systems, Version 1.0, Unnumbered report, LAP/GRAI, University of Bordeaux 1, France.

DoW (2002). Description of the Work – UEML project, IST–2001-34229, www.ueml.org.

EXTERNAL (2000). Extended Enterprise Resources, Networks and Learning, External project, IST- 1999-10091.

Gruber, T.R. (1993). A translation approach to portable ontologies. Knowledge Acquisition, vol. 5, no. 2, 199-220.

Guarino, N. (Ed.) (1998). Formal Ontology in Information Systesm. IOS Press. Amsterdam.

IFAC-IFIP Task Force (1999). GERAM: Generalised Enterprise Reference Architecture and Methodology, Version 1.6.

ISO (1995). ODP-Open Distributed Processing. ISO/IEC 10746.

ISO (1998). ISO DIS 15 704 Requirements for Enterprise Reference Architectures and Methodologies, ISO TC 184/SC5/WG1.

Jochem, R. (2002). Common representation through UEML – requirement and approach. In proceedings of ICEIMT 2002 (Kosanke K., Jochem R., Nell J., Ortiz Bas A. (Eds.)),, April 24-26 Polytechnic Univeristy of Valencia, Valencia, Spain, Kluwer. IFIP TC 5/WG5.12.

Jochem, R. Mertins, K. (1999). Quality-Oriented Design of Business Processes. Kluwer, Boston.

OMG (2002a). OCL 2.0 specification, www.omg.org.

OMG (2002b). UML 1.5 specification, www.uml.org.

Petit, M. (2002a). Some methodological clues for defining a UEML. In Proc. of ICEIMT 2002 (Kosanke K., Jochem R., Nell J., Ortiz Bas A. (Eds.)), April 24-26 Polytechnic University of Valencia, Valencia, Spain, Kluwer. IFIP TC 5/WG5.12.

Petit, M. *et al.* (2002b). D1.1: State of the Art in Enterprise Modelling, UEML-IST–2001-34229, www.ueml.org.

Petit, M. Gossenaert, J. Gruninger, M. Nell, J.G. and Vernadat, F. (1997). Formal Semantics of Enterprise Models. In K.Kosanke and J.G Nell. (Eds.), Enterprise Engineering and Integration, Springer-Verlag.

Ushlod, M. (2003) Where are the semantics in the semantic web? AI Magazine 24(3).

Vernadat, F. (2002). UEML: Towards a unified enterprise modelling language, International Journal of Production Research, 40 (17), pp. 4309-4321.

9. Connecting EAI-Domains Via SOA– Central vs. Distributed Approaches to Establish Flexible Architectures

Marten Schönherr
Technical University of Berlin
Department of Business Informatics
Competence Center EAI
[schoenherr@sysedv.tu-berlin.de]

The article defines adaptability as the main target to solve the problem of enterprise architecture sustainability. Flexibility is an important steering mechanism to develop adaptability. Organisational modularisation is used to flexibilise enterprise structures. Business processes are changing permanently according to business requirements. Unfortunately it is a matter of fact that IT is disabling this business-driven change. Integration Technology is being introduced to improve the situation. Establishing step by step a multi service integration architecture creates new issues as handling internal charging routines, service monitoring and service life cycle management. The CC for EAI at Technical University is working on an approach and prototype of a service management module adressing the mentioned issues. [46]

1. INTRODUCTION: NECESSITY OF FLEXIBLE ENTERPRISE ARCHITECTURES

The system "enterprise" is exposed to complex changes of its environment. The three essential environmental dimensions are complexity, dynamics and interdependence (Jurkovich, 1974: 380). As described in figure 1 particularly the dimensions complexity and dynamics (Krystek, 1999: 266) increased by current developments like globalisation and automation (Frese, 2000) confront companies with diverse problems.

To solve the problem they have in principle two possibilities to react (Kieser & Kubicek, 1992):

- They can take measures to influence the environment with the target to reduce dynamics and complexity by decreasing the interdependences.
- Or they can increase the adaptability of the enterprise.

In this article the second aspect – the increase in adaptability – is continued to be examined. For this the architecture components organisation and IT are brought into reference with each other after characterising the flexibility as a planning target in order to draw up an approach for the modularisation of the organisation and IT-architecture within the context of Enterprise Application Integration (EAI).

---

[46] This paper has been previosly presented in a poster session at IFSAM 2004/Goeteborg/Sweden.

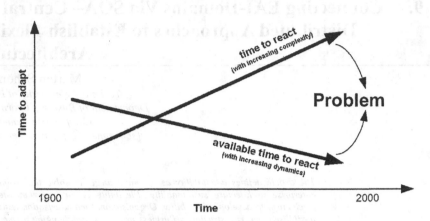

Figure 1. Complexity and dynamics as a current management problem

2. FLEXIBILITY AS AN ORGANISATIONAL TARGET

The understanding that "enterprise modification has changed or will have to change from an event, which has to be organised in longer time intervals, into a permanent state" (Krueger, 1998: 227) is accepted today to a large extent. The task is now to make companies more flexible for this change. Krueger distinguishes the change of companies depending on the requirement, willingness and capability of change. (Krueger, 1998: 227) The capability of change represents the core of the considerations of this article. In order to deal effectively with internally and externally caused requirements of change corresponding structures and organisational instances for the institutionalisation of change are to be created in the companies. In addition to institutional measures structural measures to increase the capability of change – measures for an increase in flexibility – are discussed in the following. The term of flexibility comes from Latin and means changeability, agility or ductility. A system is flexible if a requirement of change is completed by a potential of change in the system, which can be activated in an appropriate time (Kieser & Kubicek, 1992; Gronau, 2000: 125). The requirement of change includes a factual and a temporal dimension. The temporal dimension not only requires the ability to react of an enterprise but also the capability of anticipative adaptation. Hill/ Fehlbaum/ Ulrich describe this as productivity of second order (Hill, Fehlbaum & Ulrich, 1994). Kieser/ Kubicek identify the following tendencies of the structure of companies as suitable to increase their flexibility (Kieser & Kubicek, 1992):

- low specialisation on jobs and department level
- strong decentralisation
- flat hierarchies
- minimisation of strength of central supporting departments (staffs)
- simple, which means no extensive matrix structures

The mentioned tendencies aim at decoupling the structures and processes by means of a reduction of interfaces. In the following, these thoughts are further developed in order to facilitate decoupling and flexible reconfiguration through modularisation on different levels of the enterprise.

3. ORGANISATION AND IT

Today it is not sufficient to exclusively consider formal organisational planning aspects when making companies more flexible. It is rather necessary to consider the technological aspects and to combine both in an integrated architecture approach. The other way round it does not make sense to introduce or change information systems without considering the interactions with the tasks and processes, which they support (Kaib, 2002; Derszteler, 2000). EAI does not only integrate IT-systems, but is particularly the cause to plan the domain organisation and IT in an integrated way and to commonly develop them further. In the science this discussion about the mutual interdependences of IT and organisation of the enterprise has a long tradition, in which not only the technology (technological imperative), the organisation (organisational imperative) but also the complex interactions between both (emergent perspective) (Markus & Robey, 1998: 583) have been described as driving factors (Leavitt & Whisler, 1958: 41; Applegate, Cash & Miles, 1988: 128; Rockart & Short, 1989: 7; Burgfeld, 1998). Following Frese it is not assumed here that information technology and enterprise organisation are in a deterministic cause-effect context (Frese, 2000). It is rather assumed that IT represents an option which increases the scope or planning of the organizer (Frese, 2002: 191).

3.1 Modular Enterprise Architectures

In this context, enterprise architecture is understood as the interaction of technological, organisational and psycho-social aspects at the development and utilization of operational socio-technological information systems (Gronau, 2003). In the following, above all, the technological and organisational domains are considered as IT- and organisation architectures. A currently discussed method to make IT- and organisation architectures more flexible is the modularisation. By means of system perspectives in the following general characteristics of modules are described, which are taken from the organisation and IT-literature and which are equally to be valid here for both areas. A module consists of two parts, the module interface and the module trunk. The module interface contains the specification of the performances of the module, which are necessary for its environment for "utilization". The module trunk implements the specified performances. Modularisation means the structuring of a system in small, partial-autonomous subsystems. The complexity reduction in this results from the system formation within the system enterprise (Krcal, 2003: 3). The subsystem formation has a complexity-reducing effect, because on the one hand it conceals the subsystem internal complexity in the sense of the encapsulation from the system environment and on the other hand it causes a decoupling of the subsystems through the reduction to few known interfaces. The higher flexibility of the architecture results form the easier reconfiguration possibility of the decoupled modules. Thus, complexity reduction can be considered as prerequisite for making architectures more flexible. At the modularisation the following organisation targets are to be taken into consideration (Rombach, 2003: 30; Lang, 1997):

- Abstraction form implementation

- Encapsulation in the sense of concealing the internal functioning
- Exchangeability
- Reusability
- temporal validity
- Orthogonality (in the sense of "not influencing each other")
- no overlapping
- Completeness (conclusion)
- well-defined interfaces
- Interface minimalism
- Generality

By using these principles modules are created, which potentially can be combined, reused and easily changed (Rombach, 2003: 30). In the following, it is shown which approaches result from the modularisation of the organisation and IT.

3.2 Modularisation of the organisation

In the following, the modularisation of the organisation is considered on the level of the whole enterprise as macro level and the on the level of business processes as micro level (demonstrated in figure 2). By defining a macro and micro level two targets are pursued:

- Defining structures and at the same time making them more flexible
- Defining tightly structured and manageable modules

The macro level has a strongly structuring effect on the organisation. The modules of the macro level, the so-called macro modules, are to be stable over a longer period of time. By this, the macro level represents a reference system as foundation for the organisation of the modules of the micro level within the respective macro modules. The modules of the micro level, however, are to reflect the dynamics of the changing process requirements through corresponding reconfiguration. Thus it is the aim to create a reference-creating, stable organisation-invariant level – the macro level – and a dynamic, flexibly configurable level - the micro level. In the following, the characteristics of the macro and micro level are described.

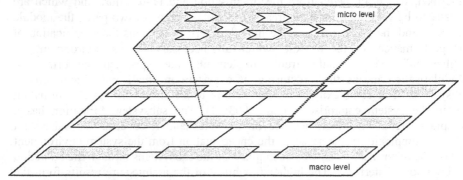

Figure 2. Modularising in macro- and micro level

Modularisation on macro level: In order to achieve a direct reproduction of strategy-oriented organisation structures on the IT-infrastructure more than the classic course- or structure-organisational points of views are required. The target of the definition of the macro modules is the consideration of the whole organisation by means of suitable criteria to be able to derive exactly this step. In the following, special criteria for the formation of macro modules are mentioned to complete the above mentioned general criteria:

- Similarity of the processes within a module
- Similarity of the necessary process know-how
- Minimisation of the external interdependences
- Maximisation of the internal coherence
- Independence of operative process adaptations
- Sector-specific best practices for the definition of modules

The enumerated criteria are partially used in the area of component ware in order to determine the granularity of software components. The analogous application on macro level at least leads to similar methods, with which later a match between macro, micro and software architecture level can be achieved. The defined macro module do not represent an instruction for a reorganisation. They rather pursue the target to guarantee a special way of looking at the organisation. This special way of looking at the organisation supports the modularisation of the respective IT-systems, which support the processes of a macro module. On the level of the macro modules structural similarities are to build the foundation for the definition. In addition to the above mentioned criteria, further criteria can be determined depending on the situation. As a rule, processes run through different modules, which means no limitation is made according to the beginning or end of process chains.

Example: In figure 3 it is shown on the basis of a classic matrix organisation, which is composed of sections and functional areas, how macro modules are formed through the above mentioned criteria. In this, the same colourfully highlighted areas respectively represent a macro module. For example, within the different sales organisations two areas were defined as macro module. The reason for this could be that the processes in both areas are very similar and require similar information for implementation.

Modularisation on micro level: Within the defined macro modules now the structural similarities are to be considered in detail. For the IT-system world particularly the business processes existing within the modules are relevant. Here all process steps have to be compared and the ones with the same structure have to be identified. In principle, a business process is composed of different steps. The probability that there are identical steps in an amount of similar process chains is very high.

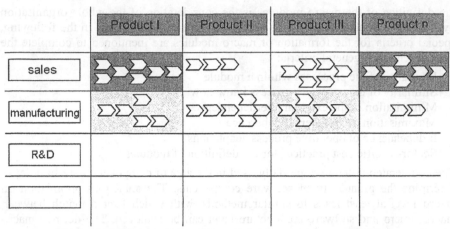

Figure 3. Definition of macro modules in a matrix organisation

Even if the partial processes are slightly different it can be considered if it is possible in the sense of a complexity reduction to modify the actual business processes, that it applies as many partial processes as possible. Gerybadze derives from the methods of modular product organisation criteria for the determination of the degree of modularising capability of processes. According to Gerybadze, there is a good modularising capability if (Gerybadze, 2003: 83):

- for each activity an exactly defined function of the overall system can be defined,
- the quality of the result of this activity can exactly be determined,
- prices or offset prices can be determined for the respective output
- the interfaces can be defined very exactly.

Apart from the general question of modularising capability the determination of the optimal module size is decisive for the successful implementation of the modularisation on process level. Module that are too big can avoid the expected flexibility of the organisation, because the modularisation is to create small, flexible fragments which can be integrated, which are more mobile than an overall monolithic structure. Modules that are too big can restrict the domination of the processes within the module due to their complexity. On the other hand, too many small modules mean a stronger division of labour and partition. The smaller the modules the more specialised the processes naturally have to be within a module. In order to define general criteria to create an optimal size of module, the characteristics of modularisation have to be considered. The minimum size of a module which has to be complied with results from the activities for a clearly definable (interim) product. The maximum module size is determined by the domination of complexity within a module. The complexity may not be so big that the responsible people are no longer able to control the module.

3.3 Implementation principles

The implementation of a modular organisation strongly depends on the prevailing conditions of the situation. Essential influence factors are the size of the organisation, its programme of supply, its internationalisation and culture, the existing structuring as well as the organisational environment (Kieser & Kubicek, 1992). In spite of this interdependence on the situation some basic implementation principles can be found:

The main principle is to build small organisational units. They have to be big enough to comprise matching processes to one object (for example, product, product group etc.). With regard to the scope and the complexity, however, they may not exceed the limits of reception and problem-solving capacities of the human being. The process orientation is expressed by the demand to permanently orient the modules to the processes to draw up performances. The process orientation shows that the modularisation does not aim at a function-oriented but an object-oriented structuring.

In close context with the process orientation is the customer or market orientation. According to this all added value activities of the modules are to be oriented to the external and/ or internal customer demands. This results from the main target of the flexible adaptation to the requirements of the market.

Integration of tasks: the processes in a module are to belong together to a great extent depending on their kind in order to guarantee the completeness of the processes concentrated in a module.

Non-hierarchical coordination forms: The coordination of autonomous activity units, which are not in a hierarchical relationship to each other any more, requires new forms of cooperation that are not based on external control any more but on self control.

3.4 Modularisation of IT

On the basis of the described modular organisation architecture adequate IT-architectures are to be defined. The coordinated organisation of strategies, processes and technical infrastructures is a classic subject of business computing science (Wall, 1996). For this different architecture concepts were developed, which have as a target a homology between organisation and IT (Krcmar, 1990: 395; Pohland, 2000; Scheer, 1991).

These concepts, however, are based on the assumption to talk about implementation and new introduction of IT. In order to create lasting architectures, if possible, the paradigm of the structural analogy between organisation and IT should be maintained as far as possible (Gronau, 2003).

EAI vs. service-oriented architecture (SOA) Service orientation and system integration have been discussed independently from each other for many years. EAI-platforms centrally integrate heterogeneous system landscapes on process, method and data level. However, the stronger integration projects are implemented with object-oriented procedures, the stronger seems the closeness to service orientation. The service-oriented application integration can be compared with the integration on interface and method level. It represents an alternative point of view, which attempts to integrate "wrapped" modules from old applications and new applications within a SOA which are already implemented in a service-oriented way. One of the most important objectives of service orientation is the re-utilization of existing components. The prerequisite for this is a central service management, which has to provide functions like service-life-cycle, service distribution and the versioning (Lublinsky & Farrel, 2003: 30). Since meanwhile the most development tools support a service orientation, applications can be developed much easier as services in the future. However, it is frequently difficult to split old applications into modules which can be defined as service. Apart from development environments which actively support a new implementation with service character in the meantime some integration suites also offer tools which make wrapping of old applications in services easier. The costs of this initial transformation can be high and often the intervention in the source code contradicts the strategic guidelines of the responsible IT-staff, who, above all, wants to leave monolithic legacies unaffected (Apicella, 2002). At the consideration of an overall IT-architecture old applications that have to be integrated and planned new implementations have to be assumed. Not only the central EAI-approach but also the decentralised SOA provide methods to implement solutions for the described area of tension. Thus, EAI and SOA are harmonising parts of an overall IT-architecture.

However, within the context of system integration it has to be considered that one of the primary targets of EAI, the reduction of the point-to-point integration scenarios is not replaced by a similar complex SOA. Two main problems at the connection of old applications are standardised interfaces and the granularity of services. At the service-oriented integration of existing applications the granularity of the functions that are packed as services is decisive. Thus, it makes sense to make those functions accessible as services which are generally necessary and thus should be reused. They should carry out a complete working unit and they should be easy to describe in their function and result (Narsu & Murphy, 2003).

Considering the organisational modularisation on the macro and micro-level there is another advantage which hasn´t been described yet. As shown in figure 4 (according to the Credit Suisse Information Bus approach) it is to recommend to implement a tight coupling of applications inside a module. The integration strategy (EAI vs.SOA) of each module depends on the individual situation of the specific module. According to the fact that there are less interactions between modules than inside the module itself with the utmost probability there will be implemented a louse coupling via service-oriented Technology.

3.5 Service Management

Establishing a SOA means managing a huge number of business and/or technical services. Web Service based SOA do not offer these features so far. The Competence Center for EAI at the technical University is working on an service management module which for instance supports the whole service life cycle; reuse and service monitoring. The following features are part of the concept developed.

- Services identification
  - Service definition
  - Definition of service features
  - Definition of service granularity
  - Defining the Service Provider
- Localise and group services
  - Functional oriented domains
  - Technology oriented domains
  - Application oriented domains

- Service capsulation
  - Service orchestration and allocation
  - Service monitoring

To proof the approach the integration lab has established an invironment with 5 commercial EAI-products and a SOA based on Open Source components. An example business case has been implemented to show how service integration can be orchestrated.

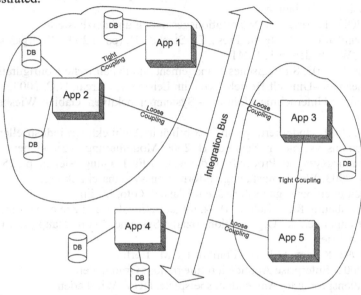

Figure 4. Tight and loose coupling inside and between modules

4. CONLUSION

Many publications, which are technically focussed, currently discuss decentralised architectures for the integration of complex IT-infrastructures. Companies again deal with the subject of re-use, although some years ago it almost completely disappeared from the experts discussion under the name of business process repositories, which were planned to implement business processes that can be reconfigured. Platforms with higher performance partly give reason for this unexpected renaissance. The subject EAI also contributes to the current discussion, which is led on an enterprise architecture level. In the first step, it deals with the technical definition of the modules (or services), only in the following, technologies for implementation will play a role. In this sense the article particularly concentrates on the modularisation on a specialist level and its transformation into a technical level.

REFERENCES

Apicella, M. 2002. Side by side in perfect harmony? InfoWorld
Applegate, L.M., Cash, J.I., Miles, D.Q. 1988. Information technology and
  tomorrow's manager. Harvard Business Review 66, p 128-136

Burgfeld, B. 1998. Organisationstheorie und Informationstechnologie. DUV, Wiesbaden

Derszteler, G. 2000. Prozessmanagement auf Basis von Workflow-Systemen. Josef Eul, Lohmar, Koeln

Erlikh, L. 2003. Integrating legacy into extended enterprise: Using web services. Relativity Technologies.

Frese, E. 2000. Grundlagen der Organisation: Konzept - Prinzipien - Strukturen. 8 edn. Gabler, Wiesbaden

Frese, E. 2002. Theorie der Organisationsgestaltung und netzbasierte Kommunikationseffekte. in: Frese, E.; Stoeber, H. (eds.) 2002. E-Organisation. Gabler, Wiesbaden, p 191-241

Gerybadze, A. 2003. Strategisches Management und dynamische Konfiguration der Unternehmens-Umwelt-Beziehungen. in: Leisten, R.; Krcal, H.C. 2003. Nachhaltige Unternehmensführung - Systemperspektiven. Gabler, Wiesbaden, p 83-100

Gronau, N. 2000. Modellierung von Flexibilität in Architekturen industrieller Informationssysteme. in: Schmidt, H. 2000. Modellierung betrieblicher Informationssysteme. Proceedings der MobIS-Fachtagung. Siegen, p 125-145

Gronau, N. 2003. Wandlungsfaehige Informationssystemarchitekturen - Nachhaltigkeit bei organisatorischem Wandel. Gito, Berlin

Hill, W., Fehlbaum, R., Ulrich, P. 1994. Organisationslehre 1: Ziele, Instrumente und Bedingungen der Organisation sozialer Systeme. 5 edn., Haupt, Bern, Stuttgart, Wien

Imai, M. 1993. Kaizen. Ullstein, Frankfurt a. M., Berlin

Kaib, M. 2002. Enterprise Application Integration: Grundlagen, Integrationsprodukte, Anwendungsbeispiele. DUV, Wiesbaden

Kieser, A., Kubicek, H. 1992. Organisation. 3 edn. De Gruyter, Berlin, New York

Krcal, H.C. 2003. Systemtheoretischer Metaansatz fuer den Umgang mit Komplexitaet und Nachhaltigkeit. in: Leisten, R.; Krcal, H.-C. 2003. Nachhaltige Unternehmensführung - Systemperspektiven. Gabler, Wiesbaden, p 3-30

Krcmar, H. 1990. Bedeutung und Ziele von Informationssystemarchitekturen. Wirtschaftsinformatik 32, S. 395-402

Krueger, W. 1998. Management permanenten Wandels. in: Glaser, H.; Schröder, E.F.; Werder, A. v. (eds.) 1998. Organisation im Wandel der Märkte. Gabler, Wiesbaden, p 227-249

Krystek, U. 1999. Vertrauen als Basis erfolgreicher strategischer Unternehmungsfuehrung. in: Hahn, D.; Taylor, B. (eds.)1999. Strategische Unternehmungsplanung - Strategische Unternehmungsführung. Physica, Heidelberg, p 266-288

Jurkovich, R. 1974. A core typology of organizational environments. Administrative Science Quarterly 19, p 380-394

Lang, K. 1997. Gestaltung von Geschaeftsprozessen mit Referenzprozessbausteinen. Gabler, Wiesbaden

Leavitt, H., Whisler, T. 1958. Management in the 1980s: New information flows cut new organization flows. Harvard Business Review 36, p 41-48

Lublinsky, B., Farrell, M. 2003. 10 misconceptions about web services. EAI Journal, p 30-33

Markus, M., Robey, D. 1988. Information technology and organizational change: Causal structure in theory and research. Management Science 34, p 583-589

Narsu, U., Murphy, P. 2003. Web services adoption outlook improves. Giga Information Group

Oesterle, H. 1995. Business Engineering: Prozess- und Systementwicklung. 2 edn., Volume 1. Springer, Berlin *et al.*

Pohland, S. 2000. Globale Unternehmensarchitekturen - Methode zur Verteilung von Informationssystemen. Weissensee-Verlag, Berlin

Rockart, J.F., Short, J.E. 1989. IT in the 90's: Managing organizational independence. Sloan Management Review, p 7-17

Rombach, D. 2003. Software nach dem Baukastenprinzip. Fraunhofer Magazin, p 30-31

Scheer, A.W.1991. Architektur integrierter Informationssysteme. Springer, Berlin

Thom, N. 1996. Betriebliches Vorschlagswesen: ein Instrument der Betriebsfuehrung und des Verbesserungsmanagements. 5 edn. Lang, Berlin

Wall, F. 1996. Organisation und betriebliche Informationssysteme - Elemente einer Konstruktionslehre. Gabler, Wiesbaden

10. A Meta-methodology for Collaborative Network Organisations: A Case Study and Reflections

Ovidiu Noran

School of Computing and Information Technology, Griffith University
E-Mail: O.Noran@Griffith.edu.au

Presently, there is a great need for methodologies and reference models to assist and guide the creation and operation of various types of Collaborative Networked Organisations (CNO). The efforts to fulfil this need can be greatly assisted by a meta-methodology integrating diverse CNO creation and operation knowledge. This paper continues previous research on the concept, requirements, design, verification and potential implementations of a CNO life cycle meta-methodology, by describing an additional case study and subsequent reflections leading to the refinement and extension of the proposed meta-methodology.

1. INTRODUCTION

In the current global market conditions, organizations worldwide often need to come together in order to bid for, and execute projects whose requirements go beyond their individual competencies. Collaborative Networked Organisations (CNO) in their various forms of manifestation (such as Virtual Organisations (VO) or Professional Virtual Communities) are recognized to offer an advantage in a competitive situation by capitalising on the overall pool of knowledge existing in the participants. However, CNO set-up and operation includes human aspects, such as establishing partner trust and a sense of community, which can only be effectively addressed in time. Similarly, CNO technical aspects such as the establishment of agreed business practices, common interoperability and distributed collaboration infrastructures require time that may not be always available. 'Business ecosystems', 'breeding / nesting environments' (Camarinha-Matos, 2002), or 'company networks' (Globemen, 2000-2002) may enable a prompt formation of CNOs; however, they need to be supported by effective reference models containing methods describing CNO set-up and operation. Unfortunately, the methodologies contained in such reference models are too generic to allow their effective use for a particular project; thus, specific methodologies have to be created for each scenario. In this sense, a *meta-methodology*[47] may be of great help in integrating knowledge relating to several CNO-type creation and operation methodologies, and thus being able to promptly suggest a suitable method for a particular CNO project.

This paper presents the application of the proposed meta-methodology to a specific scenario: the creation of a VO in the higher education sector. In this context,

---

[47] a method (here, consisting of steps and their applicability rules) on how to design a method.

the meta-methodology is used both in practice (to create a VO formation / operation method for the participating organisations) and in theory (for the triangulation[48] of the meta-methodology concept and for reflections leading to its extension), thus fulfilling the dual purpose of action research (McKay & Marshall, 2001).

2. META-METHODOLOGY PRIMER

2.1 Research Question and Strategy

The research question has asked *whether a methodology describing how to construct customised modelling methods may be built and what other factors may (positively) influence such an endeavour.* This topic has provided an opportunity to employ action research (Galliers, 1992; Wood-Harper, 1985), which allows for both practical problem solving and generating / testing theory (Eden & Chisholm, 1993; McKay & Marshall, 2001). The research strategy was based on two cycles containing lab / field testing and reflections leading to theory extension (Checkland, 1991). The cycles are entered after stating the research question, completing the research design and adopting the theoretical model. The cycles' exit conditions are based on a compromise between the time / resources available and the accuracy of the result. The design of the research strategy (design decisions, research stance justifications and methods adopted) is described in (Noran, 2001), while conceptual development of the meta-methodology and testing are contained in (Noran, 2004c).

2.2 A First Case Study

In this case study, the application of the meta-methodology has yielded a design (and partly operation) method for a Breeding Environment (BE)[49] and the Service Virtual Enterprises (SVEs) created by the BE (Hartel *et al.*, 2002). The lead partner(s)[50] wished to retain control of the identification and the concept of the SVEs created, with the rest of the SVEs' life cycle phases covered by the BE. The model audience (various levels of management and technical personnel) was partly familiar with the IDEF[51] family of languages and with the Globemen[52] reference model.

The application of the meta-methodology has resulted in a multi-level IDEF0[53] model of the design methodology for the BE and the SVEs created by it, based on the Globemen Reference Model (due to audience proficiency). The chosen language has allowed different levels of the model to simultaneously target various audiences (management, working groups, etc), while ensuring the overall model consistency. The methodology has received a positive response and is being currently used in BE creation and operation, and in SVE creation.

---

[48] triangulation is possible in the context of an previous case study (Bernus et al., 2002) and based on the assumption that while humans subjectively interpret reality, it is possible to build a descriptive and commonly agreed upon methodology model (Noran, 2004c).
[49] called Service Network Organisation (SNO) in this case study
[50] one or several partners that initiate BE creation and may influence SVEs created by the BE.
[51] Integration DEFinition, a family of languages aiming to create computer-implementable modelling methods for analysis and design (Menzel & Mayer, 1998).
[52] Global Engineering and Manufacturing in Enterprise Networks (Globemen, 2000-2002)
[53] Integration Definition for Function Modelling (NIST, 1993) - an IDEF functional language

The meta-methodology has also recommended modelling of the decisional aspect of the partners, BE and potential SVE(s) using GRAI54 Grids and applicable reference models55. This case study is described in detail in (Bernus *et al.*, 2002).

2.3 Meta-methodology Content Before the Second Case Study

In brief, before the second case study the meta-methodology comprised the following steps and associated rules of application:

- identify enterprise entities involved in the BE / CNO creation task: mandatory;
- create a business model able to express relations between life cycle phases of the identified enterprise entities (using a suitable formalism): mandatory;
- create an activity model of the BE / CNO design and operation: mandatory (main deliverable), depth level according to requirements;
- recommend additional aspects to be modelled and formalisms / tools (such as information, decision, organisation, time): project specific .

3. A SECOND CASE STUDY

This case study has been chosen due to significant differences from the first case study regarding the participant organisations' type, culture and the target CNO. This noticeable disparity has enabled the effective testing, triangulation and reflection leading to the extension of the proposed meta-methodology.

3.1 Background

Faculty FAC within university U contains several schools (A to D), with schools A and B having the same profile. School A is based in two campuses, situated at locations L1 and L2, while school B is based on single campus, situated at location L3 (AS-IS in Figure 4). Although of the same profile, and belonging to the same FAC and U, schools A and B are confronted with a lack of consistency in their products and resources, such as the programs and courses offered, allocated budget, research higher degree (RHD) scholarship number and conditions, academic profile and availability of teaching staff, etc. This situation causes negative effects, such as additional costs and difficulty in student administration and course / program design / maintenance, inter-campus competition for RHD students and staff fallacies of unequal academic and professional standing between campuses, all of which are detrimental to the Faculty as a whole.

Proposed Solution

The issues previously described could be optimally resolved by schools A and B forming a VO (called *merged school* (MS) in the TO-BE state in Figure 4) with cross-campus management and policies ensuring intrinsic consistency in the product delivered and resources allocated to the individual campuses at L1, L2 and L3. Thus, the individual campuses are set to retain much of their internal decisional and organisational structure except for the highest layer, which will be replaced by the

---

[54] Graphes avec Résultats et Activitées Interreliées (Graphs with Results and Activities Interrelated), a decisional modelling formalism / reference model (Doumeingts et al., 1998)
[55] such as the Partner-BE-SVE reference model described in (Olegario & Bernus, 2003)

VO governance structure. Campus interoperability within the VO will be assisted by a common ICT infrastructure (Camarinha-Matos & Afsarmanesh, 1999)[56] .

3.2 Specific Features of the VO Formation Scenario

In this case study, the function of Breeding Environment (BE) internal to U could be performed by the Faculty FAC, which contains several schools forming VOs as necessary. The largest school participating in an internal VO formation project (e.g. school A in Figure 2 or Figure 4) could be identified as the *lead* partner. In the current situation, partners A and B within the BE have come together at the initiative of the BE and the lead partner for the purpose of an on-going VO project; thus, the VO was set to have a long life, possibly equal to that of the BE (the FACulty). Importantly, the partners will cease to operate independently during the life of MS.

Note that, in the internal environment (U), the frequency of VO creation and thus the degree of preparedness for VO creation of the partners (schools) within the BEs (Faculties) in U, is *low*. However, it is likely that individual, or groups of schools will increasingly participate in BEs and VOs *outside* U, in which case additional *agility* (Goranson, 1999) and preparedness is required. This could be achieved by using *reference models* (reusable templates) of possible VO types, which should include VO set-up and operation methodology templates. However, the use of such reference models requires user proficiency and implies further customisation work. Thus, the proposed meta-methodology can assist in enhancing preparedness and agility by producing a methodology suited to a specific VO creation / operation task

The audience of the deliverables was to be the management of U, FAC, A and B and the academic and general staff of A and B. This audience diversity dictated that the chosen formalisms should allow effective complexity management, and that primers to the modelling methods must be included with the model(s). This has led to the choice of IDEF0 language for functional, and GRAI Grids for decisional / organisational57 modelling. A simplified version of rich pictures (Checkland & Scholes, 1990) have also been considered, in order to facilitate a common understanding of the models. In addition, it was desired that the chosen modelling tool(s) be capable of versioning and Internet publishing in order to facilitate model management and dissemination.

Documentation from a previous similar project was available, although its particular nature required additional effort for its use as a possible reference model[58].

3.3 Meta-methodology Application

The case study has been approached using the meta-methodology content described in section 2.3. As expected, the practical application of the meta-methodology to a markedly different situation has brought about some step variations and additions, leading to the enrichment and refinement of the meta-methodology.

---

[56] this requirement is partially satisfied initially, because all campuses belong to the same University U that owns (or leases) the cross-campus infrastructure .

[57] may be represented in a GRAI-Grid by assigning human resources to decision centres.

[58] e.g. filtering out noise and abstracting reusable knowledge from the available information.

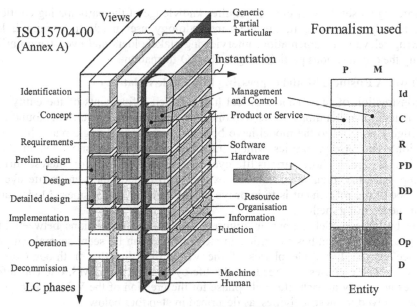

Figure 1. Model used: simplified ISO15704 Annex A

Step One: Identification of the Entities Involved.

This step has included a) deciding on the entity aspects that need to be modelled, b) the choice of a suitable entity modelling formalism, expressive enough to cover aspects decided in a), and c) constructing an initial[59] set of entities relevant to the project, in a form usable for the next step. In addition, the meta-methodology required a modelling formalism able to represent the selected entity aspects *in the context of the entity life cycle*, and capable to express the interaction of life cycle phases both within the same entity and between separate entities.

The modelling framework of ISO 15704 Annex A (GERAM[60]), namely the Generalised Enterprise Reference Architecture (GERA) contains placeholders for concepts such as life cycle phases and views; they can be used in selecting suitable architecture frameworks to create life cycle models of various entity aspects[61] at various levels of abstraction[62].

Using GERA as a checklist of possible views in this particular case, it has been initially established that the main aspects needed to be modelled during the life of the entities involved were the management, and the product / service. Subsequently, it has become clear that decisional and organisational models of the entities directly participating in the VO formation will also be necessary.

---

[59] this set may be subsequently revised as further described
[60] Generalised Enterprise Reference Architecture and Methodology (ISO/TC184, 2000)
[61] this capability is inherited from the architecture frameworks GERAM originated from, i.e. the Purdue Enterprise Reference Architecture (PERA) (Williams, 1994), the Open System Architecture for CIM (CIMOSA Association, 1996) and GRAI-GIM (Doumeingts, 1984)
[62] GERA may also be used to guide the creation of partial models, glossaries, meta-models and ontologies

Constructing a useful set of entities requires knowledge of the participating entities; if not already possessing it, the meta-methodology user needs to acquire it by reviewing relevant documentation, interviewing stakeholders and even temporarily entering the organisations participating in the VO formation .

Step Two: A Business Model expressing Life Cycle Relations

The construction of the business model has included refinements of the entity set through additional iterations of the step one. Entities whose life cycle was found not to be highly relevant to the model have been collapsed into a single phase (usually operation); conversely, entities which initially had only been modelled in their operation phase but whose various life cycle phases have become of interest during this step, have had their representation expanded to include their entire life cycle. The scope of the refinement is to improve the accuracy of the Business Model, and thus of the resulting method.

The business model (shown in Figure 2) describes the relations between life cycle phases of the entities identified in step one. As can be seen, several entities influence various life cycle phases of the VO entity, directly, or through other entities' life cycle phases. By 'reading' the life cycle diagram of MS, one can infer activities necessary at each life cycle phase for the creation of the VO and elicit the entities involved in these activities, as described in step four below.

Once again, an accurate business model demands deep knowledge of the structures and relations between participating entities. Generally, the business model may be constructed either to reflect a combined AS-IS / TO-BE view (such as used in this case study – see Figure 2), or it can be split in sub-models for the present and desired state(s); the choice of representation depends on model complexity and on the audience preferences. Once the model is constructed, it needs to be submitted for review and feedback to the stakeholders in order to improve its accuracy and achieve acceptance. *The quality of the business model is paramount to the creation of an effective VO creation / operation method.*

Step Three (Additional): Models of the AS-IS and TO-BE States

The target VO was envisaged to reuse as much as possible from the existing entities' structures and thus not to differ radically from the existing individual entities, except for the very top governance layer. In addition, there was a need to fully understand the existing entities' operation and to decide on a decisional and organisational structure of the VO. Thus, it has become necessary to undertake an additional step within the meta-methodology application, i.e. to construct models of the decisional and organisational aspects for the AS-IS[63], and *several* proposed TO-BE states.

The AS-IS model has revealed a large degree of intervention of the Planning decision centre in both the Product and the Resources decision centre groups at most horizons. This corresponds to a high degree of turbulence and a lack of clear and effective strategy within the organisation. In essence, the Head of School (HOS) in the role of Planner has to put out 'fires' (product / resource discrepancies requiring immediate reconciliation) on a short-term basis, rather than having strategies in place to avoid the *cause* of such problems. The AS-IS decisional model has also

---

[63] AS-IS modelling is beneficial if management believes that the desired TO-BE state may be achieved by *improving* the AS-IS state rather than replacing it (Uppington & Bernus, 2003)

shown a lack of sufficient financial and decisional independence of the schools A and B and a shortage of information crucial to long-term strategy making[64].

Thus, in constructing the TO-BE decisional model, attention has been paid to confine the authoritarian role of Planning to the strategic level[65], to increase the financial and decisional independence of the target VO (MS), and to provide the necessary external information to MS for the purpose of strategy making and self-governance. The strategy thus created can then be used by lower decision centres to set decision frameworks for subordinate centres in a top-down (rather than lateral) fashion, resulting in more predictable organisational behaviour and less turbulence.

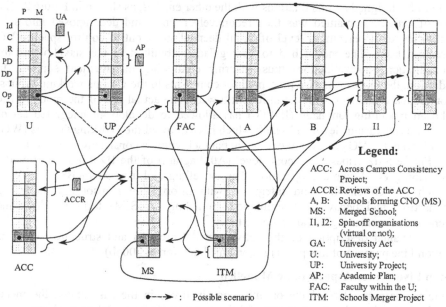

Figure 2 Business model expressing entity life cycle phases' relations

Interestingly, the resulting TO-BE decisional model has been found to match all envisaged scenarios, with differences only obvious in the *organisational* structure, i.e. the allocation of the various human resource groups to the decision centres represented in a single TO-BE decisional framework (Figure 3). Thus, the various scenarios were based on the same decisional framework, while proposing various decisional roles and memberships of these roles. For example, all scenarios agreed on the existence of a School Executive and its allocation to approximately the same decision centres, but differed on the *composition* of that body (e.g. from one person (HOS) to several (HOS plus Associate Heads of Schools)).

---

[64] e.g. the lack of information regarding generated income and projected size of allocated budget over *several* years means that the frequency of budget allocation from higher echelons (e.g. yearly) directly constrains the horizon over which effective strategy can be made
[65] thus, highly demanded resources presently allocated to Planning at lower levels (e.g. HOS, Executive) may be redistributed at higher levels for greater efficiency

Sub-step: Additional Representations

The type and cultures of the participating organisations have brought about the need for broad consultation and feedback. The diverse audience background and limited time available has necessitated the use of an additional sub-step, namely the use of rich pictures (Figure 4). They have allowed to promptly and effectively communicate the stakeholders' vision on the future VO options and have also acted as a primer to the complete, but more complex representations shown in Figure 3.

Step Four: Activity Model of the VO Design and Operation

The creation of the Activity model started by 'reading' the life cycle diagram of the VO to be designed and its relations with the other entities, as shown in Figure 2. The set of activities obtained was then recursively refined and decomposed into sub-activities, until reaching a level of detail deemed sufficient to control specific VO creation tasks. The model had to be regularly submitted to the stakeholders for consultation and validation, thus ensuring continuing management support. In addition, the model was also presented for comments to the various working groups, for feedback and to obtain commitment towards the actual use of the model. The chosen modelling language (IDEF0) has allowed to develop an integrated set of diagrams on various levels of detail, which have allowed the Working Groups (WG) involved in the VO creation to operate in parallel within a consistent framework.

Figure 5 describes the second level (A0 diagram) of the IDEF0 activity model describing VO design and operation. The modelling tool used was KBSI[66] AI0Win, which is web enabled and integrated with information / resources modelling (SmartER) and behaviour / simulation modelling (ProSIM) tools by the same vendors, thus ensuring model(s) consistency[67].

A detailed description of the activity model creation and structure (which is beyond the purpose of this paper) is contained in (Noran, 2004b).

Step Five: Other Aspects to be Modelled

Due to the particulars of the organisations involved in the case study, the meta-methodology has identified several additional aspects to be modelled, such as:

- Organisational culture and design: e.g. gap analysis, gap reduction strategies for the organisations participating in the VO, change management strategies;
- A time dimension for the set of activities in the VO creation project;
- AS-IS / TO-BE comparative costs to the University
- IS / IT infrastructure associated with production and customer service.

Modelling of such aspects must employ relevant available reference models whenever possible. For example, the first aspect has used (Keidel, 1995) to identify the organisational pattern of the schools involved, and has used gap analysis concepts from (Ansoff, 1965; Howe, 1986) and the (Kotter, 1996) eight-stage organisational change process.

---

[66] Knowledge-Based Systems, Inc. www.kbsi.com
[67] the IDEF family of languages is presently not integrated by a published metamodel (Noran, 2003); thus, consistency across models must be enforced by the modelling tools or the user.

The resulting method and its by-products have been well received, allowing working groups to refine their concepts, to develop and communicate several models and to use an ordered and consistent activity set throughout the VO creation project.

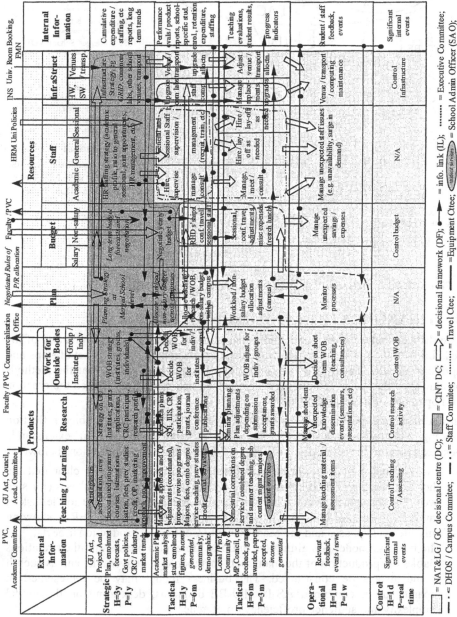

Figure 3. To-BE Decisional Model (Noran, 2004a)

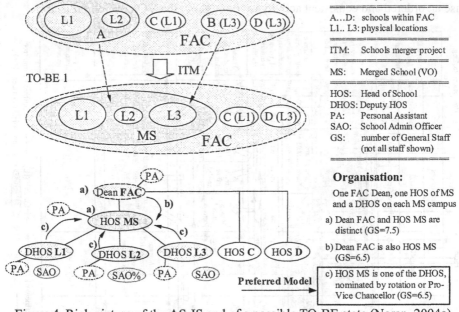

Figure 4. Rich picture of the AS-IS and of a possible TO-BE state (Noran, 2004a)

4. TRIANGULATION AND REFLECTION

The different nature of the case studies has allowed an effective triangulation of the feasibility of the meta-methodology concept and its content - i.e. the set of steps and application rules. Thus, in two different cases it has been found that the methods produced by the meta-methodology have helped and effectively guided the creation and operation of BEs and CNOs.

In addition, as expected and desired, the second case study has enabled reflection leading to the refinement and extension of the meta-methodology. This extension is best expressed in the revised set of steps and their application descriptions below:

- Entity identification should be performed in all cases. It is now clear that the chosen modelling formalism should be able to express life cycle phases (for step two). The formalism needs to be restricted to the aspects relevant to the task at hand. For example, both case studies have used a simplified GERA, showing only life cycle phases and the Management / Control vs. Production aspects;
- The business model containing the entities' life cycles phases and their relations enables the production of the main meta-methodology deliverable, and as such it should always be constructed. The business model may be represented in separate, or in combined AS-IS /TO-BE diagrams, depending on the model complexity and audience proficiency / preference;

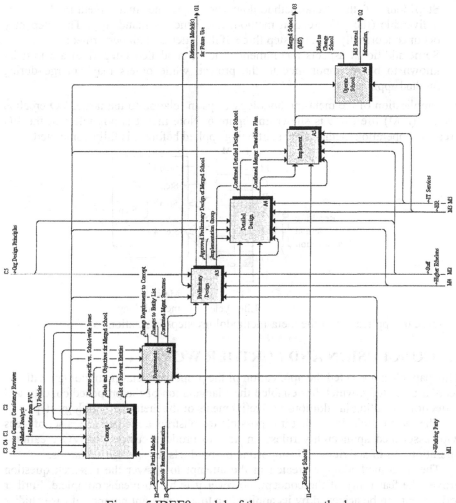

Figure 5 IDEF0 model of the merger method

- Modelling the present (AS-IS) state of the participating organisations is needed especially if it is not fully understood by stakeholders or if the future state can be obtained by *evolving* rather than replacing the AS-IS. Modelling of the future (TO-BE) state(s) is needed especially if the stakeholders have not yet agreed on the structure of the future organisation. AS-IS and TO-BE models may be required for several aspects, such as functional (and decisional), informational, organisational, resources, etc[68]. A modelling formalism is to be chosen based on audience proficiency and suitability and a primer to the chosen formalisms has to be provided. Additional representations may also be needed to assist understanding by the stakeholders and to communicate their vision to the staff;

---

[68] for example, if organisational design is needed, then modelling of the TO-BE decisional and organisational states has to be performed

- Step four of the meta-methodology produces the main meta-methodology deliverable (the VO creation method) and hence is mandatory. This step may occur concurrently with the step three, if the latter is at all performed;
- Some additional aspects recommended to be modelled (step five) are initially known to be of importance to the project, while others may emerge during method application.

The application of the meta-methodology steps in relation to the target VO creation project (ITM) life cycle is shown in Figure 6. Note that it is possible that the VO creation / operation method may start being applied before it is fully completed.

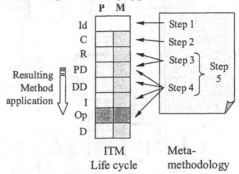

Figure 6. Application of the meta-methodology steps in relation to ITM life cycle

5. CONCLUSION AND FURTHER WORK

This paper has presented the application of the proposed meta-methodology within a second case study, which has enabled the triangulation of the proposed concept and reflections resulting in additions and refinements of the meta-methodology steps.

The main conclusion is that it is possible to construct a method (i.e. a set of steps with associated applicability rules) on how to produce customised VO creation / operation sets of activities and supporting by-products for specific scenarios.

The meta-methodology created in the attempt to answer the research question proves the feasibility of the concept, however it is by no means complete. Further refinement can be achieved by its application to a significant number of case studies and by incorporating BE, CNO and VO creation and operation knowledge from various streams of research. The open character of the meta-methodology allows its extension subject to some basic constraints, such as avoiding contradicting the existing rules (e.g. by performing a reconciliation of the existing and proposed steps / application rules, or by creating profiles for project categories) or preventing knowledge scattering over an excessive amount of steps and application rules.

The space and scope limitations of this paper do not allow the in-depth presentation of the meta-methodology; however, a full description of the findings and contributions are described in (Noran, 2004b).

REFERENCES

Ansoff, H. I. (1965). Corporate Strategy. New York: McGraw-Hill.

Bernus, P., Noran, O. and Riedlinger, J. (2002). Using the Globemen Reference Model for Virtual Enterprise Design in After Sales Service. In I. Karvoinen *et al.*

(Eds.), Global Engineering and Manufacturing in Enterprise Networks (Globemen) VTT Symposium 224. Helsinki / Finland. 71-90.

Camarinha-Matos, L. M. (2002). Foreword. In L. M. Camarinha-Matos (Ed.), Collaborative Business Ecosystems and Virtual Enterprises (Proceedings of PROVE02: 3rd IFIP Working Conference on Infrastructures for Virtual Enterprises) (Sesimbra / Portugal: Kluwer Academic Publishers.

Camarinha-Matos, L. M. and Afsarmanesh, H. (1999). Infrastructures for Virtual Enterprises: Kluwer Academic Publishers.

Checkland, P. (1991). From Framework through Experience to Learning: the Essential Nature of Action Research. In H.-E. Nissen, *et al.* (Eds.), Information Systems Research: Contemporary Approaches & Emergent Traditions. Amsterdam: Elsevier.

Checkland, P. and Scholes, J. (1990). Soft Systems Methodology in Action. New York: J. Wiley & Sons.

CIMOSA Association. (1996). CIMOSA - Open System Architecture for CIM,. Technical Baseline, ver 3.2. Private Publication.

Doumeingts, G. (1984). La Methode GRAI (PhD Thesis). Bordeaux, France: University of Bordeaux I.

Doumeingts, G., Vallespir, B. and Chen, D. (1998). GRAI Grid Decisional Modelling. In P. Bernus *et al.* (Eds.), Handbook on Architectures of Information Systems. Heidelberg: Springer Verlag. 313-339.

Eden, M. and Chisholm, R. F. (1993). Emerging Varieties of Action Research: Introduction to the Special Issue. Human Relations, 46, pp. 121-142.

Galliers, R. D. (1992). Choosing Information Systems Research Approaches. In R. Galliers (Ed.), Information Systems Research - Issues, Methods and Practical Guidelines. A.Waller Ltd. 144-162.

Globemen. (2000-2002). Global Engineering and Manufacturing in Enterprise Networks. IMS project no. 99004 / IST-1999-60002. Available: http://globemen.vtt.fi/.

Goranson, H. T. (1999). The Agile Virtual Enterprise. Westport: Quorum Books.

Hartel, I., Billinger, S., Burger, G. and Kamio, Y. (2002). Virtual Organisation of the After-sales Service in the One-of-a-kind Industry. In L. Camarinha-Matos (Ed.), Collaborative Business Ecosystems and Virtual Enterprises (Proceedings of PROVE02: 3rd IFIP Working Conference on Infrastructures for Virtual Enterprises). Sesimbra / Portugal. pp. 405-420.

Howe, W. S. (1986). Corporate Strategy. London: MacMillan Education.

ISO/TC184. (2000). Annex A: GERAM. In ISO-2000 (Ed.), ISO/IS 15704: Industrial automation systems - Requirements for enterprise-reference architectures and methodologies.

Keidel, R. W. (1995). Seeing Organizational Patterns: a New Theory and Language of Organizational Design. San Francisco: Berrett-Koehler Publishers.

Kotter, J. P. (1996). Leading Change. Boston, MA: Harvard Business School Press.

McKay, J. and Marshall, P. (2001). The Dual Imperatives of Action Research. Information Technology & People, 14(1), pp. 46-59.

Menzel, C. and Mayer, R. J. (1998). The IDEF Family of Languages. In P. Bernus *et al.* (Eds.), Handbook on Architectures of Information Systems. Heidelberg: Springer Verlag Berlin. pp. 209-241.

NIST. (1993). Integration Definition for Function Modelling (IDEF0) (Federal Information Processing Standards Publication 183): Computer Systems Laboratory, National Institute of Standards and Technology.

Noran, O. (2001). Research Design for the PhD Thesis: Enterprise Reference Architectures and Modelling Frameworks, (Report / Slides). School of CIT, Griffith University. Available: www.cit.gu.edu.au/~noran [2001, 2001].

Noran, O. (2003, Dec 2003). UML vs. IDEF: An Ontology-oriented Comparative Study in View of Business Modelling. In Proceedings of 6th International Conference on Enterprise Information Systems (ICEIS 2004), Porto (Portugal).

Noran, O. (2004a). Application of the Meta-methodology for Collaborative Networked Organisations to a University School Merger, [Report to the IT School Merger Project Working Party]. School of CIT, Griffith University.

Noran, O. (2004b). A Meta-methodology for Collaborative Network Organisations, PhD Thesis. School of CIT, Griffith University.

Noran, O. (2004c). Towards a Meta-methodology for Cooperative Networked Organisations. In Proc. of 5th IFIP Working Conference on Virtual Enterprises (PRO-VE04), Toulouse / France.

Olegario, C. and Bernus, P. (2003). Modelling the Management System. In P. Bernus, *et al.* (Eds.), Handbook on Enterprise Architecture. Heidelberg: Springer Verlag. pp. 435-500.

Uppington, G. and Bernus, P. (2003). Analysing the Present Situation and Refining Strategy. In P. Bernus *et al.* (Eds.), Handbook on Enterprise Architecture. Heidelberg: Springer Verlag. pp. 309-332.

Williams, T. J. (1994). The Purdue Enterprise Reference Architecture. Computers in Industry, 24(2-3), pp. 141-158.

Wood-Harper, A. T. (1985). Research methods in IS: Using Action Research. In E. Mumford, *et al.* (Eds.), Research Methods in Information Systems - Proceedings of the IFIP WG 8.2 Colloquium. Amsterdam: North-Holland. pp. 169-191

11. An Ontological Approach to Characterising Enterprise Architecture Frameworks

Oddrun Pauline Ohren
SINTEF, Norway. Email:[oddrun.ohren@sintef.no]

Currently, several enterprise architecture frameworks exist, and there is a need to be able to communicate about them. To this end we have proposed an Architecture Framework Ontology (AFO) providing characteristics to be assigned to the framework under consideration. AFO is then used to characterise and compare six existing frameworks, and results from this task are presented.

1. INTRODUCTION

During the last couple of decades, quite a few enterprise architecture frameworks have been presented. An architecture framework (AF) may be viewed as *a set of rules, guidelines and patterns* for describing the *architecture* of *systems*. Traditionally, in this context architecture means *information system* architecture or *software* architecture. However, a growing understanding of the importance of the context within which an application is to operate, gave rise to a "new" type of architecture, namely enterprise architecture and with it enterprise AFs. While such frameworks differ considerably in a number of respects, their application domains tend to overlap. The fact that there exist several potentially useful frameworks for any architecture effort creates the need for users to be able to assess frameworks and compare them with each other. Given their relative comprehensiveness and complexity, this is no easy task.

Although some work is performed concerning architecture framework issues (Mili, Fayad *et al.*, 2002; Martin and Robertson, 2003), we have not been able to find a simple framework for assessment and comparison of AFs.

As a first step towards such a mechanism, this paper outlines an ontology for characterising AFs. (A short version of this paper (Ohren, 2004) is presented in the poster session at the EMOI- INTEROP 2004 Workshop).

Such an ontology provides a *vocabulary*, a conceptualisation for communicating about architecture and AFs. It also supports *comparison* between AFs, as it points to distinguishing features of AFs.

Also, when embarking on an architecture project it is important to choose an AF that fits the task at hand. Although the proposed ontology is not primarily directed towards matching problems with AF, it does identify the characteristic features of a framework, which should help evaluating its suitability for the case at hand.

1.1 Related work

While some interesting work on AF issues has been published, this is as yet not a very developed research area, especially when it comes to enterprise AFs. (Dobrica and Niemelä, 2002) and (Medvidovic and Taylor, 2000) study two different aspects of *software* architecture, namely architecture analysis methods and architecture description languages, each proposing a framework for classification and comparison of analysis methods and description languages, respectively. As for enterprise architecture, there exists a few studies related to specific frameworks, for example (Cook, Kasser *et al.*, 2000) assessing the defence-oriented C4ISR[69] AF and (Goethals, 2003), studying the architecture products prescribed in several frameworks. (Martin and Robertson, 2003) performs an in-depth comparison of two distinct enterprise framework types, whereas (Mili, Fayad *et al.*, 2002) identifies some important issues related to the *management* of enterprise AF in general.

2. CHARACTERISING ARCHITECTURE FRAMEWORKS

In this chapter we propose a conceptualisation for talking about AFs, evaluating them and relating them to each other.

2.1 Ontological basis

A study of six enterprise AFs (Ohren, 2003) forms the basis for the ontology. These frameworks are: Federal Enterprise Architecture Framework (FEAF) (Chief Information Officers (CIO) Council, 1999), Department of Defense Architecture Framework (DoD AF) (Department of Defense Architecture Framework Working Group, 2003), Treasury Enterprise Architecture Framework (TEAF) (Department of the Treasury CIO Council, 2000), Zachman Framework (ZIFA), The Open Group Architectural Framework (TOGAF) (The Open Group, 2002) and Generalised Enterprise Reference Architecture and Methodology (GERAM) (IFIP-IFAC, 2001).

The proposed ontology has also been influenced by the ongoing work on MAF (Aagedal, Ohren *et al.*, 2003). MAF is a high-level *model-based* AF, implying a strong focus on models as the main formalism for describing architectures. The MAF metamodel (the model of which MAF is an instance) has greatly inspired the identification of the distinguishing features forming the framework ontology presented here.

2.2 Designing the Architecture framework ontology (AFO)

The framework ontology is realised as a class hierarchy with Architecture framework characteristic as the top node. The leaf nodes are instantiated, forming a set of concrete characteristics to be applied to the AF under consideration.

The framework ontology was designed according to the following guidelines:

- The concepts should *characterise well*, that is, refer to features that are *important* in an AF, i.e. important for deciding if and how to utilise the framework in a specific situation.

---

[69] Command, Control, Communications, Computers, Intelligence, Surveillance and Reconnaissance

- The concepts should refer to features that *discriminate* well between the AFs. However, this consideration may be over-shadowed by the *importance* requirement.
- The concepts should be clearly and thoroughly defined, with particular regard to revealing implicit similarities between the frameworks, especially in cases where the similarity is obscured by diversity in vocabulary.

The framework ontology consists of the following top level concepts, each representing a type of characteristic of AFs: Application domain characteristic, Ontological characteristic, Prescription for ADs, Methodological characteristic, and Relation to other frameworks.

Table 1 lists the classes of AFO in the left column, representing *types of characteristics* of AFs. Subclasses are indented. The corresponding instances are listed to the right, representing *concrete* characteristics to be assigned to the AFs.

Table 1 The classes and instances of AFO

Concept	Instance (individual characteristic)
Application domain characteristic	
System scope	System type in general, US Department of Defense, US Department of Treasury, US Federal Government
System type	Enterprise, Software system
Ontological characteristic	
Ontology scope	Application domain, Architecture
Ontology form	Formal concept model, Glossary of terms
Prescription for ADs	
Prescription regarding AD content	Enumeration of products, Implicit specification of products
Prescription regarding AD organisation	Architecture domain, Analytical approach,, Life-cycle phase, Stakeholder, Level of system abstraction
Prescription regarding AD representation	Formal representation, Informal representation
Methodological characteristic	AD Development process, Architecture evolution support, Principles of conformance and consistency
Tool support characteristics	Full tool support, No tool support, Some tool support.
Relation to other framework	Developed from, May be combined with, Used in, Uses

Note that AFO should not be viewed as a fixed and complete classification scheme for AFs. The currently defined instances mainly have their origin from the AFs in the study, and do not form a complete set. However, introducing additional characteristics in the ontology is not a problem; there is no requirement of the instances being semantically disjunctive. Nor does AFO require the set of instances of a class to span the class. On the other hand, defined associations between classes make it possible to relate instances to each other in various ways. For example, instances of the characteristic type System scope may be internally related by the included in relation. We do not require that a complete and disjunctive set of scopes be chosen as instances. Instead, the user is free to introduce a scope that fits the AF at hand. The important thing is that *we know it is a scope*.

2.3 The concepts of AFO

In this section we will look into the main concepts of AFO in more detail.

The application domain of an architecture framework

According to the IEEE 1471 standard (IEEE, 2000) 'architecture' is the 'fundamental organization of a *system* embodied in its components, their relationships to each other, and to the environment, and the principles guiding its design and evolution'. The term 'system' is defined as 'a collection of components organized to accomplish a specific function or set of functions'. The term 'system' being extremely generic, its range of instances might be expected to be somewhat heterogeneous. Therefore, when evaluating an AF, it is crucial to know what kind of systems it is intended to serve.   In AFO this is represented by the concepts Application domain characteristic, with subtypes System type and System scope. Examples of system types are *software system* and *enterprise*. System scope is a feature intended to restrict the System type, and may be various things like geographical scope (e.g. Norwegian), industrial branch (e.g. chemical industry) and others.

AFs as providers of conceptualisations of architecture

Most AFs use their own more or less proprietary vocabulary when describing architectures. Our study of the frameworks mentioned above shows that there is a tendency to supply commonly used words with very specific semantics. Moreover, different frameworks often use the same word in different meanings. This fact tends to obscure both differences and similarities between frameworks, especially if the frameworks do not provide explicit definitions of their key terms.

The Ontological characteristic indicates whether the framework provides ontologies   for the architecture domain and relevant aspects of the application domain, and possibly other relevant areas. Two subtypes are defined for this characteristic, Ontology scope and Ontology form. Ontology scope indicates the domains covered by the ontology (e.g. architecture), whereas the Ontology form indicates how the ontology is realised in the framework.

AFs as templates for architecture descriptions

Most AFs contain prescriptions concerning the architecture description as artifact. While their degree of specificity varies considerably, these prescriptions usually deal with what information should be included in the architecture description, how it should be structured and sometimes how it should be represented.

The Framework ontology includes four concepts to this end; Prescription for ADs and its subtypes Prescription regarding AD content, Prescription regarding AD organisation, and Prescription regarding AD representation.

AFO presently contains four instances of Prescription regarding AD organisation, each of which constituting a structuring criterion for the AD products. These are:

- Architecture domain: In the context of enterprises, it is common to recognise three or four types of architecture, each corresponding to its particular domain: Business architecture, Information system architecture (often subdivided into Data architecture and Applications architecture) and Technology architecture. Note: Architecture domain as a structuring criterion for a collection of architecture products should not be confused with the application domain of the framework as such.
- Analytical approach: Analysing a system often implies focusing on one angle at a time. This criterion orders the architecture products according to the angles applied in the various analyses producing the products. Typical examples of analytical approaches are: Functional analysis (how does the system operate?), Structural analysis (which components do the system consist of, and how are they structured?), Spatial analysis (at which locations does the system reside, how is it distributed?) and Information analysis (what information is handled in the system?).
- Stakeholder: According to IEEE 1471, any system has one or more stakeholders, each of which has certain interests in that system. In cases where stakeholders are defined solely in terms of another criterion, the two criteria will result in the same logical partitioning, although the level of granularity may differ.
- Life-cycle phase: Organises the architecture products according to the life-cycle phase (e.g. design phase) addressed by the product.
- Level of system abstraction: Organises the architecture product according to the level of abstraction at which the system in question is described in the product. Examples of system abstractions are: *Physical manifestation* of the system, *Implementation* of the system and Purpose of the system.

AFs and architecture development methodology

Describing an existing or future architecture is often a major undertaking, and it is by no means obvious how to approach such a task, even if one uses an AF which details the content of the resulting description. *Whether* and eventually *how* an AF supports the architecture description development process is an important feature of the framework.

To indicate the possible methodological support offered by a framework, the concept Methodological characteristic with subtype Tool support characteristic are used. Examples (instances) of Methodological characteristic are whether the framework specifies an AD development process, or provides any Architecture evolution support, whether it is supported by software tools, etc. Examples of Tool support characteristic are No tool support and Full tool support.

Relations between architecture frameworks

Some frameworks are related to each other in various ways. Whether factual (declared) or purely conceptual, some of these relations reveal important information about the background of the participating frameworks, in which case they should be identified and documented. Examples of instances are Used in between two frameworks of which one is used as a part of the other and Developed from between frameworks where one is part of the history of the other.

3. COMPARING SIX ENTERPRISE ARCHITECTURE FRAMEWORKS

In this chapter AFO is used to characterise the six frameworks from the study. For reasons of space, we focus on a subset of AFO, discussing only the characteristic types believed to be the most interesting or challenging. For each chosen characteristic class, all six frameworks will be assigned one or more of its instances (concrete characteristics). The findings are summarized in Table 2.

3.1 Application domain

System type and System scope

All frameworks in the study claim to be *enterprise* AFs. For all except GERAM, the interpretation of 'enterprise architecture' is 'architecture of information systems in an enterprise context' rather that 'architecture of the enterprise as such'. However, all the frameworks prescribe inclusion of key operational and organisational information; hence it is fair to assign Enterprise as System type to all of them.

As for System scope, three of the AFs are designed by and for the US Government, hence their system scope must be specified accordingly: The scope of FEAF is the US Federal Government in general, while DoD AF and TEAF have narrower scopes; US Department of Defense and US Department of Treasury, respectively.

The frameworks TOGAF, GERAM and Zachman do not specify any restrictions regarding scope, hence they are applicable to any system of the type given by their System type characteristic, meaning they will be characterised by the System scope instance called System type in general.

AFO Shortcomings

Even though FEAF, DoD AF and TEAF are designed for use within the US Government, this might be more of a *declared* restriction than a *factual* one. For example, FEAF has few, if any, features that makes it unusable in other enterprises than the US Government.   The distinction between declared and factual delimitations of the application domain is not as yet addressed in AFO.

FEAF aims to provide support for developing Federal wide and multi-departmental architectures, as well as a high level structure with which to link more specific architecture efforts within a single department. This information is not easily expressed with AFO, although part of it may be conveyed by using the association included in defined between system scopes. Thus the scopes of DoD AF and TEAF may be modelled as parts of the scope of FEAF.

3.2 Prescriptions about the architecture descriptions

Prescriptions regarding AD content

Within the six frameworks studied, there are great variations concerning the level of detail and abstraction at which the frameworks specify the content of an architecture description.

The most specific, thorough and formal frameworks in this respect are DoD AF and TEAF, both enumerating specific architecture products to be included in the architecture description, along with prescriptions for data representation. TEAF has derived its list of products from the Department of Defense's C4ISR AF (C4ISR Architecture Working Group, 1997) (not part of our study). Moreover, DoD AF is generally based on C4ISR, hence the product collections specified by TEAF and DoD AF turn out to be very similar, although TEAF naturally has had to adapt the products for use by Treasury.

FEAF also enumerates a list of architecture products to be developed, although this is less comprehensive and far less formal.

The three remaining frameworks, TOGAF, GERAM and Zachman are, as we have seen, more generic regarding application domain (System scope was set to System type in general), and are naturally less specific when listing architecture products. However, all of them specify in more or less general terms the *types* of deliverables or products that should be developed to form the architecture description. A further distinction exists between TOGAF and the others. TOGAF focuses more or less exclusively on methodology, and its specification of architecture products is to be considered more as an example than a prescription.

Prescription regarding AD organisation

With the exception of TOGAF, all the frameworks in the study employ some kind of structuring principle(s) upon the AD. For AFs offering methodological guidance, such criteria do not merely organise the architecture products, but are usually also reflected in the development process. The number of structuring principles prescribed in a single framework varies from 1 to 3. The result of applying them is a *partitioning* of the set of architecture products into subsets, each of which forming a logical unit expressing a *view* on the system in question. The purpose of splitting up the collection of architecture products is invariably to simplify and adapt the AD description to individual user groups, while internally maintaining an integrated, consistent and complete AD. Thus, capturing and visualising the whole architecture without getting lost in details is made possible.

Zachman organises the architecture products in a matrix where the rows are defined by a specific set of *stakeholders* (Planner, Owner, Designer, Builder, and Subcontractor), and the columns are called *dimensions* obtained by applying interrogatives (what, who, where...) when eliciting information about the system. Each stakeholder is defined in terms of his interest in an architecture domain, hence the AFO terms Stakeholder and Architecture domain are assigned to Zachman, even though the two characteristics are redundant. Zachman's *dimension* concept corresponds to the AFO term Analytical approach, hence this term is added to the characterisation of Zachman's structuring principles.

FEAF defines four levels at which to view the architecture. Going from level 1 to 4 implies adding detail for each level. The metaphor used to illustrate this is the observer's distance from the system. The closer one gets, the more details are visible. Note that this model does not impose an organisation of the architecture description, but a kind of zoom effect, presently not possible to represent by AFO. The actual description products are to be found at level 4. Here an adaptation of Zachman is used to organise the architecture products. The rows are still stakeholders, and the columns, although labelled as architecture domains (data, applications, technology), by inspection have more in common with Analytical approach than with Architecture domain. Hence, the structuring of architecture products in FEAF may be characterised by Stakeholder, Architecture domain and Analytical approach.

TEAF also bases its organisation of architecture products on Zachman, but has simplified it to four rows and four columns. However, the rows (in TEAF called *perspective*) still represent stakeholders defined like those in Zachman and FEAF, hence should be characterised by Stakeholder and Architecture domain. The columns (in TEAF named Functional, Information, Organisational and Infrastructure *views*) seem best characterised by Analysis approach, although the Infrastructure column (as opposed to the Technology architecture column of FEAF) is technology oriented even at the business level, and as such may be said to cover the technology domain. This represents a deviation from the original Zachman matrix, in which the technology domain is covered by the Builder's row rather than the 'where' column.

DoD AF operates with a fixed partition of three elements called *views* (Operational, System and Technical views) as the sole organising principle. The documentation of the views implies a close correspondence to the Architecture domain criterion, the DoD AF views covering the Business, Information system and the Technology domains, respectively. A comparison with TEAF, which prescribes the same products as DoD AF, confirms this. DoD AF's Operational view corresponds to the perspective of Planner and Owner (Business domain) in the TEAF matrix, while the System view roughly corresponds to the perspective of Designer and Builder. Technical view corresponds to the Infrastructure column of the Planner and Owner perspective in TEAF, hence covering the technology domain (see discussion on TEAF in the paragraph above).

GERAM uses a three-dimensional structuring principle, consisting of life-cycle phase dimension, genericity dimension (generic, partial and specific), and a third called view dimension (content view, purpose view, implementation view and physical manifestation view). Although the view dimension has similarities to the Stakeholder dimension in the Zachman varieties, we chose to assign Life-cycle phase and Level of system abstraction as the organising characteristics of GERAM. So far, the genericity dimension can not be expressed in AFO.

TOGAF does not prescribe any specific structuring of architecture products. However, TOGAF ADM (methodology part) is partly organised according to architecture domains (first build business models, then information system models, etc).

Methodological characteristics

In general, the frameworks in our study focus on the specification of the architecture products rather than how to generate them. However, it is equally true that most of them do offer *something* in the way of a method or approach.

FEAF offers guidelines for architecture development in a separate document, in which an eight step development process is outlined. Also, FEAF talks explicitly about *as-is* architecture, *target* architecture and *transitional* processes (from as-is to target), hence architecture life-cycle issues are addressed. FEAF is therefore to be characterised by the AFO instances AD Development process and Architecture evolution support.

TEAF does not include an AD development process, but encourages each bureau to supplement TEAF with a development methodology suited to its need. As for architecture life-cycle issues, TEAF does address things like Enterprise Architecture Roadmap and Enterprise Transition Strategy. TEAF is therefore to be characterised by the AFO instance Architecture evolution support.

Table 2 Summary of key findings

Characteristic	FEAF	DoD AF	TEAF	Zachman	TOGAF	GERAM
Prescription for AD content						
Explicit enumeration of products	X	X	X			
Implicit specification				X	X	X
Prescriptions for AD organisation						
Arch.itecture domain	X	X	X	X	(X)	
Analytical approach	X		X	X		
Life-cycle phase						X
Stakeholder	X		X	X		
Level of system abstraction						X
Methodological characteristics						
AD development process	X	X		X	X	
Architecture evolution support	X		X		X	
Relations to other frameworks						
Uses	Zach.		Zach. C4ISR			
Used in				FEAF TEAF		
Developed from		C4ISR				
May be combined with					any	

DoD AF offers guidance for architecture development in terms of a briefly described six step development process. Although DoD AF mentions architecture life cycle in the latest version, it is still pretty much a framework for a snapshot architecture. DoD AF is therefore to be characterised by the AFO instance AD Development process.

Zachman is a commercial product around which a number of services are offered, including architecture development guidance in the form of seminars. Zachman is therefore to be characterised by the AFO instance AD Development process

In many respects GERAM is more generic than the other frameworks. For example, its guidance concerning methodology is on a meta level compared to the others, as GERAM is mainly concerned with giving directions for methodology development. Hence, it is not evident how the existing characteristics in AFO can be assigned to GERAM and at the same time conveying the difference in abstraction level.

TOGAF has methodology as its main concern, and is definitely to be characterised by AD Development process and Architecture evolution support.

Relations between frameworks

Our study of the six AFs has indicated that several of them are related, for example by sharing each other's history. Both TEAF and DoD AF have derived their set of architecture products from C4ISR AF. Adaptations of Zachman form a major part of FEAF as well as TEAF. TEAF proclaims to be compliant with FEAF. TOGAF (ADM, the methodology part) may be used in combination with any of the others. The exact assignments of AFO terms to represent this are shown in Table 2.

4. CONCLUSION AND FURTHER WORK

Given the increasing inclination to cooperate both across enterprises and across borders within an enterprise on one hand, and the number of different enterprise AFs on the other, we have in this article argued for the need to be able to assess, compare and generally communicate about AFs as artifacts separate from the architecture descriptions that they produce. To this end we have proposed an Architecture Framework Ontology (AFO) providing *characteristics* to be assigned to the framework under consideration. AFO has been used to characterise and compare six existing frameworks. Performing this task has revealed several shortcomings in the specific AFO as well as the general approach, and as such contributed to the future research agenda.

AFO is mainly focused on identifying distinguishing features of AFs. However, while it is useful to be able to compare AFs in a systematic way, there is also a need to perform *assessment* of frameworks, e.g. evaluate the suitability of a particular framework to the problem at hand. One way of obtaining this is to extend AFO with concepts related to architectural problem characterisation.

To support interoperability *within* and *between* enterprises, we need to be able to *relate architecture descriptions* created by different frameworks, preferably with some kind of computer support, which suggests a model-based approach to architecture descriptions. In a model-based world relating *descriptions* means relating *models*, which again requires a consistent mapping between their metamodels. This is, modestly phrased, a non-trivial task, and should get a lot of attention in the years to come.

Acknowledgments

This work is sponsored by the Norwegian Defence Logistics Organisation.

REFERENCES

Aagedal, J. Ø., O. Ohren, *et al.* (2003). Model-Based Architecture Framework, v 3.0: SINTEF Telecom and Informatics. 68.
C4ISR Architecture Working Group (1997). C4ISR Architecture Framework, v. 2.0: US Department of Defense.

Chief Information Officers (CIO) Council (1999). Federal Enterprise Architecture Framework. V. 1.1. [Washington DC, USA]: Chief Information Officers Council (CIO Council).

Cook, S., J. Kasser, *et al.* (2000). Assessing the C4ISR Architecture Framework for the Military Enterprise. International Command and Control Research and Technology Symposium, 5. 24-26 October 2000, Canberra, Australia.

Department of Defense Architecture Framework Working Group (2003). DoD Architecture Framework. Vol. I-II; Deskbook. Available from: http://aitc.aitcnet.org/dodfw/

Department of the Treasury CIO Council (2000). Treasury Enterprise Architecture Framework. V.1. [Washington DC, USA]: Department of the Treasury.

Dobrica, L. and E. Niemelä (2002). "A survey on software architecture analysis methods." IEEE TRANSACTIONS ON SOFTWARE ENGINEERING 28(7): 638-653.

Goethals, F. (2003). An Overview of Enterprise Architecture Framework Deliverables; A study of existing literature on 'architectures'. http://www.econ.kuleuven.ac.be/leerstoel/SAP/downloads/Goethals%20Overvie w%20existing%20frameworks.pdf

IEEE (2000). IEEE Std 1471-2000: IEEE Recommended Practice for Architectural Description of Software-Intensive Systems: IEEE,.

IFIP-IFAC (2001). GERAM: Generalized Enterprise Architecture Architecture and Methodology. ISO 15704:2000. Industrial automation systems; Requirements for enterprise reference architecture and methodologies. Annex A. GERAM. T. C. 184: ISO.

Martin, R. and E. Robertson (2003). A comparison of frameworks for enterprise architecture modeling. Conceptual Modeling - Er 2003, Proceedings. Berlin: SPRINGER-VERLAG BERLIN. 2813. 562-564.

Medvidovic, N. and R. Taylor (2000). "A classification and comparison framework for software architecture description languages." IEEE TRANSACTIONS ON SOFTWARE ENGINEERING 26(2000)(1): 70-93.

Mili, H., M. Fayad, *et al.* (2002). "Enterprise frameworks: issues and research directions." Software-Practice & Experience 32(8): 801-831.

Ohren, O. P. (2003). Rammeverk for beskrivelse av virksomhetsarkitektur [in Norwegian]. Oslo: SINTEF. 61.

Ohren, O. P. (2004). Ontology for characterising architecture frameworks. To be presented at the INTEROP Workshop "Enterprise modelling and ontologies for interoperability" EMOI - INTEROP., Riga, Latvia.

The Open Group (2002). The Open Group arcitectural framework (TOGAF), version 8. "Enterprise edition". Reading, UK and San Francisco, USA: The Open Group. Available on the web at http://www.opengroup.org/architecture/togaf8/.

ZIFA The Zachman Institute for Framework Advancement. http://www.zifa.com

Coherent Models of Processes and Human Systems

Weston R. H.

Loughborough University Email: R.H.Weston@lboro.ac.uk

Enterprise processes are characterised to specify a conceptual model of MEs. The model is developed, with reference to processes classes, resource system types, product flows, and organisational views, so as to exemplify general interoperability needs and to highlight deficiencies in current EM and EI provision. One significant deficiency relates to modelling human resources. Hence ME enhancements are proposed centred on the coherent modeling of human systems and enterprise processes.

1. INTRODUCTION

The need for coherent models of enterprise processes and human resource systems is described with reference to a characterization of processes found in most Manufacturing Enterprises (MEs). Subsequently this characterization is used to identify means of overcoming observed deficiencies in the existing EM (Enterprise Modeling) provision.

2. CHARACTERISATION OF ME PROCESSES

2.1 Definition of Terms

According to Vernadat (1996): "processes represent the flow of control in an enterprise"; they constitute "a sequence of enterprise activities, execution of which is triggered by some event"; "most processes have a supplier of inputs and all have a customer using outputs". Scheer (1992) emphasised the dynamic nature of decision and action making about processes, with respect to (i) the need to transform material (physical) and informational (logical) entities, and (ii) resource allocation and the design of information systems. Processes are a conceptualisation of reality, not reality itself and exist over finite lifetimes; although multiple, similar process instants may be realised (Poli, 1996). Weston (1999) further explained that process models naturally define enterprise activity requirements in a reusable form; and that resource systems are needed to 'realise' those requirements within time, cost, flexibility and robustness constraints.

It follows that resource systems must possess functional abilities needed to realise instances of processes assigned to them. Functional abilities of technical resource systems (i.e. machines and software) are often referred to as 'capabilities'. Whilst functional abilities of human systems (i.e. teams, groups of people or individuals) are normally termed 'competences' (Ajaefobi 2004). Resource system organisation is achieved via both relatively enduring and short-lived structures such as methods, project plans, procedures, product structures, business rules, communication rules, role descriptions, workflow specifications, process routes, work to lists, state transition descriptions and the like.

Significant benefit can be gained by developing and reusing separate models of (1) processes and (2) candidate resource systems, with abilities to realise processes (Vernadat, 1996). Figure 1 conceptualises such a separation which is important in MEs where processes and resource systems often have distinctive life times and change requirements. For example the introduction of a new production philosophy may require a once only restructuring and re-engineering of enterprise activities, but various alternative resource systems may need to be deployed during the useful lifetime of the restructured process.

Organised Sets of Enterprise Activity

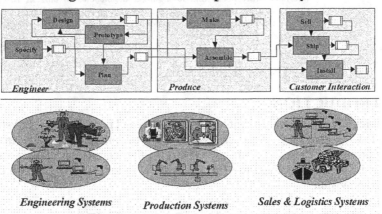

Engineering Systems   Production Systems   Sales & Logistics Systems

Human, Software & Machine Systems

Figure 1 Resourcing Processes in an Engineer-to-Order Manufacturing Enterprise
(Source: Weston 2003)

2.2 The Nature of ME Processes and Instances of ME Processes

Figure 2 shows conceptually how many MEs organise product realisation (Weston *et al* 2001). Implicit in this graphical representation are causal dependencies between products realised, processes needed, roles of resource systems and organisational boundaries

Assume that distinctive product families are realised by the alternative process flows depicted (i.e. by sequences of value adding activities) that need to be resourced for finite periods of time as needs for multiple process instants may arise. In a given ME: A1 might be a 'materials procurement' activity; A2 a 'sales order entry' activity; A8 a 'turn shaft' activity; A12 an 'assemble gear' activity; etc. Therefore with respect to individual process flows, opportunities arise to mass produce products, i.e. by involving multiple, sequential instances of the same (or a similar) process flow to produce large numbers of (similar) products. This can give rise to economies of scale, because the elemental activities that constitute a process, and their interrelationships, can over time be developed to (a) be effective and robust and (b) so that optimal resource utilisation can be achieved. It may also prove possible to invoke multiple, sequential instances of a single process flow to realise different batches of customised products, thereby giving rise to both *economies of scope* and *economies of scale*. Economies of scope may arise for similar reasons to those outlined for economies of scale but normally increased process flexibility (in terms of ability to cope with needed variations in product applications) will be needed and this can induce lead-time and cost penalties.

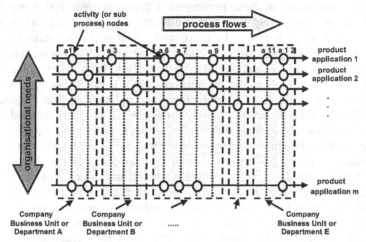

Figure 2 Process and Organisation Streams

Where products have significantly distinctive processing requirements, then distinct process flows may need to be created, also illustrated by Figure 2. New and existing process flows may share common product realising activity, but to meet differing product realisation requirements these activities may need to be linked differently, e.g. via distinctive 'physical flows' (e.g. of materials and products) and 'logical flows' (e.g. of information and control). Essentially this kind of organisation structure is both process and product oriented, so that a variety of product applications can be realised in quantities, and by due dates, required by customers. *Economies of scope* may come primarily from using a common resource set to realise sequential and concurrent instances of multiple process flows; as this increases opportunities to cost-effectively utilise available resource system capacity. But whether it arises in respect of having one or more distinctive process flows the down side is that it introduces organisational concerns: as indicated indirectly in Figure 2 by the introduction of dotted organisational boundaries. Aspects of those organisational concerns are discussed in the following.

We return to the earlier point that the notion of having process flows may constitute no more than abstract thought (Poli 2004). But real physical things (like a person supported by an order entry system; or a combination of machine operator, CNC machine, jigs and fixtures) are needed to do activities. Also real materials, sub products and products require physical movement, whilst logical entities like information need to be processed and physically moved to points of use. The human and technical resources deployed to realise activities must therefore have abilities required to do activities assigned to them. Also they must have the *capacity*, and be *available* and *willing,* to do the activities assigned to them within timeframes over which relevant process stream instances occur (Ajaefobi 2004). Indeed process flow specifications provide a time dependent organisational structure which can be referenced when synchronising individual and group resource behaviours within given time frames.

On considering the need to resource activities related to multiple process flows (and multiple instances of those processes) it becomes clear that some form of organisational boundary may be necessary to manage high levels of operational complexity, and to decide how to make both short and long term changes to processes and product applications. Clearly the size of MEs, product complexity, product variants, product volatility, production numbers, etc will determine how this might best be done, such as by forming company partnerships, business units, departments, manufacturing cells, production lines and so forth. But in general any such boundary is likely to impact (mainly for social but also for technical reasons) on process lead-times, costs, flexibility and robustness.

In many MEs a functional organisational paradigm is deployed, where cognate resource capabilities are grouped and assigned to similar activities. For example a sales office will be able to develop relevant (I) functional capabilities with regard to the human and technical resources it owns and (II) structural/organisational capabilities, in respect to the way in which it deploys its resources. By such means, for example, it will be able to schedule and seek to optimise the use of the resources it owns, so as to contribute collaboratively along with other organisational units to the execution of one or more process streams that constitute a specific ME.

Figure 2 is a simplification of the reality in actual MEs, but it does show process thinking in action: such as by conceptually defining how value is added by activities and providing a framework on which to 'anchor' specifications about needed resource capabilities, capacities and availabilities, as well as synchronisation needs, and related resource behaviours. Further cost, lead-time and other metrics can be attributed to process and product flows so as to calculate and predict revenues, etc. This in turn can lead to resource costing and efficiency calculations and help apportion appropriate budgets, costs and rewards to organisational units, or even individuals.

Over many decades the established process industry sector (populated for example by steel, petrochemical and pharmaceutical companies) has commonly organised its product realisation in a fashion characterised by Figure 2. In fact commonly they have done this from both 'logical' and 'physical' standpoints[70]: primarily because in this industry relatively large quantities of product need to be realised, with relatively little variation over relatively long time frames. In such a case it is appropriate to physically organise resource systems along process-oriented lines, as it can lead to robust, high quality, cost effective and short lead-time production and can much simplify organisational concerns. But intuitively one might expect that this kind of physical resource organisation can only be competitively applied where product variety and product volatility is relatively low.

---

[70] Here distinction is made between 'conceptualised' and 'physically realised' aspects, because in general ME processes and their underpinning resources may not adhere to both of these viewpoints.

Historically most (but by no means all) other industry sectors have preferred to physically organise their resources in (cognate) functional groupings, such as in engineering, financial, sales, machining, assembly, testing, packaging departments, sections, sub groups and teams. Despite this fact, recent literature reports how many (if not most) industries have become aware about potential benefits from process thinking. These benefits arise because a 'logical overlay' of product and process streams can be conceived, designed and mapped onto functionally organised groups of physical resources (be they people or machines). In fact the time-dependant usage of resources can then be driven largely by a logical overlay of control, data and material flows (conveyed by process thinking) despite actual resources being physically located into cognate functional groupings.

Naturally one might expect the choice of physical organisational structures in a specific ME to be influenced by (a) the stability, longevity, complexity, variability and robustness of the products and services currently realised and (b) the product and service realisation processes the ME has chosen to deploy to achieve product and service realisation. Such a theoretical stance is taken here because it is presumed that these (and possibly other) key factors will impact upon the rate at which existing and new processes need to (or have ability to) start and end, be resourced by suitable systems and be managed, maintained and changed (i.e. be reconfigured, improved, developed and/or replaced). Bearing these kinds of organisational concerns in mind, the next sub section brings out distinctions between common types of process found in MEs.

2.3 Common ME Process Types

A number of authors have classified ME processes. Table 1 compares and contrasts three such classifications developed independently by Salvendy (1992), Pandya *et al* (1997) and Chatha (2004). All three describe ME processes at a high level of abstraction.

Contrast for example MEs making computer products as opposed to MEs making roller bearing products. Clearly specific properties of instances of 'product development' and 'order fulfilment' processes needed to realise bearings will be very different to instances of the same process type used to produce computers. Whereas both ME types may usefully deploy fairly similar instances of 'business planning', 'obtain an order', and 'information management processes'. It should also be noted that the manufacturing strategies adopted by MEs will influence the nature of dependencies between process classes. For example a Make to Stock (MTS) ME is likely to have well decoupled instances of 'obtain an order', 'order fulfilment' and 'product and service development processes', each having different start times, cycle times, frequency of occurrence, etc. Whereas Engineer to Order (ETO) MEs will require occurrences of these three process types to be well synchronised.

It follows therefore that process type descriptions listed in Table 1 enable similarities and differences to be drawn between MEs. However all MEs are unique in that they:

- differently decompose process segments into organisational units
- resource processes and process segments differently

- have very different numbers and patterns of process instances, so that they can: realise large or small batches of products for customers; achieve lean, as opposed to agile manufacturing; and so forth.

Another important observation that can be drawn with reference to Table 1 is that 'operational processes' comprise those activities that should be repeated to realise products and services for customers. Whereas 'strategic processes' and 'tactical processes' should collectively ensure that all needed operational processes are specified, designed, implemented, resourced, managed, monitored, maintained, developed and changed through their lifetime, such that they continue to realise products and services of quality, on time and at an appropriate price for customers; whilst also ensuring that the ME achieves its defined purposes for stakeholders.

Table 1- Common ME Process Type Descriptions – from various authors

Salvendy (1992) Process Classification	Pandya *et al* (1997) Process Classification	Chatha (2004) Process and Activity Classification
Strategy Making Process	Generic Management Process Group, includes:

'Direction setting process'
'Business planning process'
'Direct business process' | Strategic Process: predominantly "what activities": that decide what the ME should do and develop business goals and plans to achieve the ME purposes defined |
| Product Planning & Development Process | | Tactical Process: predominantly 'how activities': that decide how segments of the business plan might best be achieved and as required specifying, designing, developing new products, processes & systems with ability to achieve business plans. |
| Manufacturing Support Process | Generic Operate Process Group, includes:

'Obtain an order process'
'Product and service development process'
'Order fulfilment process'
'Support fulfilment process' | |
| Production Operation Process | | Operational Process: predominantly 'do activities' that repetitively create products and services for customers, and thereby realise business objectives and goals |
| | Generic Support Process Group, includes:

'Human resource management process'
'Financial management process'
'Information management process'
'Marketing process'
'Technology management process' | |

Pandya *et al*'s (1997) process classification separates out a support process group, that is 'infrastructural' in nature, i.e. the purpose of this support group is to enable other process groups, rather than control or directly contribute to strategy, process, system, product or service realisation. Such a conceptual separation promotes separated execution and (re-)engineering of processes over appropriate timeframes.

Many similar, concurrent and/or sequential, instances of processes within Pandya (1997) 'generic operate group' may be required to satisfy customer demands. However the adoption of alternative manufacturing policies (such as 'make to stock', rather than 'assemble to order') and the deployment of different scheduling policies and resource configurations (such as by deploying 'synchronous dedicated production lines' as opposed to 'flexible manufacturing cells') will determine the frequency with which these processes must repeat and the variance needed between process instants so that necessary product customisation can be realised. As mentioned previously though, if similar process instances frequently occur then increased opportunity arises to continuously improve process repeatability, robustness and utilisation of resources which in turn can improve product quality, cost and lead-times. Also if the variance between process instants is well understood, and many process instants are likely to arise, then it may become appropriate to automate some (or even all) of the enterprise activities that constitute the process.

In most, possibly all MEs, it is probable that processes within Pandya *et al*'s (1997) 'generic management group' will not repeat often and will have significant variance between process instants. Indeed typically these process instants define and realise strategy making and tactical changes on project by project basis. 'Direction setting' and 'business planning' process instants may repeat annually (or possibly episodically in response to a new business threat or opportunity) but generally their constituent activities will require insight, analysis, prediction and innovative human-centred thought with respect to a new set of circumstances. Consequently opportunities to automate or continuously improve direction setting and business planning processes may not occur. Rather it will be important to ensure that the purpose, overall structural decomposition and means of managing these process types is well defined and possibly even more importantly that competent people (and teams) with sufficient motivation and time, are assigned suitable roles for enterprise activities that comprise generic management process instants. It is probable, for example, that only large-grained enterprise activity definitions can be specified and deployed, in relation to process instants of this type, and that the realisation of some of these activities will require the invocation of various 'child process instants' to which suitable human and technical resources are assigned. It follows that processes within the 'generic management group' may be recursive in nature, in as much that instances of high level processes may invoke multiple reporting instances of lower level process, but that outputs from these lower level process instances may significantly impact on the flow of higher level process instants.

In most MEs, except probably for 'technology management' process instants, instants of Pandya *et al*'s (1997) 'support process group' are likely to occur with predictable frequency and variance; albeit that they may require complex decision making and access to different data sets. Consequently multiple instants of these processes may best be realised by competent and capable resource systems comprising people whose activities are well structured by group productivity tools and well supported by personal productivity tools. Further, and possibly except for the 'technology management' process instants, support processes can be continuously improved so that they become robust, even standard.

2.4 The Need for Change Processes

Implicit within foregoing discussion is the notion that realising any change to ME processes and resource systems itself requires a process, i.e. a set of enterprise engineering activities that add value to the ME and need to be resourced by suitable human and technical resource systems. Such a *change process* can take many forms but many instants of change processes will be needed during the lifetime of any specific ME: because the environment in which it operates will change and because it will need to change itself to continue to operate competitively. At one extreme a complex change process may be needed following a merger or acquisition, which may comprise many lower level change processes, each comprising organised sets of enterprise activities. At another extreme a change process may be required to: program a production machine so that it can machine a new product; set up a production facility, so that it can manufacture a different batch of products; or deal with a known exception type and condition, e.g. as a customer modifies an order. Thus instances of change processes will range significantly as they may require large-scale, complex programmes, projects, processes and resource systems or alternatively, simple and predictable processes and resource systems. This is as one might expect because instances of all classes of process illustrated by Table 1 may need to be changed during the lifetime of MEs. We can deduce that change processes need to re-engineer fragments (or all) of MEs for some purpose and that in so doing they will require suitable: (1) change actors operating as part of an underpinning resource system and (2) models of the ME, focused primarily on issues of concern to the change actors used to realise change processes. However notwithstanding the specific aspects of concern that need to be modelled, it is important that these sub-models be appropriately positioned within the specific ME context, otherwise the changes specified, designed, implemented and maintained may not suit their intended purpose. Bearing in mind the forgoing observations about common ME process types, and the need for various change processes, in the next section common enterprise modelling and integration requirements are considered.

3. IMPLIED EM, EI AND INTEROPERABILITY REQUIREMENT

Previous sections characterise MEs from a process oriented viewpoint. However necessarily the discussion considered other viewpoints, including resource system, product flow, organisation and lifecycle views. Multi-perspective considerations were needed to cater for the high levels of complexity involved. The use of decomposition techniques and multi-perspectives is a common practice in MEs; as means of simplifying problem understanding and solution generation. Figure 3 has been constructed to illustrate common perspectives and decompositions deployed.

A process-oriented perspective usefully segments concerns about what the ME should do, by when. Day-to-day processes should add value (to materials, components and systems received from suppliers) in order to generate products and services for customers and benefits and profits for stakeholders. Whilst strategic and tactical processes should periodically plan, manage and support product and service realisation so as to renew the MEs purpose, structures, composition and behaviours over time. Such a process decomposition enables common understandings to be conceptualised and shared but as explained in section 2, in reality causal dependencies exist between process segments which will be ME specific and much complicate process interoperation and ongoing ME development and change.

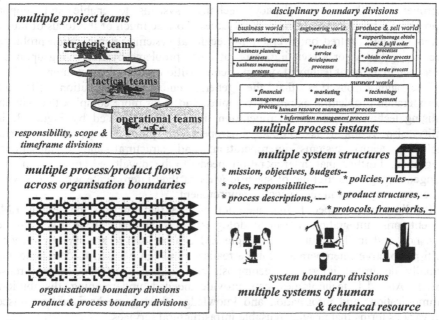

Figure 3 Conceptual Need for Integration (& Inter-Working) Across Multiple Boundaries

The reality is that various system structures need to be deployed to ensure that (human and technical) resource systems realise all ME activity requirements effectively and on time. In specific MEs structural decompositions can take various forms. Some structural elements may be: implicit within an organisational culture; be implicitly imported as part of a technical system (such as an ERP package); or be explicitly defined as part of a role, rule or procedure. The result of deploying various combinations of structural elements should be that those resources selected and assigned to enterprise activities should interoperate to realise ME goals and objectives, over both short and long time frames. It is important to stress once more the high levels of complexity involved here, one dimension of which is illustrated by the product decomposition in Figure 3. Section 2 explained the need to share a finite set of resource systems, so as to realise economies of scale and scope. It also described a common need for additional organisational decompositions to cope with product, process and system complexity and change, such as by overlaying departmental, business and company boundaries or by deploying project based organisational structures and teams.

Much of the complexity and change issues discussed in the foregoing can lie within the scope and control of a specific ME. However other issues do not. Clearly MEs must also suitably fit their environmental context which itself will change; possibly because of competitor, customer, supplier, stakeholder, legislative or political actions. Hence strategic and tactical processes and resource systems (such as teams) must operate accordingly. One general set of tactical problems can arise because proprietary technical resource systems may not readily or adequately interoperate with existing (legacy) technical systems to enable and support appropriate interoperation of human resources. To date much of the focus of the IT community has centred on: overcoming technical systems interoperation problems faced by industry; or alleviating some of those problems by improving upon the *status quo* in terms of technical system decomposition and implementations.

The author observes that the general enterprise integration (EI) and interoperation problem is extremely complex, and is far more complex than simply realising technical systems interoperation. Rather as illustrated by Figure 3, in reality also required is: process interoperation; product integration; human (including team) systems interoperation; and structural and organisational integration. Further that interoperation needs to be developed, maintained and changed through the lifetime of specific MEs as they interoperate with other MEs and complex systems in their specific environment.

It follows, with respect to interoperation in MEs, that enterprise modelling (EM) and enterprise integration (EI) methodologies and technologies have a key role to play, now and in the future. In principle EM can be used to specify how needed multi-perspective interoperation can be realised, and can be used to determine and formally document improved decompositions (in both performance and reuse terms). Also EI technologies can provide needed underpinning distribution, communication, and information and knowledge sharing mechanisms, organised into various forms that provide reusable infrastructural services.

From experience of modelling various ME process types, with reference to Figure 3 the author observes certain limitations of state of the EM and EI. Current EM methods do provide a plethora of multi-perspective modelling concepts but currently they do not adequately support the reuse of models of (i) functional, behavioural and structural aspects of human systems, (ii) causal dependencies between multiple processes and (iii) products and product instances, and organisational boundaries, and their mapping onto multiple process instants. It follows that improved EM tools are required to support the lifecycle engineering of multiple processes and associated product and resource systems; particularly with respect to enabling context dependent simulation and model enactment. The capabilities of current EI technologies have also advanced significantly in recent decades. However improved concepts, methods, architectures and tools are still needed to (1) facilitate component-based enterprise engineering and (2) enable the development of much improved infrastructural support (e.g. knowledge management, human resource management, technology management and financial management support services).

4. ON MODELLING HUMAN SYSTEMS

4.1 The Need to Model Human Systems

People are the prime resource of any ME. It is people who determine the ME's purpose, goals and objectives. It is people who conceive ME products and services and decide what, how and when product and service realising processes and systems should be deployed. People also make decisions and do many of the activities needed to realise products and services in conformance with plans. Therefore it is vital that the interoperation of ME personnel is suitably systemised and co-ordinated.

Different enterprises will vary significantly in the number of people they deploy and the roles those people play but the need for their systematic and co-ordinated working transcends the vast number of ME instances in operation globally. The high levels of complexity involved in satisfying this need necessitates the use of systemic problem decomposition, so that well structured solution realisation can be achieved. But that systemisation must be cognitive of factors that impact on people motivation, innovation and the like. Co-ordination is needed because of the time invariant nature of ME activities and the need to for interoperability amongst many resource systems used to realise ME activities. Some MEs and their operating environments may be subject to frequent change, e.g. to products and product realising processes. This in turn can require frequent change to the assignment of people roles, and dependencies between people roles.

Evidently therefore there is a general requirement within MEs to develop models of human resource systems (be they models of individuals, groups of people or teams of people) so that they can be used coherently in conjunction with developed models of ME processes to specify, implement and maintain timely and cost effective interworking of human resources through the lifetime of specific MEs. However, it is known that developing general purpose models of people (and systems of people) *per se* is impossible to achieve. Consequently before embarking on such a task the purpose for which human resource models will be used needs to be well specified; and even then it is understood that care needs to be exercised in respect to the use of derived models because of complex behavioural, motivational and cultural factors that impact on humans in the workplace.

4.2 Common Uses and Needed Attributes of Human System Models

Kosanke (2003) and Noran (2003) argue that in spite of progress made by ISO, IEC and CEN in regard to Enterprise Modelling (EM) more work is needed; especially on human related aspects like model representation of human roles, skills and their organisational authorities and responsibilities. With a view to addressing this need, Weston *et al* (2001), Ajaefobi and Weston (2003) and Ajaefobi (2004) built upon findings from previous human systems modelling studies to determine ways of classifying and modelling human competencies, (workload) capacities and the assignment of human roles and responsibilities. Bearing in mind the need to utilise human systems to resource multiple, dependent instances of the ME process types characterised by Table 1, the following generic modelling requirements were observed:

- Enterprise Activities (EAs) need to be modelled in the context in which they are to be realised, thereby providing a formal description of key structural and co-ordination aspects of processes that can readily be overlaid onto candidate human systems that need to interoperate with other enterprise resource systems to realise defined goals.
- Models of enterprise activities need to be explicitly characterised in terms of competency, capability and capacity requirements that must be satisfied by candidate systems of (human and technical) resource.
- Candidate human systems need to be modelled with respect to their potential to bring to bear competency types and competency attributes that suitably match activity requirements.
- Both long lived (static) and short lived (dynamic) structural aspects of human systems need to be modelled, including descriptions of: functional roles and responsibilities for groups of enterprise activities; and related causal dependencies between activity groupings and their associated information, material, product and control flows.
- Key behavioural aspects (such as reachable states and state transitions and associated performance levels) of unified process and human system models need to be usefully represented so that the operation and interoperation of candidate human and technical resource systems can be: simulated, and their performance predicted with respect to lead-time, cost and robustness; and enacted, via the use of suitable workflow technology.

Thus the need for separate but coherent models of context dependent processes and candidate human systems was observed as being needed to realise the following benefits:

- Ability in a given ME context to choose between alternative candidate human systems that satisfy requirements of activities (individual or grouped) from a functional viewpoint, namely in terms of their relative (a) competencies and (b) workload capacities.
- A systemic facilitation of process design, redesign and ongoing improvement (with reference to suitable candidate human systems) either at process, sub process or activity levels of granularity based on dual criteria of (a) performance lead times and (b) labour costs.
- The systemic attribution of values (e.g. as part of a knowledge capitalisation project) to processes and their elemental activities; here capital value can be placed upon human and structural assets of an enterprise by attributing to them (1) competency, capacity and structural attributes of assigned human systems and (2) appropriate business metrics.

4.3 Enhanced MPM Enabled Modelling of Human Systems

Suitable means of realising the modelling requirements described in 3.2 needed to be determined. Here the Multi-Process Modelling (MPM) method (Monfared *et al*, 2002) and its EM constructs and tools was selected as a baseline. MPM itself extends the use of CIMOSA modelling constructs and targets CIMOSA model enactment on (1) dynamic systems modelling and simulation and (2) workflow modelling, control and management. However by developing and incorporating into MPM a suitable set of human systems modelling tools an Enhanced Multi-Process

Modelling (Enhanced MPM) method was created and its development is reported in the PhD thesis of Ajaefobi (2004).

'Enhanced MPM' development centred on conceiving and testing a generalised methodology for selecting from amongst candidate human systems (namely individuals, groups of people and teams possibly supported by technical system elements); such that selected systems possess needed abilities to realise specific cases of the process types described in section 2. Here it was presumed that an initial match should be made between (i) *competency requirements* (identified as being necessary to realise a specific process and its elemental activities) and (ii) *competencies possessed* by alternative human systems. It was also presumed that a secondary matching would be necessary between (a) *capacity requirements* (identified in respect of specific processes) and (b) *capacity availability* vested in alternative human systems. Here it was understood that more than one viable candidate human system might possess competencies needed (to realise a specific enterprise activity or group of enterprise activities) but that the achievable performance of viable alternatives might differ significantly and/or they may vary significantly in their susceptibility to mental and/or physical workload stressors.

If those presumptions hold true then implicitly there is a need to achieve both a static and dynamic match between coherent models of processes and human systems. Here static competency matching should enable selection on the basis of relatively enduring abilities of candidates, which might be 'functional', 'behavioural' or 'organisational' in nature. Whereas the time variant (dynamic) nature of processes, process instants and process loading (in terms of product flows, information flows and the like) will in general impose workload variations on viable candidates (who pass static matching criteria) and their relative ability to cope with predicted load variations should usefully inform human system selection; as this choice could have process performance implications, e.g. on process lead-times, cost, repeatability, robustness and flexibility.

Hence to facilitate 'Enhanced MPM' development a set of generic and semi-generic 'competency', 'performance' and 'workload' modelling constructs was defined that can coherently be attributed to static and dynamic models of enterprise processes and candidate human (resource) systems. Figure 4 shows conceptually how such an attribution was designed to semantically enrich process and resource system models, related properties of which can be analysed to achieve static and dynamic matching of human systems to specified segments of processes. Thereby process costs, lead-times, etc, can be predicted before actual processes and process instants are implemented and run (so as to avoid future process loading problems. Further it was intended that the semantically enriched process and human resource system models would be mapped onto computer executable workflow models (run in proprietary workflow management tools) so as to support aspects of workflow design, execution and performance monitoring.

It was observed that competencies of human systems can be modelled at alternative levels of 'genericity', namely by defining: (1) generic competency classes pertaining to different process types discussed in Chapter 2; (2) semi-generic competency types, pertaining to common functional, behavioural or organisational competencies needed by processes operating in different domains (such as sales order processing, product engineering, manufacturing, logistical and project engineering domains); and (3) particular (functional, behavioural and organisational) competencies needed in respect of specific processes and process instants. To facilitate the application of Enhanced MPM, generic competency classes were defined to enable their use as reference models which can be particularised (in domain or specific cases) and incorporated into semantically enriched process and resource system models. Table 2 lists four generic competency classes so defined. This table also shows three generic performance classifications developed, which can also be particularised and attributed to (domain and specific) process and resource models. The use of these generic performance classes has proven useful in industrial case testing and has supported the second stage human systems selection, where process behaviours and performance are predicted should alternative viable human resource systems be deployed.

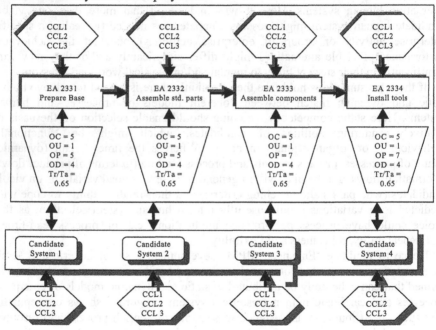

Figure 4 Example *Semantically Enriched Process Model Segment*

Existing human factors literature on workload stresses was reviewed to determine suitable means of modelling capacity attributes of processes and human (resource) systems. Here, in relation to human executors, it was observed that enterprise activities can generally be associated with some prerequisite level of workload that must be satisfied to guarantee adequate performance.

Workload was also observed to be multifaceted, involving mental, physical and organisational aspects. Because industry research sponsor interest was so oriented, emphasis during Enhanced MPM development was on mental aspects of workload. Another simplifying assumption made was that workload stressors lead mainly to two dimensional effects on people (assigned to execute enterprise activities); namely in terms of time stress and sensory modalities/effectors. The PhD thesis of Ajaefobi

(2004) describes the rationale, development and initial testing of Enhanced MPM workload modelling, while Table 2 lists some of the workload modelling construct conceived that were found to be particularly useful in support of human systems selection and process behaviours and performance prediction.

Table 2 Example Enhanced MPM constructs defined and used to match human systems to enterprise processes and to predict process performance

Generic Competency Classes (used as a reference)	Generic Performance Classes (used as a reference)	Mental Workload Modelling Constructs (common constructs used during Enhanced MPM simulation modelling)
CCL1: Competency to execute defined set of general operations based on specified methods, procedures and order. Here activities are essentially routine and results are predictable.	Level 1: People at this level are generally competent and active so they achieve satisfactory quality and timeliness of performance but have low degree of autonomy, low level of flexibility and are not conversant with the operational environment.	Operation Criticality (OC) Operational Uncertainty (OU) Operation Precision (OP) Time Ration (Tr/Ta): where Tr is the time required and TA is the time available. Auditory Demand (AD) Visual Demand (VD) Cognitive Demand (CD) Psychomotor Demand (PD)
CCL2: Competency to understand, interpret and implement concepts, designs, and operation plans linked to specific product realization and to apply them in solving practical problems e.g. system installation, operation and maintenance.	Level 2: People at this level are competent, resourceful, 'reflective', with a higher degree of autonomy and flexibility and hence can do things alternatively if need be. Further more, they are amiliar with tools, operating procedures, and technology and therefore can be trusted to deliver expected output even under critical situation.	
CCL3: Competency needed to translate abstract concepts into shared realities in the form of product designs, process specifications, operation procedures, budgeting and resource specifications	Level 3: People at this level have versatile experience and consequently are very proactive, innovative and creative. They have long outstanding years of confirmed experiences in solving problems in their areas and therefore can predict and effectively manage system behaviours in their area of expertise.	Note: During Enhanced MPM simulations, typically these constructs are assigned integer values in the range 1 (low) to 7 (high).
CCL4: Competency needed to formulating high level business goals, mission, policies, strategies and innovative ideas		

During subsequent simulation studies it was observed that an ordinal scale of 1 to 7 can usefully quantify those workflow constructs attributed. Also usefully incorporated into human systems selection methods has been sensory modality conflict theory of North *et al* (1989) and visual, auditory, cognitive and psychomotor (VACP) workload models proposed by Aldrich *et al* (1989).

The human resource modelling constructs developed to underpin the realisation of Enhanced MPM methodologies can also be used to formally describe important organisationally related attributes of human systems. For example 'role' modelling constructs have been defined to explicitly attribute to human systems, various responsibilities for process segments and process instants. By combining role and process definitions, key structural aspects of human systems can be formally specified from both individual and collective viewpoints.

Another thread of ongoing MSI research into human systems modelling concerns that of modelling behavioural competencies of human systems, and more particularly formally describing, predicting and monitoring key aspects of 'team member selection' and 'team working development'. Here it was observed that improved synergy and performance in teams is key in many MEs and commonly emerges from behavioural interaction between team members. Often also this behavioural interaction occurs in parallel with task execution, and constitutes a reflective process that leads to (a) improved team performance (in terms of task realisation) and (b) ongoing change in team roles, and hence team organisation. Thus Enhanced MPM modelling constructs were conceived and deployed that formally describe and predict the impact of team working behaviours {Byer 2004).

Thus it is concluded that the modelling of human systems, as executors of enterprise processes, is a fruitful area of research study which requires significant new efforts from enterprise modellers so as to meet industry needs and further the utility and applicability of state-of-the-art EM and EI.

5. CONCLUSION

General deficiencies in current EM and EI provision have been identified which point to a current unsatisfied need for:

*   Enriched understanding about: ME process types; long and short lived dependencies between those process types, and between their derivative process instances and related physical and logical flows; long and short lived operation and interoperation requirements of resource systems that collectively have capabilities to realise ME goals within defined constraints.
*   Improved means of developing and reusing coherent and context dependent models of ME processes and resource systems in support of large and small scale enterprise (re)design, engineering and change. The reader is referred to the PhD thesis of Chatha (2004) for a detailed description of this research need.
*   New human systems modelling concepts that: usefully underpin the attribution of individuals, and groups and teams of people, to activity elements of ME processes; provide means of analysing the performance of alternative candidate systems; and provide a basis for capitalising intellectual capital in MEs. This research need is reported in detail in PhD theses of Ajaefobi (2004) and Byer (2004).

This paper reports some progress towards addressing these needs, with particular emphasis on human systems modelling. New modelling concepts reported in this paper have been partially tested in support of the lifecycle of selected processes and human systems used by a global consortium of companies operating in the automotive sector. However a much broader base of industrial evaluation work is ongoing which is jointly funded by the UK government and by small and medium sized MEs operating in furniture, leisure, electronics and aerospace industry sectors.

REFERENCES

Ajaefobi, J.O. (2004) Human systems modelling in support of enhanced process realisation. PhD Thesis, Loughborough University, Leics., UK.

Ajaefobi, J O, Weston, R H, (2003) An Approach to Modelling and Matching Human System Competencies to Engineering Processes, IFIP WG 5.7 Int. Working Conf. 2003, Karlsruhe. Integrating Human Aspects in Production Management.

Ajaefobi, J.O., Weston, R.H. and Chatha, K.A.(2004) Enhanced Multi-Process Modelling (E-MPM): in support of selecting suitable human systems. Submitted to Proc. of I.Mech.E. Part B: J. of Engineering Manufacture.

Aldrich *et al* (1989) Aldrich, T.B., Szabo, S.M. and Bierbaum, C.R., 1989, The Development and Application of models To predict Operator Workload During System Design, *Applications of Human Performance Models To System Design*, Grant R. McMillan *et al* (eds), (London: Plenum Press).

Byer, N.A., (2004) Team systems development and the role of enterprise modelling technology. PhD Thesis, Loughborough University, Leics., UK.

Chatha, K.A. (2004) Multi process modelling approach to complex organisation design. PhD Thesis, Loughborough University, Leics., UK.

Kosanke (2003) Standardisation in enterprise inter- and intra- organisational integration, In Proc. of 10th ISPE Int. Conf. C.E.: Research and Applications, Madeira, Portugal, 873-878. R. Jardim-Goncalves *et al* (eds).

Monfared, R.P., West, A.A., Harrison, R., Weston, R.H. (2002) An implementation of the business process modelling approach in the automotive industry. In Proc. of the Institution of Mech. Engineers, Part B: J. of Engineering Manufacture, v 216, n 11, 1413-1428.

Noran, O. (2003), A mapping of Individual Architecture Frameworks (GRAI, PERA, C4ISR, CIMOSA, ZACHMAN, IRIS) onto GERAM, *Handbook on Enterprise Architecture*, Bernus, P. *et al* (eds), 65-210, (Springer, Berlin).

North, R.A. and Riley, A.A. (1989) W/INDEX: A Predictive Model of Operator Workload, *Applications of Human Performance Models To System Design*, Edited by Grant R. McMillan *et al*, (London: Plenum Press).

Pandya, K.V., Karlsson, A., Sega, S., and Carrie, A. (1997) Towards the manufacturing enterprises of the future, *International Journal of Operations & Production Management*, Vol.17, No.5, 502-521.

Poli, R., (1996) Ontology for Knowledge Organisation, R.Green (ed.) Knowledge Organisation and Change. Frankfurt / Main, IDEKS Verlag. 313-319.

Salvendy, G. (1992) 'Handbook of Industrial Engineering', A Wiley-Interscience Publications, John Wiley & Sons, Inc., USA, ISBN: 0-471-50276-6.

Scheer, A.W. (1992), Architecture of Integrated Information Systems, Springer-Verlag, Berlin, ISBN 3-540-55131-X.

Vernadat, F.B., (1996) Enterprise Modelling and Integration: Principles and
    Applications, (Chapman & Hall, London)
Weston R.H. (1999) A model-driven, component-based approach to reconfiguring
    manufacturing software systems., Special issue of *Int. J. of Operations & Prod.
    Management*, Responsiveness in Manufacturing. B MacCarthy & D McFarlane
    (eds). 19(8), 834-855.
Weston, R.H., Clements, P.E., Shorter, D.N., Carrott, A.J., Hodgson, A. and West,
    A.A. (2001) On the Explicit Modelling of Systems of Human Resources. In
    Special Issue on Modelling, Specification and Analysis of Manufacturing
    Systems, Eds. Francois Vernadat and Xiaolan Xie, *Int. J of Prod Research*, Vol.
    39, No. 2, 185-204.

13.   If Business Models Could Speak! Efficient: a Framework for Appraisal, Design and Simulation of Electronic Business Transactions

Michael Schmitt, Bertrand Grégoire,
Christophe Incoul, Sophie Ramel, Pierre Brimont and Eric Dubois[1]

1 Centre de Recherche Public Henri Tudor
Av. John F. Kennedy, L-1855 Luxembourg
Email:{ Michael.Schmitt, Bertrand.Gregoire, Christophe.Incoul,
Sophie.Ramel, Pierre.Brimont, Eric Dubois}@tudor.lu

In this paper we investigate the development of an appropriate business model associated with B2B transactions, designed according to the newly introduced ebXML standards. We explain the added value of such business model in complement to the more technical models defined by ebXML. In particular we explain the importance of achieving a better definition of the economic value associated with a B2B transaction. Together with the proposed business model ontology we also introduce a tool for supporting its management as well as a simulation tool for supporting decision making between different models.

1. INTRODUCTION

For more than 25 years, heavy and complicated standards such as UN/EDIFACT and ANSI X12 are dominating the field of electronic data interchange. They define an industry specific set of electronic messages that are the counterparts of the non-electronic document types that facilitate the business transactions. Several problems have led to a limitation in the spread of such technology. One problem is that grammars describing the syntax of the business documents are often complex and in some cases ambiguous. Specialized IT experts and a high level of communication are hence required for message implementation. Another problem is the message-oriented view of EDI standards. There is a need for a global view of the business transaction that would include their governing rules and alternative possible scenarios of execution easily. The application of EDI has, therefore, been limited to the big players with static transactions, and seemed not feasible for SMEs.

To overcome such problems, the ebXML initiative, launched by UN/CEFACT and OASIS, aims at working out XML based specifications for the facilitation of electronic document interchange. Along with the use of XML, a transaction-based view is suggested that caters for the needs of the whole business transaction. Together with recommendations of the XML definition of messages ebXML also define how to specify a transaction through a set of UML models associated with the flow of messages.

In section 2 we introduce the results regarding the development of the Efficient toolset supporting the design and the animation of a transaction.

In the ebXML proposal little is said about the analysis of the economic value associated with a B2B transaction. The core of this paper is related to this issue. In section 3 we first provide a rapid overview of the academic research conducted for the past few years in the business-modeling field. Then in section 4 we propose our ontology of concepts to capture the business value of a transaction as well as a supporting tool for its management. Finally in section 5 we explain how a business simulation tool can be used for supporting decision making among different business transactions proposals.

2. THE EFFICIENT PROJECT

This paper presents the work carried out within the framework of the research project Efficient. Efficient (eBusiness Framework For an effIcient Capture and Implementation of ENd-to-end Transactions) proposes an integrated tool set that supports the design, modeling and validation of ebXML based business transactions.

The tool set consists of an extension of a commercial UML-based CASE tool that supports the modelling of ebXML business transactions, and an animator tool that supports the execution of the above UML models, based upon a workflow engine. The animator allows business experts to cooperatively validate transaction models at the time they are built, before their implementation has started. Rather than simulation, we prefer to use the word 'animation' since the validation is done in an interactive way, each business expert playing the role of a business actor and participating in the execution of the transaction by receiving messages and sending answers. By doing this, business experts can validate the transaction by playing different possible scenarios that include different messages.

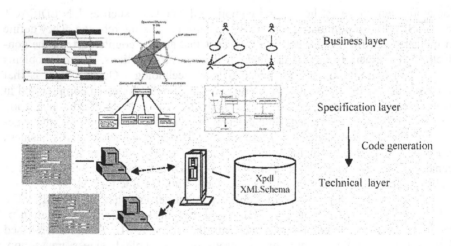

Figure 1 multi-layered approach of the Efficient project

In our project, we use a three-layered approach depicted in Figure 1. The *business layer* supports the appraisal and the design of the planned transaction from an economic point of view. Among the topics addressed at this layer is the model's value potential along with a managerial view on the ingredients needed for its implementation. It further details the model adding typical business activities and the players involved. The information content that accompanies the execution of the business transaction is structured into what we call the business domain.

The *specification layer* adds the sequence of activities and the flow of information (documents) that form the base of the transaction. Passing through an automated generation process, this layer feeds into the business simulator that enables XML based message exchange for an effective *simulation and validation* of the transaction, its activities, documents and flows. Governing the message exchange, business rules can be specified in natural language that control behavior and content during each step of the execution. The models employed and details as to the implementation are discussed in (Eshuis *et al.*, 2003).

The practical choices shaping the two last layers were guided by the maxims to use *open source software* and *follow standards* whenever possible, to guarantee the independency of our proposals against proprietary solutions and the ease of development. The considered standards include UML from the model point of view, and XML, Xlinkit and web services from the implementation point of view.

While ebXML introduces as a first level in modeling the business domain and process discovery, e.g. in its UMM methodology (UN/CEFACT, 2001), it does not highlight the importance of taking into account the economic context of a transaction. A business process emerges directly from strategic objectives of satisfying customer demand, and hence needs to be embedded in its economic environment that is, a sound business model.

2. RECENT BUSINESS MODELLING WORK

2.1 Business modelling objectives

Many people talk about business models today, and it seems there are as many different meanings of the term. A linking element seems to be the underlying motive to model a business in order to better understand the reasons that make some firms prosper while others have dropped out of the market. Unlike with business process models where the interest is mainly on transparency and efficiency of the operational processes, we consider business models as a more general, managerial view of a business that details the nature of the underlying business case, that is, it provides at least a description of what the company offers to the market, how it differs from its competitors and what core ingredients (partners, activities, resources, competencies) it employs to provide its offering.

2.2 Theoretical foundation

We follow (Gordijn *et al*, 2000) in their argumentation that while a process viewpoint on a firm may be suited to explain *how* a business case is or should be carried out, it seems not feasible to reason about the business itself. According to them, a business model details which actor provides what object or service of value to which others and what benefits he expects in return. A business model hence describes the way economic value is created and consumed along the chain of activities among its participants.

An important aspect of this definition is the idea of reciprocity of economic exchange (see (McCarthy, 2000)). Each service or good provided by an economic actor must be complemented by a reward or incentive flowing in the opposite direction. This entails that the profitability and sustainability of a business model depend not only on its value creation potential but also on the attractiveness of the benefits and incentives it offers to its participants. (Wise & Morrison, 2000) e.g. refer to a lack of attractive benefits in their explanation why many of the electronic marketplace providers were not able to sustain their initial success.

So far, we have identified the creation of economic value and the benefits structure as core elements in the notion of business models, but we have not explained how a business model differentiates from competition, nor how the firm plans to reach its customers and on which cost and revenue models it plans to earn money from value creation. In this respect, (Timmers, 1999) complements our definition by taking into account the potential sources of revenue. He considers a business models as "architecture for product, service and information flows including a description of the various business actors and their roles, along with a description of potential benefits for the actors and a description of the sources of revenue". However, such as point of view focuses on elements internal to the value creation network and does not discuss the various relationships and dependencies that hold with the external world: customer segments and market segmentation, promotion and customer care, law compliancy and the structure of competition.

Most of these missing elements are covered by the definition provided by (Afuah & Tucci, 2001), who point out that a business model need to answer such questions as what value is offered by the firm and which customers it provides the value to, how the value proposition is priced compared to the offerings of its competitors, what is needed to provide the value proposition conceived and what strategies it identifies to sustain any competitive advantage derived from its activities. While the answers to these questions may give us a grasp of a firm's business case, we suggest to add two more requirements in order for us to be able to exploit and capitalize on the information its contains, that means, as the title of our article suggests, to make the business model "speak":

A business model should serve a good starting point for business simulation, in that it helps to determine possible indicators of performance.

A business model should be represented formally so that it can be compared to others and evaluated to reveal strengths and weaknesses hence can feed a subsequent business simulation with valuable input.

2.3 The BML framework

We have chosen to implement the modeling framework proposed by (Pigneur, 2002) and (Osterwalder, 2004) as their approach seems comprehensive with regards to the above modeling goals and it is formal enough to allow computer-based evaluation. The core of their model consists of modeling language ontology as illustrated in Figure 2.

The *customer relationship pillar* details the market segments addressed by the business model, the distribution channels and promotional means to reach each of the segments. Starting with the customers and identifying their demands, the *product pillar* models the value proposition the firm provides in order to respond to that demand. The *infrastructure pillar* reveals the key capabilities, resources and strategic alliances that are at the heart of the business structure, and without which the value proposition could not be furnished. Finally, the *financial pillar* ties the other pillars together by aligning resources, capabilities and commercial activities with their respective costs and by opposing them with potential sources of revenue.

The shape of their ontology was motivated by the work of Kaplan and Norton (Kaplan, 1996) on performance measurement and seems well suited to support the identification of KPI and measures for business appraisal: The product pillar permits the firm to assess the innovative character of their offering, which links to the innovation and learning perspective of the Balanced Scorecard (BSC), infrastructure management corresponds to business process perspective. Financial aspects and customer interface finally refer to the equivalent perspectives of a BSC.

A formal syntactical framework alone, however, does not shape or limit the form and content of business cases modeled using the framework. In other words, a model designer still can model business ideas that probably won't be successful and that contain major conceptual flaws. In order to minimize the potential of such failure and to further research the factors that impact on success or failure, we shall discuss in the next section some approaches towards value creation that have shaped our specific implementation of the semantics of the modeling framework.

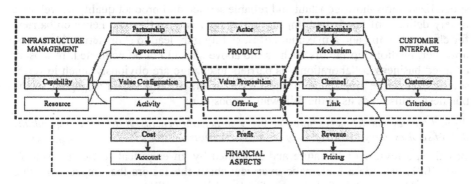

Figure 2 BML modeling architecture

3. BUSINESS MODELLING IN EFFICIENT

3.1 Investigating semantics constraints

The BML proposition discussed above is mostly a syntactical framework that needs to be completed to make sure the described business models create economic value. We describe below some paths we followed in extending the proposition of (Osterwalder, 2004) in that sense.

Economic success can be considered as a function of economic value drawn from business. (Porter, 2001) explains economic value in terms of profit level, as the difference between revenue and cost. Hence, in order to maximize value creation, companies can pursue either of two strategies. They can innovate in order to design a unique offering that earns a higher price or they can strive for operational effectiveness in order to reduce their costs. With regards to the requirements a business model must fulfill, such considerations require a sound financial model at the heart of the business model. This leads us to our first proposition:

A good business model explains how a firm plans to earn money. A description of its innovative character and its pricing are required to position a firm's value proposition with regards to the competition.   *proposition 1*

Porter further comments that improving operational effectiveness needs to be accompanied by a constant seek to improve and extend one's strategic positioning. A key factor for success is hence the steady adaptation of both the value proposition and infrastructure to match the changing requirements of the market. This includes make-or-buy decisions for missing competencies as well as the integration of the customer into the business model to maximize the strategic fit of the offering and the demand. (Timmers, 1999) argues that such flexibility favors the creation of loose business networks, which leads us to suggest the following proposition:

A good business model does support the business manager with a means to flexibly adapt his offering portfolio to the market needs. At the same time, it emphasizes the costs and benefits of such change.   *proposition 2*

Though flexibility appears highly desirable, there's also another side to the coin. A stable business relationship, for instance, usually comes with efficient process cycle times, reliable and error-free collaborative value creation processes. It is trust, specialized know-how, constant and reliable service and product quality as well as a timely delivery of goods that impact on customer satisfaction and hence on barriers for change. Pricing and a maximum of flexibility may lead to short-term advantages, however, relationship factors such as the above should not be neglected. In loosely coupled business collaborations, as barriers to change are obviously much lower, it seems even more important to stress each actors incentives for engaging in the business. This leads us to the following proposition:

A good business model makes sure that every participant benefits. It depicts along with the flow of services and goods the flow of rewards or benefits.   *proposition 3*

Based on a review of literature and supported by an empirical research, (Amit & Zott, 2001) provide a systematic overview about the factors that impact on value creation. They identify four types of value drivers: *Efficiency*, which implies the costs of carrying out a transaction, *Complementarities*, which refers to bundling effects when a product bundle is perceived more valuable than each of its parts, *Lock-In*, i.e. any kind of a barrier to change or an incentive that results in increased customer loyalty, and *novelty*, which is associated with innovation. While transaction efficiency and novelty can be associated with the Porterian view of value creation, the other two value drivers lead us to suggest the following proposition:

A good business model is a canvas that permits to exploit the shift of value levels resulting from product bundling. It further encourages the designer to integrate measures for achieving economic benefits from customer loyalty.   *proposition 4*

Another interesting work on business value drivers comes from (Hlupic & Qureshi, 2003) who examine the organizational and technical prerequisites of value creation. They consider value creation a positive function of a firm's intellectual capital, team productivity, collaboration, the task-technology fit and its social intellectual capital.

Intellectual capital refers both a firm's human capital, i.e. to the skills and knowledge of the individuals and to company values and culture, and to the structural capital of the firm, the knowledge associated at the company level: databases, software, patents, copyrights. *Team productivity* is important as it may limit the capability of people to reason, to take actions or to assimilate new knowledge. *Collaboration* relates to effective use of collaborative technologies for business management, including message systems, shared calendars and file systems or a common customer database are necessary to create a shared understanding of the business and to make sure information is synchronized among business partners. *Task-Technology* fit measures the effective use of collaborative technology and suggests that value creation is affected by the extent to which a fit can be achieved between a group's task and the technology employed. *Social intellectual capital* finally at the individual level raises the ability of people to effectively engage in communication and negotiation. At group level, a shared understanding about the purposes of the business and its functioning as well as a congruence of goals are necessary prerequisites for value creation in a collaborative environment.

Many of value drivers discussed in their paper refer to what is known as intangible assets of a firm. This had led us to suggest the following propositions:

A good business model points out the importance of intangible assets for value creation. This includes know-how, corporate culture, communication and technical skills as well as the ability to work in a team.   *proposition 5*

A good business model supports the choice of technology that fits a specific commercial task.
    proposition 6

The next section presents our specific design of the business modeling ontology taking into account the semantic constraints discussed.

3.2 Introducing new business concepts

The need to flexibly adapt the value offering portfolio to changing market needs has led us to incorporate an element in the customer relationship pillar, defined by BML, which describes the various customer needs. A *customer demand* (see Figure) in our ontology is a bundle of functional and non-functional requirements each of which is assigned a priority tag. A firm's value proposition may meet all or only some of the requirements of a customer segment.

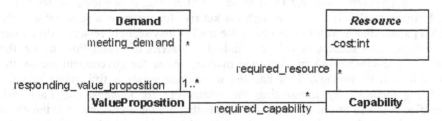

Figure 3: Calculation of the cost of a satisfying a customer *demand.*

The need to measure a business partner's incentives for engaging in the business has led us to add a benefit element to the infrastructure management pillar. Benefits may be either tangible assets such as money in return for a service, or they may be intangible such as an increase of market knowledge, a repartition of the economic risks involved or a maximum utilization of resources. We call these benefits *compensations*, as illustrated in Figure 4.

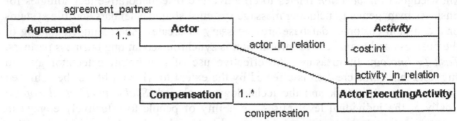

Figure 4: An actor receives a *compensation* for the activities he participates in.

In order to track the costs of changes by adapting either the value proposition or business infrastructure we have decided to associate the resources, both tangible and intangible, with a *cost* (per unit, per time) and to link that cost to a *cost account* in the financial model. As there are resources at the base of the capabilities essential in providing a value proposition, this permits us to measure the cost of each offering of the value proposition. Mapping the offerings to the customer demand means that costs can be tracked throughout the model giving us an estimate of the total cost of fulfilling part of the customer demand. Workload was also observed to be multifaceted, involving mental, physical and organisational aspects. Because industry research sponsor interest was so oriented, emphasis during Enhanced MPM development was on mental aspects of workload. Another simplifying assumption made was that workload stressors lead mainly to two dimensional effects on people (assigned to execute enterprise activities); namely in terms of time stress and sensory modalities/effectors. The PhD thesis of Ajaefobi (2004) describes the rationale, development and initial testing of Enhanced MPM workload modelling, while Table 2 lists some of the workload modelling construct conceived that were found to be particularly useful in support of human systems selection and process behaviours and performance prediction.Figure3 gives a more formal view about the relationship between demand and cost.

Critical and costly resources, changes of customer need and a high degree of dynamism represent risks that need to be identified and, if possible, catered for in the business model. If there is a high market risk for instance, a strategic alliance with a partner that is well introduced in the market may seem necessary. Also, some business model's success of failure is linked to a series of assumptions as e.g. the trade volume achieved in an electronic marketplace or the government aid for the research of a technology. This has led us to incorporate a fifth pillar into the modeling ontology by introducing the notion of *risk management*. Each risk identified is linked to one or several elements that are threatened by or at the cause of the it, these links are illustrated in Figure 5. Risk management impacts also on the financial results as additional resources may need to be provided and financial reserves need to be built for the case of loss.

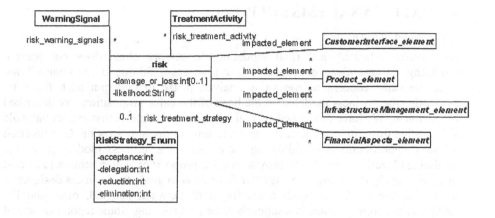

Figure 5: Risk management: impacts, warning and typology

Figure 6 gives a top-level view of our business modeling ontology. The tool that we use for the implementation, Protégé, is an open-source ontology editor from Stanford University that provides an extensible architecture for the creation of customized knowledge-based applications. It comes with a rich set of available plug-ins, one of which is a Prolog-tab for logic based knowledge extraction, which we use to exploit the information contained in a business model.

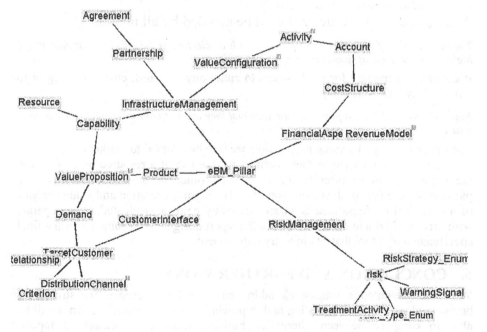

Figure 6: An implementation of the business modelling ontology based on Protégé

4. VALUE ANALYSIS SUPPORT

4.1 Report generation

As already indicated, a formal modelling framework alone does not prevent modeling business ideas that probably won't be successful due to conceptual flaws. It is the main concern of the value analysis phase to underpin such flaws by evaluating the business modeled on the basis of the value propositions we described in section 4. We have developed several value analysis reports that extract valuable information from the content of the model to support decision-making. One function of such reports consists in validating the content of a business model against the modeling objectives, such as to provide a value proposition with a minimum of cost or to effectively use a resource. Another function is to guide the business designer to improving the model to reach a maximum fit between demand, offer and the infrastructure configuration that supports value provisioning. Some reports are stated below:

A critical resource, for instance, is defined as a resource being consumed by one or several activities, which are essential for providing some part the value proposition that is especially valuable to the customer. A resource is considered as critical if a single external actor, who could not be easily replaced, provides it.

A critical resource should be replaced, as soon as possible, with another less critical one. A workaround for this thread may be the internal development of the required skill of stuff.

A risky business relationship provides critical resources and is associated with a low level of trust and a low degree of substitutability.

A single sourcing strategy should be avoided by all means.

The most costly offering of a value proposition is a selection of the offering that induces the highest cost in terms of resources, out of all such elements.

It could be interesting for the business to either buy this most costly offering or to replace it by a substitution product.

Non-competitive value propositions are such that they are not innovative and rather high in cost.

The pricing policy of a value proposition needs to be adapted to market conditions.

All the elements used in these definitions (the use of a resource by an activity, the importance of an offering for a target customer, or the substitutability of a partnership) are part of the business model. The report generation and value analysis of a described business idea is an area currently being investigated. We cooperate with private SME's in order to enhance the report design and to come up with a final specification of a tool that fits industry requirements.

5. CONCLUSION AND FURTHER WORK

We have presented a framework addressed to business experts for structuring business ideas, evaluating, testing and improving them. The model designer will be able to compare between alternative business models, by means of reports highlighting their respective strengths and weaknesses. Current efforts include the enhancement of reports that form the value analysis of a business model by studying some real-world business cases. Also, at the current stage, the strategic layer and the transaction layer remain largely unconnected. Future research will focus on a methodology that takes a promising business model through a series of (semi-) automated steps that yield one or more transaction models, which are inline with the business strategy. Other research includes an investigation in expressiveness to improve the specification formalisms and to work on formal validation applications. More information can be found at our website, http://efficient.citi.tudor.lu.

REFERENCES

Afuah A., Tucci C. L. (2001) *Internet business models and strategies*, Boston: McGraw-Hill.

Amit R., Zott C. (2001) *Value Creation in e-Business*, Strategic Management Journal No. 22, 493-520

Eshuis R., Brimont P., Dubois E., Grégoire B., Ramel S. (2003) *Animating ebXML Transactions with a Workflow Engine*, In Proc. CoopIS 2003, volume 2888 of LNCS, Springer, 2003.

Gordijn J., Akkermans H., van Vliet H. (2000) *Business Modelling Is Not Process Modelling*, ER Workshops 2000, 40-51

Hlupic V., Quereshi S. (2003) *What causes value to be created when it did not exist before?* In: Proc. 36th Hawaii International Conference on System Sciences, 2003 IEEE

Kaplan R., Norton D.P. (1996) *The Balanced Scorecard – translating strategy into action*, Harvard Business School Press, Boston, Massachusetts

McCarthy W. E., Geerts G. L. (2000) *The ontological foundation of REA Enterprise Information Systems*, working paper, Michigan state university, August 2000

Osterwalder A. (2004) *The Business Model Ontology* , Ph.D. thesis 2004, HEC Lausanne.

Pigneur Y., Osterwalder A. (2002) *An e-business model ontology for modeling e-business*, 15th Bled Electronic Commerce Conference, Bled, Slovenia

Porter M. (2001) *Strategy and the Internet*, Harvard Business Review, 63-78

Ross R. (1997) The Business Rule Book: Classifying, Defining and Modeling Rules, Second Edition, ISBN: 0941049035

Timmers P. (1999) Electronic Commerce: Strategies and Models for Business-to-Business Trading, New-York: John Wiley & Sons

UN/CEFACT (2001) UN/CEFACT's Modelling Methodolgy, TMWG, N090R10.

Wise R., Morisson D. (2000) *Beyond the Exchange: The future of B2B*, Harvard Business Review, November – December 2000, 86-96

14.   Building a Better Team

Jason Mausberg
President IDS Scheer Canada
j.mausberg@ids-scheer.ca

Building a better team focuses on the need for efficient teamwork in order to maximize project or business process success. The paper first investigates what constitutes a team environment, and then, puts forward an educational model/ framework in order to better foster a team environment. By using the ARIS tool, businesses and/ or project teams can develop evaluative and visual aids to build a more efficient team.

1. INTRODUCTION

Teamwork within and across organizations and cultures is a key criterion in the success of the development and implementation of new business modelling structures, data standards, and business process improvement initiatives. A lack of change management and cooperation at a team level is a primary reason for the failure of these initiatives. The inability for organizations, inter-organizations, and development groups to work as an efficiently functioning team can lead to failure or less than ideal results.

In this paper I propose to attempt the following tasks: first, to increase awareness of what best constitutes a team environment, and secondly, to provide suggestions as to how to create the skills at an individual and team level necessary to foster this kind of environment. To better understand what constitutes a team environment, various examples/ levels of team environments will be examined. The ARIS tool will be discussed as a platform and a central framework around which workshops can be based to help the individual and the team develop what I refer to as "team thinking" skills

The paper will be divided into three major sections:

- Exploring what constitutes a team environment
- Examining a process to help build skills at the individual level and the team level that will increase team work
- Brief discussion of how to utilize the ARIS toolset in the team building exercise

2. UNDERSTANDING THE TEAM ENVIRONMENT

Teamwork can be defined as joint work towards a common end or goal performed by a group of people organized to work together. This definition of teamwork can be interpreted to allow for the possibility of individuals working on individual tasks with a common goal.

Individual organizations, development symposiums, and cooperative development groups often define their working environment as team oriented (Hammer, 2003). Common to mission statements of such groups is the definition of goals around which periodic meetings are held to update other team members on progress. In contrast, mountain emergency rescue teams work in a much more integrated manner. In fact lives depend on the ability for rescue workers to support each other. Within the emergency rescue environment, the helicopter pilot is dependent upon the evacuation team to determine a suitable area for pick up. An error at the top of a mountain in a wind or snowstorm in terms of wind direction and proximity to the mountain peaks could result in the death of the pilot and or team. Similarly ropes and rescue attempts are most often performed by two to three individuals working as one unit with one person holding or securing the ropes/ safety systems and the others securing the victim and transporting the victim to safety. Clearly there are various levels in which a team functions. In order to assist and develop the skills required to foster a team environment, it is critical for the individuals to understand the differences amongst various "team" environments.

I would like to use the comparison of America's pastime, baseball, to the global sport, soccer, to gain a better understanding of teamwork. Prior to examining these two sports, it is important to understand that a broader definition of a team environment in sports could include social interaction, commitment to the team, and a sense of belonging. However the primary goal in professional sports is to be victorious. Victory is achieved through on field processes which are fostered through morale and "off field" bonding. Ultimately it is the processes on the playing field that illustrate the results of teamwork and showcase teamwork in action. For this reason I will focus and discuss on field activity.

When examining different levels of teamwork, a framework of questions needs to be developed. These questions could include:

- What level of interaction do team members have with each other?
- How often does this level of interaction occur?
- How dependent are individuals on each other in order to achieve a positive outcome?
- How fluid/ dynamic is the environment in which the team performs?

2.1 Soccer or Baseball, which is more team oriented?

In examining the team concept within the sports of baseball and soccer, attention will be focused only on the field of play (the actual process of competing within a game). Based on the questions above, soccer would be considered a sport requiring greater interaction amongst the players necessary for a positive outcome. In relation to the question above regarding interaction, frequency, dependency, and fluidity, consider the following points in regards to this argument:

- In soccer there are 10 players and a goalie. Within any given game it is common for any player to pass the ball to another player including the goalie. Within baseball the interaction of players is limited defensively to only certain players (for example a left fielder would never throw the ball to a right fielder). In addition in baseball the offensive interaction, which is not a pass/ receive relationship but more a succession of individual offensive attempts at bat is limited to only the players in close proximity within the batting line up.

- The level of interaction in soccer is continuous. Players are constantly running and readjusting their positions based on ball position and other player positioning (from their own team and the competition). Baseball does involve readjustment in positioning based on the same factors however the level of readjustment is much more restricted from a distance perspective and much more limited due to the fluidity of soccer versus baseball. In terms of interaction from a passing perspective, soccer revolves around passing where in baseball passing is limited to the throwing that occurs in an one dimensional basis on a single play.
- In soccer passing/ball movement is fundamental to achieving a positive outcome of scoring a goal or defending the other team. In baseball individual success can dictate an entire game for example a home run with no one on or a perfectly pitched game can dictate the entire outcome of a game.
- The fluid/ continuous nature of soccer requires significantly more coordination as well as infinite passing possibilities compared to the stop and go timing of baseball. Within the stop and go flow of the game of baseball, decisions can be made at an individual level to change strategy, introduce other players, rearrange defences, etc. In soccer with a continuous flow the players must instinctively work as a unit to achieve success.

It is important to understand the subtle and not so subtle differences in team interaction in order to apply these differences to the more complex business environment. Whether you are in agreement or not with the comparisons between baseball and soccer, the important aspect is to analyse the differences and thus gain a better understanding of what constitutes teamwork. Sport can be considered a less complex environment because the focus is well understood and universally shared (that of winning) and the result very well defined for each competition (the score).

2.2 Strategies to create teamwork

There are many variables that contribute to developing an environment that will foster teamwork. Some of these traditional variables include: remuneration plans, leadership styles, motivational tools, corporate structures, methods of communication, methods of problem solving and idea generation.

In this paper I will not discuss how the above variables contribute and enhance a team environment. Rather, I will focus on the ways in which a team applies its strengths and weaknesses in the midst of changing environmental factors to perform business practices. The purpose of this paper is not to develop a training schedule or plan but to put forth an educational training model that would lead to an increased awareness of what constitutes a team environment. This educational model is built with the following principles in mind:

- Team skills are learned first at an individual level and then refined and improved in a group environment. Using the soccer analogy passing and kicking a ball are both skills that are paramount in the play of a team. These skills when combined together are defined as team play, however these skills occur at an individual level as part of the team process. Similarly in business, individuals must first be educated in the skills necessary to function as an efficient team member prior to becoming immersed in collective processes.

- Processes/ functions of a team occur at the individual level. So once again process knowledge must first be grasped on an individual basis before the individual can interact with the team. Using the soccer metaphor, player positioning is based on ball position and other player's movements. A player must understand moving without the ball prior to interacting with other players.
- Team skills and creating the most efficient team environment are both highly complex areas and can be improved on an ongoing basis. The framework that I am suggesting can be simply used as a starting point. The most important principle is the ongoing discussion and analysis of what is responsible for fostering the team environment.
- The business environment is ever changing. An optimal process employing highly aggressive sales tactics in a growing industry such as high tech in the late nineties may be entirely inappropriate today. Today a more customer centric process involving increased customer support and education may be more appropriate for this industry.

3. BUILDING A TEAM ENVIRONMENT

The following educational model and steps could be used to foster an increased teamwork environment:

- Minimize the focus on the individual self.
- Determine and document individual strengths and weaknesses.
- Compare strengths and weaknesses with others within the immediate team.
- Examine and document processes that need to be executed in order to achieve success.
- Map individual strengths to the process model. Determine primary and secondary roles of individuals in relation to the processes.
- Identify business environment variables that affect these processes, which change over time.
- Discuss optimal applications of people's skills to the processes within changing environments.

Within each step different workshops and/ or teachings methods could be used to educate the participants. The paper will not discuss how each workshop should occur and should be structured. The paper will focus on the key objectives of each step and the importance of each step within the framework of the ARIS software tool.

4. USING ARIS TO FOSTER TEAM BUILDING

The ARIS Toolset enables the enterprise and inter-enterprise wide design of business processes, as well as their analysis and optimisation. In addition ARIS allows for the identification and graphic application of individual skill sets. The output of the ARIS toolset is multi level/ intersecting process modelling across data, process, and people plains. The ARIS tool can be used to systematically go through the steps identified in section 3 outlining how to build a team environment (with the exception of "self negation"). See Figure 1 on the following page for a graphical depiction of the steps in ARIS.

4.1 Self negation / relational conception of the self

The first task in building a cohesive team is to convince all team members that a team functions as a whole and that success is ultimately measured at a team level not an individual level. This task will be difficult to accomplish, as individuals are motivated based on their own personal success. The Western mindset typically revolves around promotion, individual bonuses, and self-advancement. Much has been written on the subject of team dynamics and motivation (for a more detailed discussion on this see Chris Harris and/ or Daniel Levi).

In an effort to promote the team concept, a more detailed discussion of an individual versus a relational conception of the self should be discussed which would highlight the array of dependencies that exist within daily life and the interdependencies that are necessary to achieve business success. This objective would not be mapped in the ARIS tool other than highlighted in an initial overview diagram.

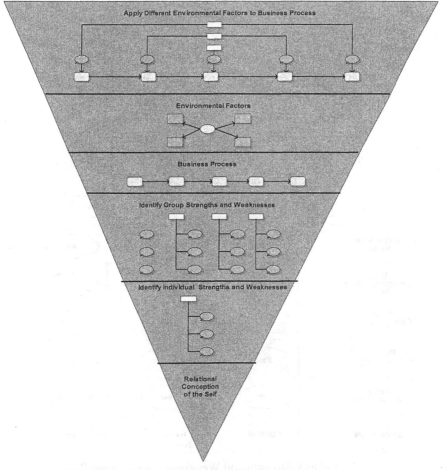

Figure 1. Team Building in ARIS

Figure 1, which is an overview of the Team Building model, illustrates the levels / steps in the education process. The model moves from the bottom upwards starting with the individual and building upwards with team members, process, and the business environment.

Figure 2 represents the 2nd level of information in ARIS. The idea is that in ARIS you can "drill down" from one level to the next. In this case, the 2nd level diagrams illustrate "identifying individual skills" and "analysing skills on a group basis".

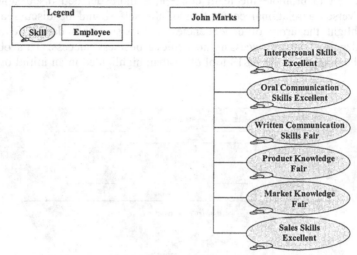

Figure 2a. Identify Individual skills (Self Assessment)

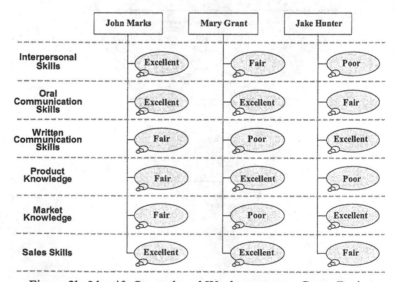

Figure 2b. Identify Strength and Weaknesses on a Group Basis

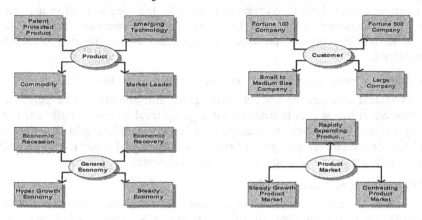

Identify Environmental Factors

Team Processes

Sales Lead Generation

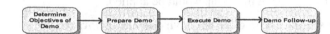

Product Demonstration

Sales Lead Generation
Rapidly Expanding Market

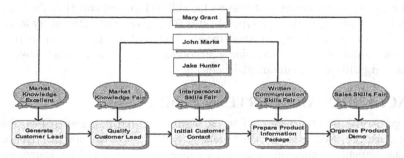

Figure 3. Second level "drill downs" from Figure 1.

4.2 Identify individual strengths and weaknesses (figure 2)

The objective of this level is to create awareness within the individual to identify their own strengths and weaknesses, as well as the strengths and weaknesses of others. The tendency in performing this task is to determine with precision exact strengths and weaknesses of the individual. While this accuracy is desirable, the goal in performing this analysis is to create awareness for the need to identify individual skills so as to maximize a team's ability to best perform processes. As progression through this model continues, individual strengths and weaknesses can be revisited and redefined.

4.3 Compare strengths and weaknesses across the group (figure 2)

Once individual assessments are complete, group assessments can be performed. The objective in this step is to determine at a group level where strengths exist at a primary and secondary level. By attempting to define where strengths lie within the group, the team members will gain a better understanding of when it is appropriate to use such strengths and how to minimize any weaknesses.

4.4 Define key team processes (figure 3)

Key business, project, or development processes should be documented to act as a framework for the application of individual strengths. Using ARIS modelling techniques, processes should be discussed and documented.

4.5 Define business environmental factors (figure 3)

Define environmental factors that are relevant to the business environment. These factors would typically be categorized into four areas product, customer, market maturity, and economically related. It is not essential to have an exhaustive list. The important part of the modelling exercise is to understand that under different circumstances different strengths of the group would be applied.

4.6 Analyse the processes based on different environment factors (figure 3)

Within this step, first examine a process without considering environmental factors. For example, if the typical sales process entails an initial site visit, followed by a technical presentation, and subsequently presentation of a contract one week later, then examine this sales process. Do not examine variations at this point. Apply an individual name (s) to processes that possess the skill (s) to best complete the task. There may be discussion as the best type of skills for a given task. Do not try to be overly precise in the mapping, instead record a second name (s) if required. Following the initial exercise of examining a process in its' basic element, rework the process using different environmental elements.

5. CONCLUSION AND FURTHER WORK

As teams go out to face business challenges, within and across organizations, the ARIS tool could be used as a basis for process modelling, teamwork facilitation, change management, simulation, and performance management. The teamwork model that has been put forth starts with a single task of re-examining the self from a relational perspective and ends up with a complex matrix involving people, process, and environment, which in turn circles back to a reflection on the individual.

Similarly the ARIS tool encompasses the complete lifecycle of team/ project mission statement development, strategic direction, project execution, performance monitoring, and readjustment of goals.

Globalisation, end consumer sophistication, and advances in technology, have contributed to dramatic increases in the capabilities of software packages and business modelling tools. The level of sophistication both from a process, business, and technology perspective is increasing. Despite the increase in sophistication of tools and project approach, the ability to manage change is still a key factor in determining success. At the core of change management is teamwork- teamwork within a company, teamwork across organizations, and teamwork within development and standards bodies.

Much has been studied and written about change management. I believe that organizations that function in a team environment are most able to adapt to change. The question becomes what is a team environment and how does an organization create a team environment? This question is not easily addressed and the answer can always be improved upon. In this paper I have set forth a basic educational model built upon the ARIS tool that focuses on the complexities and intersectionality of people's skill sets, business processes, and changing environmental factors. By using ARIS to apply this model, an increased awareness of individual's roles and intersectionality to a team process can be better understood, resulting in a more efficient performance for the team.

REFERENCES

August-Wilhelm Scheer, Ferri Abolhassan, Wolfram Jost, Mathias Kirchmer (2003) Business Process Change Management ARIS in Practice, foreword by Michael Hammer . Berlin, Heidelberg, New York: Springer-Verlag

August- Wilhelm Scheer (1994) Business Process Engineering. Berlin, Heidelberg, New York, Tokyo: Springer-Verlag

Chris Harris (2003) Building Innovative Teams. New York: Palgrave Macmillan

Daniel Levi (2001) Group Dynamics for Teams. Thousand Oaks, London, New Delhi: Sage Publications

James T. Scarnati (2001) Team Performance Management. Bradford:2001.Vol. 7, Iss. ½:pg.5

Joannie M. Schrof (1996) US News & World Report. August 5, 1996 v121 n5 pg 53

Joni Daniels (2004) The Collaborative Experience. Industrial Management, Norcross: May/Jun 2004. Vol.46, Iss. 3; pg 27

Melissa C. Thomas-Hunt, Katherine W. Phillips (2003) Leading and Managing People in the Dynamic Organization, edited by Randall S. Peterson and Elizabeth A. Mannix. Managing Teams in the Dynamic Organization: The Effects of Revolving Membership and Changing Task Demands on Expertise and Status in Groups, Lawrence Erlbaum Associates Inc.

Patricia K. Felkins, B.J. Chakiris, Kenneth N. Chakiris (1993) Change Management, A Model for Effective Organizational Performance. Teamwork as a Structure for Change,. Kraus Organization Limited, New York

15. A Reference Model for Human Supply Chain Integration: an Interdisciplinary Approach

S. T. Ponis[1] and E. Koronis[2]

1 National Technical University of Athens Email: [staponis@central.ntua.gr]
2 University of Warwick Email: [ep.koronis@proel.gr]

The focus of this paper is to adopt an interdisciplinary approach of the education system's strategic planning process, by drawing insights and critically evaluating the possibility of applying a mechanistic view of the work force development inspired from the vast and numerous literature of supply chain management. The outcome of the study is a proposed high-level reference model for Human Supply Chain (HSC) integration.

1. INTRODUCTION

This paper is mostly a "thought provoking" paper and its main concept emerged as a result of discussions that have overwhelmed national media, about the increasing numbers of unemployment, especially in the "warmware" (skilled labour) area. The aim of this study is to adopt an interdisciplinary approach of the national education system's strategic planning by applying a mechanistic view of the workforce development processes. This effort relies on an attempt to stress similarities between the discrete planning phases of educational planning and the traditional supply chain steps (plan, source, make deliver, return) as described in the relative bibliography. In this way, the education process is decomposed and conceptualized as a mechanism for educating individuals and empowering them with skills and knowledge which meet the labour demand within an open market. Such notion would add to current occupational forecasting projections the functional aspects of streamlining processes and planning time and available resources.

In doing so, a study of the analogical reasoning methods was carried out, especially in the field of metaphorical viewing of social phenomena. It is true that while analogies can become excellent carriers of explanatory messages and provide useful tools for abstraction and inspiration, extreme caution should be taken when they become tools for analogical reasoning or assumptions. In addition, Morgan (1986) suggests that metaphors may end up becoming erroneous expressions of a false analogical thinking. Despite these impedimenta, it is our opinion that Supply Chain Management (SCM) methodologies, tools and techniques may provide education analysts and government services a source of valuable and creative ideas towards an effective process for education planning. This paper aims at bridging the raising demand for efficient education planning and the solid and well accepted fields of supply chain methodology, thus producing for education specialists and labour economists a positivist and normative approach on education planning.

To support this point of view, this study is proposing a high level reference model. The logic of the proposed model is based mostly on SCOR (Supply Chain Operation Reference Model) and is fathoming two levels of abstraction (strategy and tactical); the long term goal of these efforts is the elaboration of a full scale and operational model that will be able to support educational planning in such a way to streamline employees' sourcing with labour demand and increase the overall Human Supply Chain effectiveness and integration.

The paper is organized in four Sections including this introduction, which keeps the place of Section 1. Section 2 reviews the literature concerning Supply Chain Management and provides an analysis of the problem of Labour Market imbalances as well as a presentation of past and modern approaches addressing the problem. In Section 3, the core of our rationale is presented; Human Supply Chain is defined and the first two levels of the corresponding reference model are presented. Finally, in Section 4 the limitations of the research are recited followed by the prospects for further research efforts and model development.

2. LITERATURE

2.1 Supply Chain Management

The concept of Supply Chain Management has its origins to Forrester (Forrester, 1958; 1961; 1968), who identified the pattern of response to changes in demand in supply chain situations. A supply chain situation suggested the existence of a network of organizations connected to each other, through upstream and downstream linkages, carrying out in collaboration different processes and activities that produce value in the forms of products and services in the hands of the ultimate customer (Christofer, 1998). Croom *et al.* (2000) argue that an antecedent of Forrester's ideas can be found in the Total Cost approach to distribution and logistics (Heckert and Miner, 1940; Lewis, 1956). Both these approaches show that focusing on a single element in the chain can not assure the effectiveness of the whole system.

Hau *et al.* (1997), indicate the phenomenon of distortion in demand patterns created by the dynamic complexity present in transferring demand from end users along a chain of supply to manufacturers and material suppliers. It has been the identification of this kind of distortion and inefficiency along with the realisation of managers that actions taken by one member of the chain can influence the profitability of all the other chain members that have driven many organizations to managing their whole supply chain instead of short-sightedly focusing on their own organization (Lee *et al.*, 1997). Since then, SCM remains a topic of considerable interest among supply chain academicians and practitioners from both large and small companies as they strive for better quality and higher customer satisfaction (Larson and Halldorsson, 2002, Mentzer *et al.*, 2000; Chopra and Meindle, 2001).

Opportunities for companies to use supply chain management principles to improve their competitive position are well documented in the literature (Davis 1993; Cooper and Ellram 1993; Gattorna, 1996). Successful implementation of supply chain management creates a number of benefits, these being cost deterioration (Mainardi *et al.* 1999; Cooper, R and Yoshikawa, T., 1994), technological innovation (Hult *et al.* 2000), increased profitability and productivity

(Gryna 2001), risk reduction (Chase *et al.* 2000), and improved organizational competitiveness (Fisher 1997; Christopher 2000). While supply chain management principles derive from a particular settings of problems and address a dialogue concerning these main issues, it may be argued that supply chain might offer an interesting paradigmatic view capable of providing new ideas for different fields of scientific thought.

2.2 Structural Unemployment and Labour Market Imbalances

There is a growing discussion about structural unemployment and the critical question of matching labour supply with labour demand. Inevitably, the complexity of social structuring and the transformation of social systems into dynamic entities (Castells, 2000) magnified the importance of accurate human development planning. Analysts often argue about the need for institutional adaptation and education reforms which must provide society with the appropriate labour supply. As Manacorda & Petrongolo (1999:182) suggest, "any increase in the relative demand for skilled labour would not cause major labour market problems if it were matched by a parallel adjustment of supply". Best (2001) also argues that the availability of the adequate skill base and the matching of supply and demand of technical skills become crucial factors for societal success and a critical question for innovation achievement.

Neo-classical models have ignored the importance of structural unemployment and the need for planning, arguing that labour markets are adjusting any imbalances in their own accord. But this is hardly the reality; inflexibilities and internal problems in structuring of education and training policies do exist. It has therefore been early stated that planning efforts must take place (Willems, 1996) in order to streamline education and labour market needs. In the past, the concept of manpower planning (Ahamad & Blaug, 1973), which consists of a solid methodological toolset for prediction and planning of labour demand, has been used. It was then occupational forecasting which boosted the efforts towards predicting and forecasting the jobs within the labour market in the long run (Johansen, 1960) and although occupational forecasting studies they require long effort and adequate funding, several surveys of that kind are still being published (Hughes, 1993; 1994).

It is arguable though, that connecting a forecasted labour demand with supply requires a great number of institutional arrangements and socio-economical reforms that must accompany a labour supply reform (ibid; Corrales, 1997; Whitley, 2001; Lloyd & Payne, 2003). It is also true that even in today's manpower forecasting efforts the study of the supply side of the labour market remains relatively under-explored (Willems, 1996:1). In other words, we need to integrate an effective prediction process with the adequate institutional measures and educational policy in order to come up with a realistic and effective plan of educational action.

In this paper we argue that the prevention or resolution of labour market imbalances may be resulted by integrating the educational and vocational training systems with the labour market, the same way that in the enterprise world, companies are integrating procurement, production and delivery systems for planning, interoperability and efficiency purposes. It is the aim of this study, to propose an alternative holistic view of the educational and labour market infrastructure and processes, taking advantage of the well proven methods picked out of the enterprise reality, more specifically Supply Chain Management. Education specialists and governmental initiatives, we argue, may draw useful insights and ideas from a model-based view of education. The first step of these efforts will be presented in this paper in the form of a high level reference model, stating the analogy and creating the sub base for further research that will hopefully result into a fully operating human supply chain management model.

3. UTILISING THE ANALOGIES: THE HUMAN SUPPLY CHAIN REFERENCE MODEL

The sciences do not develop in complete isolation. On the contrary there is influence between the disciplines (Thoben, 1982). One way of achieving such as an interdisciplinary approach is through identifying analogies. This metaphorical way of thinking is a natural cognitive process primarily met in Aristotelian thinking. Recent literature provides us with many examples of utilising analogical reasoning techniques and metaphorical viewing in training and learning (Gregan Paxton & John, 1997), in manufacturing (Mill & Sherlock, 2000), for the theorizing of the firm and the organization (Penrose, 1952; Alchian, 1953; Keely, 1980; Morgan 1986; Oswick et. Al, 2002), for sociology and economics, a scientific field extensively viewed under a mechanistic and organistic view (Thoben, 1982). Successful analogical reasoning is possible when it assimilates a process of transferring an explanatory structure from the source domain to the target domain (Tsoukas, 1993:337; Oswick et. Al, 2002). In other words, we suggest that an analogy provides a solid methodological ground for research when it is the result of an effort to 'compare' instead of 'assimilate' two different systems (Morgan, 1986).

This is exactly the aim of this paper, to use analogical thinking and transfer principles and perspectives of Supply Chain Management into the problematic areas of labour economics and structural unemployment. In this paper it is suggested that the power of the SCM concepts may be found in their simplicity; an abstract view of a supply chain may be adopted, providing useful thinking for educational planning.

The Supply Chain Operations Reference Model (SCOR) is a tool for representing, analyzing and configuring supply chains. It was developed by the Supply Chain Council (SCC) and the consulting firm Pittiglio Rabin Todd & McGrath (PRTM), as well as over 65 major companies (Supply Chain Council, 2002a). Unlike optimising models, no mathematical formal description of a supply chain or heuristic methods for solving a problem are given. Instead, terminology and processes are standardized, enabling a general description of the supply chain under study. In this case, SCOR will be used as the basis for establishing a first abstract view of what we call the Human Supply Chain Reference model.

The first step, towards the elaboration of the primary two upper levels of the Human Supply Chain Reference Model (HSCRM) was the determination of the Supply Chain it self, the actors and the alternative pathways a human can follow in transforming himself from an unskilled worker to a ready to occupy a job position employee. A schematic time-oriented representation of the Human Supply Chain is shown in Figure 1.

The next step in the proposed approach was the application of analogical thinking in order to establish the necessary semantic bridges into initiating an interacting connection between the two different scientific research fields, these being Supply Chain Management and Educational & Occupational Planning.

Enterprise Integration (EI) has grown in the past ten years at a pace where there is an obvious need for a more frequent forum where these strategic discussions can be continued bringing together leading thinkers of industry, defence and research.

As soon as the analogies were created, the actual core processes could be identified along with the actors that initiate and utilize them. This is shown in Figure 2, in the form of a UML (Unified Modelling Language) Business Use Case Diagram.

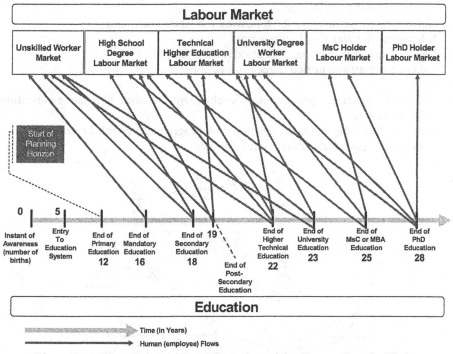

Figure 1: A Time Oriented Representation of the Human Supply Chain

The final step of the proposed approach was the application of the utilized analogies and the elaboration of the first two levels of the Human Supply Chain Reference model. According to the Supply Chain Council (2001, 2002b) Level 1 consists of five elementary distinct management processes which in our case have the following characteristics that are described below:

- Plan- Its scope includes the following planning activities:
- Developing and calculating all the necessary projections after processing the available data and estimations.
- Balancing resources with requirements and establish/communicate plans for the whole supply chain,
- Transition management, planning configuration, institutional arrangements and regulatory framework requirements and compliance.

- Align the supply chain unit plan with the financial plan.
- Source: Its scope includes the following sourcing activities:
- Educational system infrastructure management and development
- Graduating policies assessment and update.
- Demographic policy management.
- Admission and quotas framework.
- Make: Its scope includes the following activities:
- Educational institution establishment and function management.
- Syllabus management.
- Academia and labour market relationship management.
- Education staff procurement.
- Deliver: Its scope includes the following job market activities
- Job market functionality improvement.
- National and private career services development.
- Labour equality and justice framework establishment.
- Return:
- Continuing education programmes development including alternative education training schemes.
- Vocational training and skills development programmes establishment.
- Executive postgraduate schemes development.

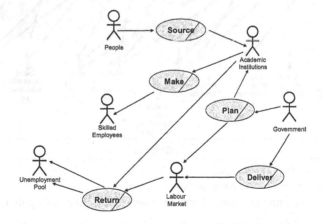

Figure 2: Human Supply Chain UML Business Use Case Diagram

The five distinct management processes described above are further decomposed into 30 process categories. At this level, typical redundancies of established business, such as overlapping planning processes and duplicate or unneeded sourcing activities can be identified. Each process category belongs to one of the types: planning, executing and enabling.

More specifically:

Planning (decomposition of Level 1 Plan process): Process categories of this type support the allocation of resources (educational or other) to the expected demand. They incorporate balancing of supply and demand in an adequate planning horizon. These processes are executed in a periodical manner and they directly influence the supply chain's flexibility in rearranging it self when demand changes. For example a new emerging technology that is going to dominate the business environment will subsequently create a demand for employees capable of using this new technology. Planning is responsible for reallocating existing (e.g. change of the current syllabus of related courses to include this new technology) or establishing new resources (constitute new educational establishments or initiate training programs) in order to reschedule the human supply chain, thus enabling it to provide the labour market with the employees that will match in the best possible way the specific market needs at the time created.

Executing (decomposition of Level 1 Make process): These are processes that are triggered by planned or current demand. Process categories of the type executing, directly influence the time interval between incoming orders and delivery. They depict the core processes of a supply chain, which are responsible for the implementation of the orders and the resource and time constraints that rule them, as dispatched by the planning processes, in the strategy level. The process types source, make and deliver are divided with respect to the seven different human categories (see Figure 3) corresponding to the level of their education.

Enabling (decomposition of Level 1 Source, Deliver and Return process): This type of processes support the later two process types. They prepare, preserve and control the flow of information and the relations between the other two types of processes.

In the next Section, a set of limitations and constraints of this study will be presented along with the further research efforts to overcome them.

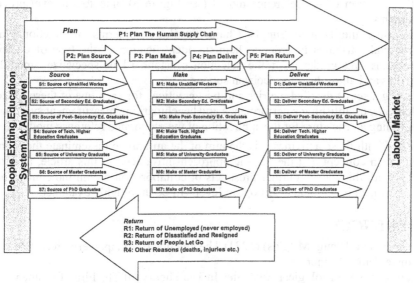

Figure 3: The Human Supply Chain Reference Model – Level 2

4. LIMITATIONS AND PROSPECTS FOR FURTHER RESEARCH

Supply Chain management is a flushing field of exploration for both researchers and practitioners. Major international consulting firms, academic institutions and enterprise R&D departments are developing large practices in the supply chain field, and the number of related research papers is growing rapidly. Despite these facts, there are no efforts spotted in the international bibliography that try to connect supply chain disciplines with human chains and educational planning.

It should be stated at this point, that research proposed in this paper is limited by certain constraints which are resulted by the social nature of the problems itself. It could be suggested that the model under consideration requires a high level of abstraction from the complexity of reality; such complexity is magnified by the fact that materials flowing within the systemic boundaries refer to people who maintain their own beliefs, expectations and decisions. In addition to that, education choices are not only a question of personal choice or institutional strategy but also a social process often presented as a 'social right' (Corrales, 1999: 78).

Nevertheless, we argue that a supply chain perspective on human resources planning within a social system may provide policy makers the missing link between forecasted jobs and the decisions related to the structuring of education, thus improving the overall human supply chain efficiency. In doing so, several research efforts should take place in the future, as described below:

- The further decomposition of the process categories identified in the second level of abstraction of the reference model (see Figure 3), into fourth level process elements.
- Finalising the Human Supply Chain Typology in terms of functional and structural attributes in order to help the identification of the type of decision problems and guidance of the selection of standard or specialized modules, models and algorithms for decision support.
- Issuing a set of detailed metrics and best practices for each one of these process elements, establishing a performance and benchmarking measurement system.
- Outlining a procedure for the application of the elaborated reference model.

In parallel with these activities it is the aim of the authors to disseminate the human supply chain concept and the holistic approach introduced in this article, an effort initiated with this paper. Further prospects of collaboration within the context of a research project in national or European level will be thoroughly examined.

REFERENCES

Ahamad, B. and Blaug, M. (Eds) (1973) The Practice of Manpower Forecasting. Amsterdam: Elsevier.

Alchian, A. (1953) Biological Analogies In The Theory Of The Firm: Comment, American Economic Review, Vol. 43, Issue 4, pp 600-604

Best, M. (2001) The New Competitive Advantage, Oxford University Press, New York.

Castells, M. (2000) The Rise of the Network Society, 2d edition, London: Blackwell Editions

Chase, R.B., NJ. Aquilano, and R.E Jacobs (2000) Operations Management for Competitive Advantage, Irwin Publishing Co., Chicago, IL, pp. 30-45.

Chopra, S. and P. Meindle (2001) Supply Chain Management: Strategy, Planning, and Operation, Prentice Hall, Inc., Upper Saddle River, NJ, pp. 1-24.

Christopher, M. (1998) Logistics and Supply Chain Management – Strategies for reducing cost and improving service, 2nd ed., London *et al.*

Christopher, M. (2000) Managing the Global Supply Chain in anUncertain World, www.indianifoline.com, February 23, 2000.

Cooper, M.C., Ellram, L.M., Gardner, J.T., Hanks, A.M. (1992) Meshing Multiple Alliances, Journal of Business Logistics 18 (1), pp 67-88.

Cooper, R., Yoshikawa, T. (1994) Inter-organisational cost management systems: the case of the Tokyo-Yokohama-Kamakura supplier chain. International Journal of Production Economics 37, pp 51-62.

Corrales, Javier (1999) The Politics of Education Reform: Bolstering The Supply And Demand; Overcoming Institutional Blocks, The Education Reform And Management Series, Vol. II, No 1

Croom, S., Romano, P., and Giannakis, M. (2000). Supply chain management: An analytical framework for critical review. European Journal of Purchasing and Supply Management, 6, 67-83.

Davis, T. (1993) Effective Supply Chain Management, Sloan Management Review, Summer 1993, pp. 35-46.

Fisher, M. (1997) What is the Right Supply Chain for Your Product?, Harvard Business Review, March/April 1997, pp. 105-116.

Forrester, J. (1961). Industrial Dynamics. Cambridge, MA. MIT Press.

Forrester, J. (1968). Principles of Systems. Cambridge, MA. MIT Press.

Forrester, J.W. (1958) Industrial Dynamics: A major breakthrough for decision makers, Harvard Business Review, Vol.36, No. 4, 37-66.

Gattorna, J.L., Walters, D.W. (1996) Managing the Supply Chain. A Strategic Perspective. Macmillan, New York.

Gregan-Paxton, J., John, D. R. (1997) Consumer Learning by Analogy: A Model of Internal Knowledge Transfer, Journal of Consumer Research, Vol. 24, Issue 3, pp 266-285.

Gryna, F. (2001) Supply Chain Management, Quality Planning & Analysis, (Chapter 15), The McGraw-Hill Companies, Inc., New York, NY, pp. 403-432.

Hau L. Lee, Padmanabhan, P., and Whang, S. (1997) Information Distortion In A Supply Chain: The Bullwhip Effect, Management Science (43:4), pp. 546-558.

Heckert, J.B., Miner, R.B., 1940. Distribution Costs. The Ronald Press Company, New York.

Hughes, G. (1993), Projecting the Occupational Structure of Employment in OECD Countries, OECD, Paris, Labour Market and Social Policy Occasional Papers. No. 10.

Hughes, G. (1994), An Overview of Occupational Forecasting in OECD Countries, Paper for the ECE/Eurostat Joint Work Session on Demographic Projections, Monsdorf-les-Bains, Luxembourg, June 1-4.

Hult, G., M. Thomas, E. Nichols, and L.C. Giunipero, Jr. (2000) Global Organizational Learning in the Supply Chain: A Low versus High Learning Study, Journal of International Marketing, (8:3), 2000, pp. 61-83.

Johansen, L. (1960). A multi-sectoral study of economic growth. Amsterdam: North-Holland.

Keely M. (1980) Organizational Analogy: A Comparison Of Organismic And Social Contract Models, Administrative Science Quarterly, Vol. 25, Is. 2, pp 337-363

Larson P. and Halldorsson A., (2002) What is SCM? And, Where is it?, vol. 38, no. 4 (Fall 2002), pp. 36-44.

Lee, H., Padmanabhan, V., and Whang, S. (1997). Information distortion in a supply chain: The bullwhip effect. Management Science, 43:546.

Lewis, H.T., (1956) The Role of Air Freight in Physical Distribution. Graduate School of Business .

Lloyd, C., Payne, J. (2003) The Political Economy Of Skill And The Limits Of Educational Policy, Journal of Educational Policy, Vol. 18, Issue 1, pp 85-107.

Mainardi, C.A., M. Salva, and M. Sanderson (1999) Label of Origin: Made on Earth, Strategy Management Competition, 2nd Quarter 1999, pp. 20-28.Administration, Division of Research. Harvard University, Boston.

Manacorda, M., Petrongolo, B. (1999) Skill Mismatch And Unemployment in OECD Countries, Economica, Vol. 66, pp. 181-207

Mentzer, J.T., J.H. Foggin, and S.L. Golicic (2000) Collaboration, Supply Chain Management Review, (4:4), pp. 52-60.

Mill, F., Sherlock, A. (2000) Biological Analogies In Manufacturing, Computers in Industry, Vol. 43, Issue 2, pp 153-161

Morgan, G. (1986) Images of Organization, London: Sage Publications

Oswick, C., Keenoy, T., Grant, D., (2002) Metaphor And Analogical Reasoning In Organization Theory: Beyond Orthodoxy, Academy Of Management Review, Vol. 27, No 2, pp 294-303

Penrose, E. T. (1952) Biological Analogies in the Theory of the Firm, American Economic Review, Vol. 42, Issue 5, pp 804-820

Supply Chain Council (2001) The Supply Chain Council and the Supply Chain Operations Reference Model, Pittsburgh, URL: http://supply-chain.nidhog.com/slides/slide.asp.

Supply Chain Council (2002a and 2002b) Homepage, URL: http://supply-chain.nidhog.com/slides/slide.asp.

Thoben, H. (1982) Mechanistic And Organistic Analogies In Economics Reconsidered, KYKLOS, Vol. 35, pp 292-306

Tsoukas, H. (1993) Analogical Reasoning And Knowledge Generation In Organization Theory, Organization Studies, Vol. 14/3, pp.323-346

Whitley R. (1999) Divergent Capitalisms: The Social Structuring And Change Of Business Systems, Oxford: Oxford University Press

Willems, Ed (1996) Manpower Forecasting and Modelling Replacement Demand: An Overview, Research Centre for Education and the Labour Market, Maastricht, September 1996

16.　　A Proposition for Risk Analysis in Manufacturing and Enterprise Modeling

Vincent Chapurlat[1], Jacky Montmain[2], Djamel Gharbi[1]

1 - Laboratoire de Génie Informatique et d'Ingénierie de Production - LGI2P - site EERIE de l'Ecole des Mines d'Alès - Parc Scientifique Georges Besse - F30035 Nîmes cedex 5 - Tel : (+33) 4 66 38 70 65 - Fax : (+33) 4 66 38 70 74
email : Vincent.Chapurlat@ema.fr
2 – Unité de Recherche sur la Complexité URC CEA/EMA - site EERIE de l'Ecole des Mines d'Alès - Parc Scientifique Georges Besse - F30035 Nîmes cedex 5

This article presents a work in progress, which aims at associating a systemic reference modeling approach with formal verification concepts in order to improve the user's toolbox concerning risk analysis. This approach is here applied to a manufacturing process.

1. INTRODUCTION

A system is a composite set of people and components (plant, hardware, software), which are organized in an environment in order to perform a mission and attain objectives. Each system, whatever its nature, is said to be in **danger** when the occurrence of interdependent events puts the system in a situation where it can possibly be irreversibly damaged. A **risk** is thus commonly defined as the possible occurrence of damage resulting from exposure to a dangerous situation. The system is therefore unable to reach its objectives, less efficient or unable to execute its mission. The causes may be human errors, technical failures, environmental and financial malfunctions, and so on. For example, a manufacturing system must be stopped when a major breakdown occurs. The **damage**, reversible when repairable, can be associated to a rapidly decreasing productivity rate as long as the situation remains the same.

It remains difficult for a system designer to foresee all the possible effects and identify their causes in order to circumvent them, especially when they have never been identified in the past. The work in progress described in this article proposes a set of innovative concepts and tools and adds new tools to the risk assessment toolbox. These concepts are partially applied to a manufacturing process example.

2. RISK ANALYSIS

Risk analysis approaches are commonly based on the following sequential process:
- The identification of risks consists in describing the system and identifying dangerous phenomena and/or situations.
- The evaluation of risks, in a qualitative and/or quantitative way, consists in taking into account their possible occurrence rate, the gravity of their effects and the critical situations they potentially induce on the system, the vulnerability of

the system regarding the existing mechanisms protecting it against the undesired effects. A risk hierarchy can then be built.

- The reduction of risks consists in solving separately the potential problems causing the identified and evaluated risks until an acceptable level of system performance is achieved.

A list of 62 risk analysis methods is presented in (Tixier *et al.*, 2000). They are classified into three main clusters of approaches, which each offer their own advantages:

- The first ones enable the risk to be studied in a qualitative, quantitative, deterministic or probabilistic way. Risk occurrence and relevance can then be rationally evaluated, assuming, however, the availability of experiments, data and information about the pre existing system behavior.
- Systemic approaches such as MADS and MOSAR (Perilhon, 2003) enable the capture of risk and danger representations, but do not really describe the system itself. They use a common set of limited concepts and risk reference models that improve the user's knowledge and the relevance of the models obtained through the approach.
- Cyndinic approaches focus on a theoretical representation of situations based on a language of risk modeling but remain difficult to use in practice (Kervern, 1994).

In each case, the user manipulates several modeling languages and methods. Doubt may therefore be cast on their relevance, depending on their ability to take into account different levels of details and assumptions, simultaneously different points of view and investigation fields such as human, financial, technical or others. The verification ('is the model correctly built?') and the validation ('is the model correct with regard to the actual system?') may give some responses to achieve a satisfying level of trust in these representations but remain unknown. The goal of this work is to use:

- A system modeling approach respecting systemic concepts inspired by SAGACE (CEA, 1998; Penalva, 1994, 1997; Chatel *et al.*, 2004).
- A set of V&V concepts and mechanisms enabling: firstly the verification of the system model in order to be sure of its correctness, consistency and so on; secondly attempted validation of the model in order to achieve some of the objectives of risk analysis (identification and evaluation at least). In fact, each potential piece of damage induces the modification or non-predictable emergence of several properties in the system characterizing the system's efficiency, stability and integrity. The idea therefore consists in detecting when, under what conditions, how and in which way (event, situation, state of the system, combination of these, etc.) the truth of a property can change revealing possible problems and may be considered as a risk. As used in many works such as (Manna, 1992; NASA, 1998; Lamine, 2001; Lamboley, 2001), this research will focus on a formal property proof in a model verification and/or validation (V&V) perspective.

3. MODELING APPROACH

The designer has to build his or her own representation of the system to be analyzed. The result is a set of representations, susceptible to interpretations and critical examinations, but which still remains a source of knowledge for the user. We propose using a systemic reference approach, guiding the user to build a model that will become a representation that is sharable with other designers.

Before introducing the modeling approach itself, it is important to note that it is necessary to set up a unique and commonly defined vocabulary throughout the approach. This will unify and define the common sense meaning of each concept that will be used during the modeling and V&V phases. This is achieved through the conception and building of a system's domain ontology (Ushold, 1996). In this ontology, all the concepts required to describe the system to the user are itemized, together with the relations between these concepts, in the environment of the system at a given level of detail. This task has to be carried out by experts in the required domains. For example, a version of a vocabulary framework inspired from existing ontologies such as PSL (NIST, 2002) and dedicated to industrial processes in enterprise modeling (Vernadat, 1996; Bernus *et al.*, 2003) has been defined (Chapurlat *et al.*, 2003).

The modeling approach, SAGACE, considers a system from functional, organic and operational standpoints. The functional view is an external view of the phenomenon as a system open to its environment. The organic view is an internal view of the system as a network of interrelations and interactions among operative, logistic and auxiliary components. The operational (or teleological) view seeks to clarify the decision-making competencies involved in accomplishing the objective (control and management). A more refined typology may be obtained by combining the three views from the perspective of examining certain expected system properties: performance, stability and integrity. The combination of the three views and three perspectives determines nine system viewpoints identified in the SAGACE matrix summarized in Figure 1. The knowledge representation proposed by the SAGACE method involves a projection on the nine-viewpoint matrix to assess the complexity of the system by distributing the knowledge and questions over the viewpoints. This knowledge representation method has been described in detail in (Penalva 1997). The purpose of this system modeling approach is to produce a representation constituting a structured medium for information of different types from a variety of sources, a basis for collective discussion and argumentation, the concrete expression of shared knowledge of the operating situation.

Each viewpoint in the matrix may be defined as follows (this is a generic definition, and must be customized for each project and each system according to the nature of the problem and the type of model to be developed):

- The goal viewpoint describes the aim and functions of the system independently from their physical implementation.
- The processes and activities viewpoint describes how the functions are assumed by a set of activities.
- The resource viewpoint defines the supporting resources that are chosen in order to support the system activities.
- The resources organization viewpoint describes how these resources are really used, their allocation and the corresponding emerging network of resources needed by the system.
- The scenarios and modes viewpoints define the different possible situations of the system and the conditions enabling transition from one situation to another.

- The three last control viewpoints respectively enable description of the system management rules for adjusting, keeping stable and anticipating some situations.

The resulting models thus provide structured and univocal knowledge elements to describe the system as exhaustively as possible.

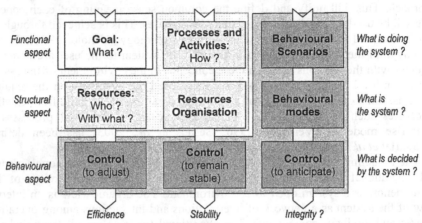

Figure 1: The SAGACE generic matrix

In its original version, SAGACE provides a unique graphical language designed to facilitate representation of a point of view, allow communication among the relevant players and materialize the shared knowledge of a subject (CEA 1998). This language enables the relations between three classes of entities to be represented: processor, flow and observer through transactions, interactions or coupling phenomena. While this language seems to be effective, it remains hard to use for a non-specialist and need to be reinterpreted when the user changes from one point of view to another. The proposed approach defines a dedicated modeling language for each point of view. It relies on the idea that dedicated modeling languages have already been developed for each of the SAGACE viewpoints and are nowadays in common use. A formal semantic must then be established between the different formalisms used in order to achieve the consistency of the different points of view as proposed in (Feliot, 2000).

For example, the selected languages used to describe a manufacturing process are shown in Figure 2. The goals are defined by forcing the user to clearly define the aims of the system (this description step remains informal because expressed in natural language) and IDEF1 to build the corresponding functional (Menzel, 1998). Processes are represented using a process modeling language defined in (Lamine, 2001). Resources are conceptually expressed by using object class diagrams issued from UML (Booch, 1998) allowing, if necessary, the concepts and relations defined in the ontology above to be refined. A database description is also necessary here in order to arrange and to manage all the data and information about the real system. A flow chart diagram is used in order to describe the implantation of physical resources and the transactions (matter, energy, information) between the system and its environment. Finally, an automata model based modeling language - such as a Petri Net - is used to capture the dynamic behavior of the system.

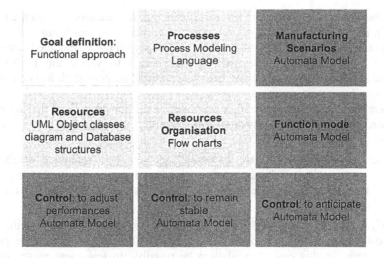

Figure 2: The *SAGACE* matrix for Manufacturing systems and its associated modeling languages

4. ANALYSIS APPROACH: VERIFICATION AND VALIDATION

The analysis approach is based on:

- A property model named CRED presented in (Lamine, 2001) and completed in (Chapurlat *et al.*, 2003).
- A reference properties database as proposed in (Chapurlat *et al.*, 2002).
- A formal verification tool introducing Conceptual Graphs (Kamsu *et al.*, 2003).

The property model allows all the properties that are expected to govern the system to be described. A property is thus a formal representation of an expectation, a need or a characteristic of a real system, which may be described as a causal structure. This is a qualitative description of the effect or influence that system entities (the causes) have on other entities (the effects). A property is thus modeled as a composite entity that consists of a set of causes (denoted C) linked up with a set of effects (denoted E) via a parameterized relation (denoted R). This relation can capture different interpretations of causality:

- Logical: the occurrence of a set of causes implies or is equivalent to the occurrence of a set of effects.
- Temporal: the occurrence of a set of causes strictly happens after the occurrence of the set of effects.
- Emerging: the set of causes describes how different objects can interact in order to bring out a set of effects which may be observable at a lower level of abstraction but not directly deducible from the causes.
- Influence: causality means variation influence and the corresponding relationships between causes and effects are interpreted as beneficial or harmful.
- All the concepts and relations defined in the unique vocabulary are then used to describe any usual (or common sense) properties governing the system. As for the vocabulary definition, a set of experts defines and classifies a set of:

- *Axiomatic properties*: properties describing natural phenomena (such as PV = nRT), rules and laws (such as *an operator cannot work more than 8 hours per day*) and norms that indisputably have to be respected by the modeled system.
- *Model properties*: properties which are needed to verify the model itself (syntactic ones – not considered here - and semantic ones such as *each activity necessarily obeys a constraint input*)
- *System properties*: properties that characterize the functional and non-functional constraints governing the system (for example, *each machine has an energy input*).

All these generic properties are gathered together in a reference database of properties, and specific mechanisms are implemented in a support tool (Chapurlat *et al.*, 2004) in order to manipulate them. The user can specify what properties must be proved in order to:

- Verify the model, consisting in proving the model has been correctly built.
- Validate the model, that is to say make sure of its accuracy regarding the pointed out system. In this case, it will then be possible to test some more complex propositions (modeling them by a (set of) properties) which, if they cannot be established and proved, seem to be the cause of a problem.

A first version of this database has been constructed for models of industrial processes. The risk reference list proposed in MADS MOSAR will extend this version to risk assessment in industrial plants.

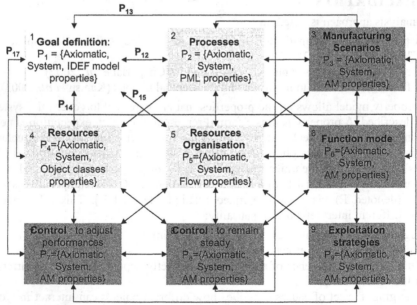

Figure 3: Point of view properties

In the proposed approach, the database allows the user to describe:

- The properties needed to explain each viewpoint more and more, independently of the other eight. These properties, inspired from Axiomatic, System and Model properties, ensure the user verifies and if possible validates the contents of each viewpoint.

- The properties needed to take into account normative rules and interconnection rules between the viewpoints as proposed in (Chatel, 2004). These properties ensure the user achieves the consistency of the whole representation. Nevertheless, it supposes that a formal semantic between the modeling languages used in the viewpoints has been established.

Figure 3 shows the properties that are used for the *Goal* viewpoint. It is composed of the properties P_1 taking into account the viewpoint itself. It is an IDEF1 model so these properties must help the user to verify and to validate the contents of this functional model. If it is not the case, a modeling problem is detected.

On the other hand, the properties gathered in the P_{15} set define the properties describing all the connecting and normative rules that need to be verified to ensure viewpoint *Goal* and viewpoint Resource organization are consistent. As soon as a property is not verified, the modeler can then investigate the database in further detail to isolate the origin of the problem. All the P_i sets are under specification at this stage of the approach development.

When all properties have been specified, there may be several properties sharing the same causes or the same effects. The causal structure of properties is thus a directed and acyclic graph where the nodes are cause or effect entities and the arcs support the relationships. This graph is translated into a conceptual graph as proposed by (Kamsu *et al.*, 2003). It allows the existence of these relations to be analyzed and proved. Finally, each property must be proved using other possible mechanisms such as model checker or theorem prover, if they exist for the chosen modeling language.

5. APPLICATION

The following example is inspired by the pedagogic literature. It is a manufacturing system shown in Figure 4 composed of manual and automated working stations. The objective is to produce three kinds of electrical devices by transforming, assembling and testing the resulting product. The resources are human operators, an automatic pallet transportation system, three dedicated assembly machines called A, B and C, a control station D in charge of electrical tests and several areas for stocking material and pallets. These resources are organized all around the transport system. On each pallet different products are installed depending on the daily customer orders. Each working station must be autonomous.

First of all, the ontology defining the common vocabulary details the different concepts that will need to be manipulated and the different relations to take into account. The concepts of Device, Actor, Activity and the relations ActivityType, ActivityDuration and so on are then refined from common sense definitions to the manufacturing domain. Second, the user must use the chosen modeling languages in order to describe all the viewpoints of this system. Each viewpoint shares or refines, as shown in Figure 5 some information with the other viewpoints.

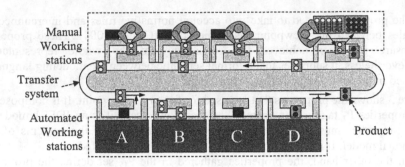

Manual Working stations

Transfer system

Automated Working stations

A   B   C   D

Product

Figure 4: Example of manufacturing system

The following shows a very simple example of a property describing a link between the resources network and the processes viewpoint. The operator involved in the control activity must have some particular skills concerning the test tasks to ensure electrical devices:

{ ∀t, [∀a ∈SetOfActivity / ActivityType(a)='control'] and [∀act ∈SetOfActor and TypeOfActor(a)='operator'], (SetOfResourcesOf(a,t) ⊃act}
⇒ {SkillOfActor(act) ⊃'Electrical test of devices'}]

If this is not the case, there is a possible risk concerning the production quality and production rate of the system.

6. CONCLUSION AND PERSPECTIVES

This article presents a work in progress based on several concepts from different cultures: enterprise modeling, systemic, risk assessment and formal verification. Our approach takes advantage of this variety and these complementarities to provide an original risk analysis method. The global proposition consists in modeling the system, verifying the resulting models, which must respect certain properties leading to possible damage or dangerous situations, and then modeling the origin of the emerging problem to provide the most relevant solution to the identified risk. The modeling approach uses the SAGACE approach. A verification approach implemented in a working platform called LUSP (French acronym of Unified Properties Modeling Language) is supported by a set of software tools (Chapurlat *et al.*, 2004, Chapurlat *et al.*, 2000).

A set of mechanisms enabling resolution of the highlighted problems at the origin of the risks is now under development. This part, not presented in this paper, is inspired by a TRIZ (Mann, 2002) analogy as proposed in (Rushti *et al.*, 2001) for business management systems and modeling tools as proposed in (Gharbi *et al.*, 2003).

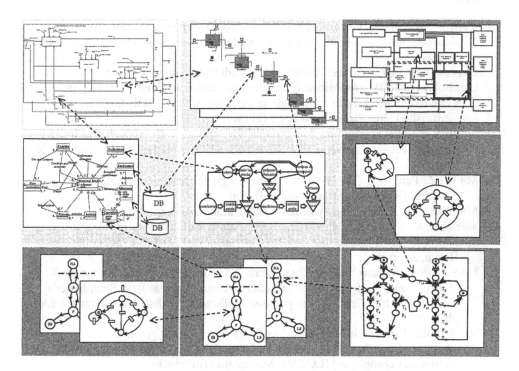

Figure 5.: Example of a SAGACE matrix

REFERENCES

Bernus P., Mertins K., Schmidt G. (2003) Handbook on architectures of information systems, Springer

Booch G., Rumbaugh J., Jacobson I. (1998), The Unified Modeling Language User Guide, Eddison Wesley

CEA (1998) SAGACE: le systémographe CEA Ed. (in French).

Chapurlat V., Kamsu Foguem B., Prunet F. (2003), Enterprise model verification and validation: an approach, Annual Review in Control, IFAC Journal

Chapurlat V., Lambolais T., Benaben F., Antoine C. (2004) Unified Properties Specification Language: a framework in Preprints of INCOM'04 congress

Chapurlat V., B.Kamsu-Foguem, F.Prunet (2002), A Property Relevance Model and associated Tools For System Life-Cycle Management in 15th IFAC World Congress on Automation Control (B'02), Barcelona

Chapurlat V., Lamine V., Magnier J. (2000) Unified Property Specification Language for industrial systems analysis: LUSP, MCPL'2000, Grenoble, France

Chatel V., Feliot C. (2004) Principe de conception système certifiée par la preuve Journées Francophones des Langages Applicatifs, JFLA 2004 (in French)

Feliot C. (2000) Modélisation systémique et techniques de la preuve de programmes pour l'analyse et la validation de spécifications systèmes, ICSSEA 2000

Gharbi D., Chapurlat V., Montmain J., Grevy G., Dusserre G. (2003) Une approche composite d'analyse de risque : identification et résolution, Congrès de Génie Industriel, Québec, Canada (in french)

Kamsu-Foguem, B., V.Chapurlat, F.Prunet (2003) Complex System Properties Representation and Reasoning by using the Conceptual Graphs, CIMCA 2003, Vienna, Austria

Kervern G.Y. (1994) Latest Advances in Cyndinics. Economica Paris

Lamboley P. (2001) Proposition d'une méthode formelle d'automatisation de systèmes de production à l'aide de la méthode B, PhD Thesis (in french) Université Henri Poincaré Nancy I (in French)

Lamine, E. (2001) Définition d'un modèle de propriété et proposition d'un langage de spécification associé : LUSP, PhD Thesis from Montpellier II University (in French)

Manna Z., Pnuelli P. (1992) The Temporal Logic of Reactive and Concurrent Systems, Editions Springer-Verlag, Berlin

Mann D., (2002) Hands on systematic Innovation, CREAX

Menzel C.P., Mayer R.J. (1998) The IDEF Family of Languages in Handbook on architectures of information systems, Bernus P., Mertins K. et Schmidt G. ed., Berlin, Springer

NASA (1998) Formal Methods Specification and Analysis Guidebook for the Verification of Software and Computer Systems, Volume II: A Practitioner's Companion, http://eis.jpl.nasa.gov/quality/Formal_Methods/document/NASA gb2.pdf

NIST (2002) Process Specification Language http://ats.nist.gov/psl/

Penalva J.M., Page E. (1994) SAGACE: La modélisation des systèmes dont la maîtrise est complexes, ILCE'94, Montpellier (in french)

Penalva, J-M. (1997) La modélisation par les systèmes en situations complexes. Ph.D. thesis, Université de Paris XI–Orsay (in French)

Perilhon P. (2003) MOSAR: Présentation de la méthode, Techniques de l'Ingénieur, traité Sécurité et gestion des risques (in French)

Ruchti B., Livotov P. (2001) TRIZ-based Innovation Principles and a Process for Problem Solving in Business and Management, proc. of European TRIZ Association

Tixier J., G.Dusserre (2000). Review of 62 risk analysis methodologies of industrial plants. Journal of Loss Prevention in the Process Industries

Uschold M., Gruninger M. (1996) Ontologies: Principles, Methods and Applications' Knowledge Engineering Review, vol.11:2, pp. 93-136

Vernadat F.B. (1996) Enterprise Modeling and Integration: Principles and Applications Chapman & Hall

17. An Ontology for Static Knowledge Provenance

Mark S. Fox and Jingwei Huang
Enterprise Integration Laboratory, University of Toronto
{msf, jingwei}@eil.utoronto.ca

Knowledge Provenance (KP) is proposed to address the problem about how to determine the validity and origin of information/knowledge on the web by means of modeling and maintaining information sources and dependencies as well as trust structures. Four levels of KP are introduced: Static, Dynamic, Uncertain, and Judgmental. In order to give a formal and explicit specification for the fundamental concepts of KP, a static KP ontology is defined in this paper.

1. INTRODUCTION

With the widespread use of Internet and telecommunication technologies that make information globally accessible, knowledge/information validity becomes a crucial factor for enterprise integration, as well as knowledge management within or across enterprises. The validation of parts catalogue information, product requirements, financial information, etc. can be quite costly. For example, an aerospace company designed a device without knowing the NASA approved parts catalogue it was using had been replaced by a newer version, thereby forcing a redesign, delay in delivery and cost overrun.

Knowledge Provenance (hereafter, referred as KP) has been proposed to create an approach to determining the origin and validity of web information by means of modeling and maintaining information sources and dependencies, as well as trust structures. The major questions KP attempts to answer include: Can this information be believed to be true? Who created it? Can its creator be trusted? What does it depend on? Can the information it depends on be believed to be true? This proposed approach could be used to help people and web software agents to determine the validity of web information.

Philosophically, we believe the web will always be a morass of uncertain and incomplete information. But we also believe that it is possible to annotate web content to create islands of certainty. Towards this end, we introduce 4 levels of provenance that range from strong provenance (corresponding to high certainty) to weak provenance (corresponding to high uncertainty). Level 1 (static KP (Fox&Huang2003)) develops the fundamental concepts for KP, and focuses on provenance of static and certain information; Level 2 (Dynamic KP (Huang&Fox2003B)) considers how the validity of information may change over time; Level 3 (Uncertainty-oriented KP (Huang&Fox2004)) considers uncertain truth value and uncertain trust relationships; Level 4 (Judgment-based KP) focuses on social processes necessary to support provenance. Since static KP is the foundation to develop other levels of KP, an explicit formal description is expected.

This paper defines static KP ontology in First Order Logic. Following the ontology development methodology of Gruninger & Fox (1995), we specify static KP ontology in 4 steps: (i) provide a motivating scenario; (ii) define informal competency questions for which the ontology must be able to derive answers; (iii) define the terminology (i.e., predicates); (iv) define the axioms (i.e., semantics).

This paper is organized as follows: Section 2 introduces related research. Section 3,4,5 and 6 define a static KP ontology in 4 steps as stated above. Section 7 introduces implementation. Section 8 gives a summary and a view on future work.

2. RELATED RESEARCH

Interest in addressing the issue of web information trustworthiness has appeared under the umbrella of the "Web of Trust" that is identified as the top layer of The Semantic Web (see (Berners-Lee 2003) slide 26, 27). Digital signature and digital certification ((Simon *et al*, 2001)) play important roles in "Web of Trust". However, they only provide an approach to certify an individual's identification and information integrity, and they do not determine whether this individual could be trusted. Trustworthiness of the individual is supposed to be evaluated by each application. For the purpose of secure web access control, Blaze *et al*, (1996) first introduced "decentralized trust management" to separate trust management from applications. Since then, trust management has grown from web access control to more general trust concerns in various web applications. Although tightly related, in the context of knowledge provenance, trust management only considers direct and indirect trust relationships to information creators but does not consider the dependencies among information units. KP addresses both trust relationships and the dependencies among information units. In addition, coming from an automated reasoning perspective, "Inference Web (IW)" (McGuinness&Silva2003) enables information creators to register proofs with provenance information in IW, and then IW is able to explain the provenance of a piece of requested knowledge. IW provides provenance information (registered by creators) for users to support them deciding by themselves to trust or not trust the requested knowledge.

In addition, information source evaluation criteria, such as, Authority, Accuracy, Objectivity, Currency and Coverage, have been developed in library and information science, and have been extended to online information (Alexander,1999).

Finally, the technologies developed in Semantic Web (Berners-Lee,2001) provide an approach to the web implementation of KP. Many technologies developed in AI, such as Truth Maintenance System (de Kleer, etc, 1989), etc., provide a basic approach for knowledge representation and reasoning in KP.

3. MOTIVATING SCENARIO

In the following, the underlying concepts of Static KP are explored in the context of two case studies.

Case 1: Asserted Information

Consider a proposition in a document found on the intranet in an enterprise. This proposition states that "a delay of more than one minute in answering a phone call may cause the customer to be unsatisfied." From a provenance perspective, there are three questions that have to be answered: 1) What is the truth value of this proposition? 2) Who asserted this proposition? 3) Should we believe the person or organization that asserted it? In this example, a further examination of the text of the web document provides the answers: It can be believed as a true proposition, asserted by a retired customer service manager, who most people in the company believe is an authority on the subject. Questions are: (1) what is the basis for us to believe this proposition as true? (2) how can the provenance process be formalized?

Case 2: Dependent Information

Consider the following proposition found in another web document: "This new approach to reduce response-delay to less than one minute may increase customer loyalty, because a delay of more than one minute in answering a phone call may cause the customer to be unsatisfied." This is actually two propositions composed of a premise, " a delay of more than one minute in answering a phone call may cause the customer to be unsatisfied." and a conclusion, " This new approach to reduce response-delay to less than one minute may increase customer loyalty." Just as in the previous case, the same questions need to be answered for each proposition. What makes this case more interesting is that answering these questions is dependent upon propositions found in other web pages. There are two types of dependency occurring. First, the truth of the premise is dependent on the truth of the proposition found in another web document. Second, the truth of the conclusion depends on the truth of the premise and upon some hidden reasoning that led to the deduction. These types of propositions are called "dependent propositions" in KP.

It is common to find information in one document that is reproduced in another. The reproduction of a proposition in a second document leads to an equivalence relationship between the two propositions, i.e., the truth values of the two propositions are equivalent. However, the relationship is also asymmetric; one proposition is a copy of the other. The copy of one proposition is classified as "equivalent information". Furthermore, a proposition can be derived using logical deduction. Hence, the truth value of the derived proposition depends on the truth values of its antecedent propositions. This type of derived proposition is classified as "derived information".

Returning to the example, determining the provenance of the premise requires that we link, in some way, the premise to the proposition in the other web document from which it is copied. The same is true of the conclusion. Minimally, we should link it to its premise, maximally we should link it to the axioms that justify its derivation. These links will also require some type of certification so that we know who created it and whether it is to be trusted.

From these two cases, a number of concepts required for reasoning about provenance emerge:

- Text is divided into propositions. Once so designated, they are assumed to be indivisible.
- An proposition must have a digital signature.
- An assertion is believed to be true, if the information user trusts the person or organization that created the assertion in the corresponding topic.
- As propositions are reused across the web, a link between where it is used and where it came from must be maintained. These links, or dependencies, must be included in the digital signatures with propositions.

- Dependencies can be simple copies, or can be the result of a reasoning process. If the latter, then axioms used in the reasoning should also be identified and signed by an acceptable organization.

4. INFORMAL COMPETENCY QUESTIONS

What static KP needs to answer, called informal competency questions, are identified as follows. These questions define the requirements to Static KP.

- Is this proposition true, false, or unknown?
- Who created this proposition?
- What is the digital signature verification status?
- Which knowledge fields does this proposition belong to?
- In these fields, can the information creator be trusted?
- Does the truth of this proposition depend on any other propositions? If so, which ones?

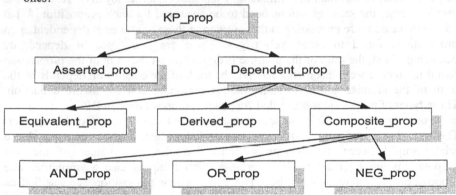

Figure 1. Proposition Taxonomy in Knowledge Provenance

5. TERMINOLOGY

There are five main classes in the static KP ontology: Propositions, Documents, Information Sources, Trust Relationships and Signature Status.

Propositions

The basic information unit in KP is a proposition. KP-Prop is the most general concept used to represent propositions in a document. From our motivating scenario in section 3 and the natures of propositions, we prefer to depict the taxonomy of propositions in KP as shown in Figure 1. An Asserted_prop is an assertion that is not dependent on any other propositions; a Dependent_prop is a proposition which truth is dependent on other propositions; an Equivalent_prop is a quotation that is a copy and its truth value is the same as the proposition it depends on; a Derived_prop is a derived conclusion based on some premises; a Composite_prop could be the "and"/ "or" / "negation" of other proposition(s).

Table I defines the predicates for depicting a KP proposition and its attributes.

Table I. Predicates depicting a KP proposition and its attributes

Predicate	Description
type(x, "KP_prop")	x is defined to be a proposition, signified by being of type KP_prop.
proposition_content(x,s)	s is the content of the proposition x. In html files, the content of a proposition usually is a string; in xml files, the content of a proposition can be an xml element.
assigned_truth_value (x,v)	Proposition x has a truth value v assigned by proposition creator.
trusted_truth_value(a,x,v)	Agent a trusts that proposition x has a truth value v. v may be one of "True", "False", or "Unknown".
type(x, "asserted_prop")	x is an assertion and does not depend upon any other proposition.
type(x, "dependent_prop")	x is a proposition whose truth value is dependent upon another proposition. Dependent-prop class is further divided into 3 subclasses: equivalent-prop, derived-prop, and composite-prop.
type(x, "equivalent_prop")	An equivalent-prop is a copy of and its truth value is the same as the proposition it depends on.
type(x, "composite_prop")	Composite-prop is defined to be the logical combination of its constituent propositions. A composite-prop is divided into 3 subclasses: neg-prop, and-prop, and or-prop.
type(x, "derived_prop")	A derived-prop indicates that the proposition is a derived conclusion based on some premises. For example, derived-prop B has dependency-link pointing to composite-prop A, meaning that A is a premise of B.
is_dependent_on(x, y)	Proposition x is dependent on proposition y. x is called dependent proposition, and y is called support proposition.
has_same_content(x,y)	Proposition x has the same proposition content as y.

Documents

To facilitate the determination of the provenance of a proposition, properties of the document in which it appears may need to be considered. For example, knowing who created the document may be important in determining the validity of a proposition within. A document can be any type of file. For the purposes of this paper, we restrict our attention to standard web files such as: html files, xml files, and xhtml files. Following are document related KP predicates:

Predicate	Definition
type(x, "document")	x is defined to be a KP document.
in_document(y,d)	Proposition y is contained in document d.

Information Source and Signature

For any document and proposition its creator can be defined. Along with it can be defined a digital signature and the verification status of the signature. Assume that a digital signature validation software provides the result of signature verification.

Predicate	Description
has_infoCreator(x,c)	KP-prop or Document x has infoCreator c. Here, infoCreator may be either creator or publisher.
has_signature(x, s)	The proposition or document x has a signature s.
has_sig_status(x, v)	The digital signature verification status of x is v, where v may be one of three status: "Verified"-- the signature is verified successfully; "Failed"-- the signature verification is failed; and "NoSignature"-- do not have digital signature.

Trust Relationships

In section 3 we stated that KP is context sensitive, where the context is largely associated with trust relationships that define the provenance requester trusts whose propositions in what topics. A trust relationship in KP is defined as a of triple (a, c, f) where the provenance requester (information receiver) a "trusts" information creator c in a topic or a specific knowledge field f, here, "trust" means that a believes any proposition created by c in field f to be true. (Note: The mathematical definition of a trust relation should be a set of triples $\{(a, c, f)\}$. A triple (a, c, f) is called a trust relationship in this paper). The following defines the trust related predicates:

Predicate	Description
trusted_in(a, c, f)	Provenance requester a trusts information creator c in knowledge field f.
trusted(x, a)	Proposition x is trusted by agent a. That means its information creator is trusted by a in one of the fields which proposition x belongs to.
in_field(x,f)	Proposition x belongs to knowledge field f.
subfieldOf(x,y)	Knowledge field x is a sub-field of knowledge field y

6. AXIOMS

In the following, a set of axioms is defined to specify truth conditions of KP-props. Basically, the truth value of an asserted proposition depends on if the proposition is "trusted"; the truth value of an equivalent proposition depends on the truth value of the proposition that this equivalent proposition points to by its dependency-link; the truth value of a derived proposition depends on if the proposition is "trusted" and if its support KP-prop is true. In addition, a KP-prop is "trusted", if the creator or publisher of the proposition is trusted in one of the fields of the proposition, and the digital signature verification status is "Verified". Finally, note that the "close world assumption" is applied to handle "not" in this paper.

Asserted Propositions

An asserted-prop is trusted to have its truth value as assigned, if the asserted-prop is trusted by the provenance requester.

Axiom SKP-1:
  for-all (a,x,v)
  ((type(x, "asserted_prop") ^ trusted(x, a) ^ assigned_truth_value(x, v))
  →   *trusted_truth_value(a, x, v)).*

A KP-prop is "trusted", if the creator or publisher of the proposition is "trusted" in one of the fields of the proposition, and the digital signature verification status is "Verified".

Axiom SKP-2:
  for-all (a,x,f,c,w)
  ((type(x, "KP-prop") ^ has_sig_status(x, "Verified") ^ has_infoCreator(x, c)
  ^ in_field(x, f) ^ trusted_in(a, c, w) ^ subfield_of(f, w))
  →   *trusted(x, a)).*

For a KP-prop that has no creator specified, the creator of the document is the default creator of the KP-prop.

Axiom SKP-3:
  for-all (x, d, c)((type(x, "KP-prop")
  ^ (not(exist (c2) has_creator(x, c2)))
  ^ in_document(x, d) ^ has_creator(d, c))
  →   *has_creator(x, c)).*

If a proposition does not have a creator, then the digital signature verification status of the KP-prop is determined by the digital signature verification status of the document.

Axiom SKP-4:
  for-all (x, d, c, v)((type(x, "KP-prop")
  ^ (not (exist (c2) has_creator(x, c2)))
  ^ in_document(x, d) ^ has_creator(d, c) ^ has_sig_status(d, v))
  →   *has_sig_status(x, v)).*

Equivalent Propositions

The trusted truth value of an equivalent-prop is the same as the trusted truth value of its support proposition, if this equivalent-prop exactly has the same proposition-content as its support proposition has.

Axiom SKP-5:
  for-all (a, x, y, v) ((type(x, "equivalent_prop")
  ^ is_dependent_on(x, y) ^has_same_content(x,y)
  ^ trusted_truth_value(a, y, v))
  →   *trusted_truth_value(a, x, v)).*

Composite Propositions

The trusted truth value of a neg-prop is the negation of the trusted truth value of the KP-prop it is dependent on.

Axiom SKP-6:
> *for-all (a, x, y)((type(x, "neg_prop")*
> *∧ is_dependent_on(x, y) ∧ trusted_truth_value(a, y, "True"))*
> → *trusted_truth_value(a, x, "False")).*

Axiom SKP-7:
> *for-all (a, x, y)((type(x, "neg_prop")*
> *∧ is_dependent_on(x, y) ∧ trusted_truth_value(a, y, "False"))*
> → *trusted_truth_value(a, x, "True")).*

The trusted truth value of an and-prop is "True" if all its support KP-props are "True"; and the trusted truth value of an and-prop is "False" if at least one of its support KP-props is "False".

Axiom SKP-8:
> *for-all(a, x)((type(x, "and_prop")*
> *∧ for-all (y) (is_dependent_on(x, y) -> trusted_truth_value(a, y, "True")))*
> → *trusted_truth_value(a, x, "True")).*

Axiom SKP-9:
> *for-all(a, x)((type(x, "and_prop")*
> *∧(exist(y) (is_dependent_on(x, y) ∧ trusted_truth_value(a, n, "False"))))*
> → *trusted_truth_value(a, x, "False").*

The trusted truth value of an or-prop is "True" if at least one of its support KP-props is "True"; and the trusted truth value of an or-prop is "False" if all its support KP-props are "False".

Axiom SKP-10:
> *for-all(a, x)((type(x, "or_prop")*
> *∧ (exist (y) (is_dependent_on(x, y) ∧ trusted_truth_value(a, y, "True"))))*
> → *trusted_truth_value(a, x, "True")).*

Axiom SKP-11:
> *for-all(a, x)((type(x, "or_prop")*
> *∧ (for-all (y) (is_dependent_on(x, y) ∧ trusted_truth_value(a, y, "False"))))*
> → *trusted_truth_value(a, x, "False")).*

Derived Propositions

The trusted truth value of a derived proposition is "True" or "False" as specified, if it is "trusted" and its support KP-prop (condition) is "True". Note that the axioms used to derive the truth value do not have to be included as part of the dependency.

Axiom SKP-12:
> *for-all (a, x, y, v)((type(x, "derived_prop")*
> *∧ trusted(x, a) ∧ assigned_truth_value(x, v)*
> *∧ is_dependent_on(x, y) ∧ trusted_truth_value(a, y, "True"))*
> → *trusted_truth_value(a, x, v)).*

Default assigned_truth value

The default truth value of an asserted or derived proposition assigned by the proposition creator is "True".

Axiom SKP-13:

for-all (a, x, y, v)((type(x, "Asserted_prop")^ type(x, "Derived_prop")
^ triple(x, assigned_ truth_value, v))
→   assigned_truth_value(a, x, v)).
for-all (a, x, y, v)((type(x, "Asserted_prop")^ type(x, "Derived_prop")
^ not (triple(x, assigned_ truth_value, v)))
→   assigned_truth_value(a, x, "True")).

Default trusted_truth value

The default trusted truth value of a proposition is "Unknown".

Axiom SKP-14:

for-all (a, x, v)((type(x, "KP_prop")
^ not (trusted_truth_value(a, x, "True"))
^ not (trusted_truth_value(a, x, "False")))
→   trusted_truth_value(a, x, "Unkown")).

7. IMPLEMENTATION

To apply KP in practice, information creators need to annotate web documents with KP metadata, users (provenance requesters) need to define their trust relationships, and a KP reasoner conducts provenance reasoning on annotated web documents.

In order to facilitate the annotation of web documents with KP metadata and define trust relationships, we have defined a KP markup language in RDFS (Resource Description Framework Schema) (Brickley&Guha, 2004). The following is a piece of example containing only one proposition in a web document annotated with kp metadata. An entire annotation example can be found in (Fox&Huang2003).

```
<kp:Derived_prop rdf:id="ReduceDelay"
    is_dependent_on="#ProblemOfDelay"
    creator ="Tim Levy"
    in_field ="Custom Relationship Management">
            The new approach to reduce response-delay to
            less than one minute may increase customer loyalty.
</kp:Derived_prop>
```

We have implemented the KP reasoner with Prolog. The system can infer the truth of any KP-prop. In the above example, assume that information user A, who requests the provenance of this proposition, trusts Tim Levy in field "Custom Relationship Management", and the digital signature verification status of the proposition is "verified", to determine the trusted truth value of this derived proposition, the system applies axiom SKP-12, and then the main goal *trusted_truth_value(A, "ReduceDelay",v)* is divided into several sub-goals: *trusted("ReduceDelay",A)* is solved by applying axiom SKP-2; *assigned_truth_value("ReduceDelay",v)* is solved by applying axiom SKP-13, and *v* is bound as *"True"*; from kp metadata, *is_dependent_on("ReduceDelay", "ProblemOfDelay")* is true; this leads to solve sub-goal *trusted_truth_value(A, "ProblemOfDelay", "True"))* by applying axiom

SKP-1. The process to solve this subgoal is similar. If this last sub-goal is solved, then the main goal is solved and the returned trusted truth value v is "True"; otherwise, the main goal is failed, the system returns trusted truth value of "*Unknown*" by applying axiom SKP-14.

8. DISCUSSION AND FURTHER WORK

Knowledge Provenance is an approach to determining the validity and origin of information/knowledge by means of modelling and maintaining information source and dependencies, as well as trust structures. Four levels of KP are introduced: Static, Dynamic, Uncertain, and Judgmental. In order to give a formal and explicit specification for the fundamental concepts of KP, a static KP ontology was defined in this paper. A KP markup language was designed with RDFS; a KP reasoner that traces the web documents annotated with kp metadata and deduces the origin and validity of requested information was implemented in Prolog.

Based on this formal static KP model, we have developed a dynamic KP model (Huang&Fox2003) that determines the validity of information in a world where the validity of a proposition and trust relationships are changing over time. Furthermore, an uncertainty-oriented KP model that introduces "trust degree" to represent uncertain trust relationships and "certainty degree" to represent uncertain truth value has been developed (Huang&Fox2004).

We will continue our work towards judgment-based KP to develop a formal "social" process representing trust propagation in social networks. In addition, a web based KP reasoner will be implemented to deduce the origin and validity of requested web information by tracing web documents across the web.

As stated earlier in the paper: "we believe the web will always be a morass of uncertain and incomplete information". The challenge therefore is to create models and processes that will enable the validation of as much information as possible.

This research was supported, in part, by Bell University Laboratory and Novator Systems, Ltd.

REFERENCES

Alexander, J. E., and Tate, M.A., (1999), Web Wisdom: how to evaluate and create information quality on the web, Lawrence Erlbaum Associates Publishers.

Berners-Lee, T., (1998), Semamtic Web Road Map,
http://www.w3.org/DesignIssues/Semantic.html

Berners-Lee, T., Hendler, J., and Lassila, O., (2001), "The Semantic Web", Scientific American, May 2001.

Berners-Lee, T., (2003), The Semantic Web and Challenges,.
http://www.w3.org/2003/Talks/01-sweb-tbl/

Blaze, M., Feigenbaum, J. and Lacy, J., (1996), Decentralized Trust Management, Proceedings of IEEE Conference on Security and Privacy, May, 1996.

Brickley, D. and Guha, R.V., RDF Vocabulary Description Language 1.0: RDF Schema, W3C Recommendation 10 February 2004,
http://www.w3.org/TR/2004/REC-rdf-schema-20040210/

de Kleer, J., Forbus, K., McAllester,D., Truth Maintenance Systems (Tutorial SA5), Int. Joint Conference on Artificial Intelligence, SA5-182~225, 1989.

Fox, M. S., and Huang, J., (2003), "Knowledge Provenance: An Approach to Modeling and Maintaining the Evolution and Validity of Knowledge", EIL Technical Report, University of Toronto.
http://www.eil.utoronto.ca/km/papers/fox-kp1.pdf

Gruninger, M., and Fox, M.S., (1995), "Methodology for the Design and Evaluation of Ontologies", Workshop on Basic Ontological Issues in Knowledge Sharing, IJCAI-95, Montreal.

Huang, J. and Fox, M. S., (2003B), " Dynamic Knowledge Provenance ", EIL Technical Report, University of Toronto.
http://www.eil.utoronto.ca/km/papers/kp2-TR03.pdf

Huang, J., and Fox, M.S., (2004), "Uncertainty in Knowledge Provenance", in Bussler, C. Davies J., Fensel D., Studer R. (eds.) The Semantic Web:Research and Applications, Lecture Notes in Computer Science 3053, Springer, 2004, PP.372-387.

Huhns, M.H., Buell, D. A., (2002), Trusted Autonomy, IEEE Internet Computing, May. June 2002.

Khare, R., and Rifkin, A., (1997), "Weaving and Web of Trust", World Wide Web Journal, V.2, pp.77-112.

Simon, E., Madsen, P., Adams, C., (2001), An Introduction to XML Digital Signatures, Aug., 2001. http://www.xml.com/pub/a/2001/08/08/xmldsig.html

18. Object Model for Planning and Scheduling Integration in Discrete Manufacturing Enterprises

Yasuyuki Nishioka[1]

1 Hosei University, Email: nishioka@k.hosei.ac.jp

This paper proposes an object model for planning and scheduling integration in system development on discrete manufacturing. The model can deal with frequent changes of the market much more agile than the traditional models for production management. In addition, the object model can be translated to another model, which additionally has table, property and data type objects for automatic translation to a RDB schema. A modification procedure to adjust it to each case in industries is also described.

1. INTRODUCTION

Production management in manufacturing enterprises had been completely divided into two layers: an enterprise management layer in which order management and production/inventory planning are executed from a view of enterprise, and a shop floor management layer where production scheduling, quality control and maintenance activities are carried on in a decentralized basis. Facing to the current market environment, this division becomes an obstacle to achieve agile manufacturing that can synchronously respond to the market by means of dynamic change of operational decisions in shop floor management. Sales and services required to the enterprise should be responded to their customers eliminating the barrier between the two layers. Methods and models for such integrated systems are strongly required.

There were many investigations on this issue to integrate manufacturing systems as a whole enterprise model. For example, Purdue reference model for CIM was proposed for developing general model for manufacturing enterprises with respect to computer integrated systems (Williams, 1989). On the other hand, CIMOSA specification (ESPRIT, 1993) was proposed by an ESPRIT research project as a solution for enterprise integration supporting both model engineering and model execution. A framework of the Generalized Enterprise Reference Architecture (GERAM) proposed by the IFAC/IFIP task force was also a model considering the integration issue (IFIP–IFAC, 1999). As ontology, TOVE project proposed ontologies for enterprise modeling (Fox, 1998), which try to represent common vocabulary for modeling. In accordance with the Purdue model, ISA provided an industrial standard referred to as S95, which will also be an IEC/ISO international standard (ISA, 2000).

With respect to production planning and scheduling, all above approaches describe separate functionality of each module and very simple interfaces between them. Considering a production planning system and a detail scheduling system, information on the interface may have production orders sending to shop floors and their production results. However, those are too simplified in order to catch up the market changes. For example, because of a lack of data connection between production planning and detail scheduling, enterprises are due to have more inventory and production capacity than the necessary. Therefore, we are trying to make a new integration framework for production planning and scheduling (Nishioka, 2004).

Since traditional MRP/MRP II based planning systems have not been success to achieve such elaborate integration, advanced planning and scheduling (APS) architecture that deals with detail scheduling in a decision process of business layer is introduced. However, APS gets a limited success because it does not have a well-designed integration model. A required integration model should deal with different granularities, different time spans and different focus points both on planning and scheduling.

This paper proposes an object model for planning and scheduling integration in manufacturing enterprises. The model is an extended version of the PSLX specifications published by PSLX consortium, which is a Japan based non profit organization for research and dissemination of general models and standards for next generation manufacturing industries (PSLX 2003). The PSLX specifications describe a decentralized architecture using a concept of autonomous agents, and define some ontologies used in their communication messages (Nishioka 2004).

The models in this paper are defined for discrete manufacturing where variety of parts and materials are composed into a final product in order to fulfill customer orders within shorter response time and less inventory. Many kinds of order processing such as make-to-order, configure-to-order, make-to-stock, and many kinds of production categories such as one-of-a-kind production, lot-production and repetitive production are in our scope.

This paper also discusses implementation issues on each manufacturer. To reduce the cost and the period of time for developing new/revised information systems, relationship between RDB schema and the proposed object model is discussed. A procedure to create an enterprise-specific RDB schema from a template of object models is also described. These procedural relationships are very important not only for the cost and time reduction but also for data federation or integration among different enterprises on a supply chain.

2. ONTOLOGIES

This section defines several important terms used in the object models. Since the terms have a variety of semantics depending on contexts of use, industrial categories, culture of shop floors and so on, the definitions are performing as ontology, which gives a meaning of itself providing a domain world without miss communications. Our approach to define ontology is to describe its intuitive explanations rather than precise specifications. However, they can be applied correctly to each particular situation, considering their structures and the relations to the expected decision processes.

2.1 The primary terms

The most primitive and important terms for planning and scheduling in a discrete manufacturing management are *order*, *operation*, *item* and *resource*. First, *order* is a term to represent an actual requirement of activities that are expected to do or already done on a certain period of time. There are a lot of order categories such as customer orders, prospective orders, purchase orders, production orders and work orders. Second, *operation* is a term to represent a class of specific activity for manufacturing in general. Activities for manufacturing generally include production operations, inspection operations, transportation operations and inventory operations.

Regarding that production operations usually produce and consume something goods or materials, *item* is a term to represent the something that is produced or consumed by an operation. Finally, *resource* is a term to represent an item that is necessary to carry out an operation but never consumed after its completion. Total capacity of a resource for each period of time can be shared by several operations that require the resource during their execution. Equipments, personnel and tools are typical classes of resources.

2.2 Other useful terms for manufacturing

There are several useful terms in addition to the primary terms. These are important for discussing production management not only from a local view of shop floor operations, but also from a view of enterprise-wide management. First, *area* is a term to represent an aggregation of resources, corresponding to a local decision making unit for manufacturing management. In addition, *product* is a term for a special item that is dealt with in business decisions. On the other hand, *process* is a term to represent an aggregation of operations that produce and/or consume products.

Special cases of resources can be defined as storage or route. Regarding that there are inventory operations and transportation operations, those operations require special classes of resources: storages and routes respectively. Storages are located at particular address and used to keep items for a particular period of time, while routes are connected to those storages and take account the traffic availability on each path. Finally, the term party is used to identify a border of an enterprise. Outside of the enterprise can be represented using customer and supplier that are sub classes of party.

Terms for relational definitions

Using the primitive terms described above, there are also definitions of several terms that represent relations between two of them. First, *source* is a term for a relation between product and area, representing possibility to supply the product. A relation between process and area is defined by a term *loading*, where the process needs to use capacity of the area during its execution period.

For operation, there are two terms for relative definitions. One is assignment, which represents a relation between operation and resource if the operation uses the resource. The other is material use, which represents a relation between operation and item that is consumed by the operation. In addition, there are other relations that define for two elements of a same kind. A term item structure presents a part-of relation between two items, if one item contains the other as a part or a material. On the other hand, precedence is a term that represents predecessor or successor of operations.

2.3 Terms for production management

When a scheduler or planner is making a decision of production management, there are some special terms useful to understand the content of management. The followings are created every time to manage actual manufacturing processes. First of all, *inventory* and *capacity* should be addressed as very important terms. Inventory is a temporal status of storages or areas with respect to the volume of items or products respectively. Similar to this definition, capacity is a term for a temporal status of resources or areas to represent availability in making reservations for the current or future orders. Calendar is a special case of capacity, because it generally shows availability to work for each day.

In traditional production management, Master Production Schedule (MPS) and Material Requirement Plan (MRP) are very important terms. This paper also makes definitions for them as follows: *MPS* is a term to represent orders that actually request to produce some products within each period of time. On the other hand, *MRP* is a term to represent actual requirements of materials that are needed for accomplishing production requests in MPS. Each requirement in the MRP corresponds to a purchase order, or other MPS if the material can be produced as a product in the same enterprise.

Finally, some additional technical terms for production management are defined using *lot tracking* and *order pegging*. Both of them are used to represent a relation between two orders. Lot tracking represents a connection from one order to another with respect to dependency of their actual production lot. This shows that a production lot produced by one work order is consumed by the other work order. Order pegging, on the other hand, represents a relation that is made for customer order fulfilment, connecting orders from a shop floor level to a business level.

2.4 Definition of planning and scheduling

Planning and scheduling is very general terms, thus we have to define their meanings to avoid any confusions. In order to do this, a concept of time, which is continuously going from the past to the future in the real world, is very important to distinguish them. With respect to time in this definition, scheduling can be defined as a term to represent a decision of time-dependent problems. The results of a decision contain parameters of time instances. On the other hand, planning is a term for decision making relate to activities in general. The definition of planning basically includes scheduling as a part. However, this paper discusses production planning as a time-independent problem, and considers that it occasionally communicates with scheduling for its time-dependent aspects.

Generally, production planning seems to have a time-dependent aspect. We think that most of all such time-dependent aspect can represent without continuous time horizon, using a concept of time bucket that represents a series of fixed-length periods of time. In a problem within a single time bucket, temporal parameters can be removed. A problem that lies across several time buckets can represent their time aspects using any other time-independent constraints.

Our definition requires that problems for some decision makings which contain time instances in its result should be solved by scheduling. In such cases, scheduling can be performed individually or as a portion of a planning problem. For example, a problem of order allocation to resources is categorized to a scheduling problem if it decides a particular start time and/or end time to reserve appropriate capacity of resources.

Since planning and scheduling have a lot of variety, the rest of this paper only focuses on production planning and production scheduling. In our model, production planning mainly deals with production order creation for many kinds of requests in manufacturers. Planning does not deal with continuous time horizon, but uses time buckets for some parameters if it needs temporal aspects. Production scheduling, then, mainly deals with work order assignment in a shop floor on a continuous time horizon.

3. PLANNING AND SCHEDULING INTEGRATION

This section briefly presents an integrated architecture of production planning and production scheduling. The architecture is based on the object model in which each object corresponds to the ontology defined in the previous section.

3.1 Order management

Before discussing the issue of integration, we have to clarify several groups of orders in manufacturing enterprises, because orders are very important information that performs in the integrated architecture. Especially, in a business layer, there are several types of orders, which are described as prospective orders, customer orders, purchase orders and production orders.

Prospective orders are created inside of an enterprise with respect to the future demand. Customer orders, on the other hand, are provided by customers, and will be connected to the prospective orders in cases where order-to-shipping lead time should be less than production lead time. For parts or materials that cannot be made in the enterprise, purchase orders are created in a procurement division. Finally, production orders are actual requirements of production activities that are aggregated for each production area. Production orders are created in a business layer, and also in a shop floor layer if necessary.

In a shop floor layer, there is another type of orders, which are referred to as work orders that deal with detail level of production due to the other orders such as production orders. A work order corresponds to particular activity that actually performs in a shop floor level. A work order can have time instances for its start time and end time. Therefore, scheduling deals with work orders that are producing a particular item in a shop floor. A work order can also represents any results of the activity.

As a special feature of our framework, production orders and purchase orders can be created and modified in shop floors. In other words, production planning and production scheduling communicate through these orders. This two-way information flow allows manufacturers to perform agility by means of collaboration of business divisions in an office and manufacturing divisions in distributed shop floors.

3.2 Objects for planning and scheduling

Regarding a manufacturing enterprise organization, we have been discussing two different layers: a business layer, which faces to customers of the enterprise, and a shop floor layer, which faces to reality of the shop floor controls. In terms of production management, functions of business divisions and functions of shop floors almost correspond to production planning and production scheduling respectively, because, business divisions are responsible to customer orders and make production orders for the customers, whereas shop floor managers try to fulfill the production orders by making a schedule.

Figure 1 illustrates our object model clarifying the border line of planning and scheduling. The outer area restricted by the dashed line represents a planning decision, while the area within the inner dashed line indicates objects related only to scheduling. Product, process and area are the planning objects as well as customer order and prospective order. The figure also shows that item, operation, resource and work order belong to scheduling objects. Production order and purchase order are on the border line, depending on the fact that they can be created by either planning or scheduling.

Relations between two of those objects in Figure 1 can also be pointed out. Some of them are defined in the previous section. For example, the relation between process and area represents loading. The relation between operation and resource represents assignment, and so on. In Figure 1, there are also self-referring relations such as precedence, item structure, order tracking and order pegging.

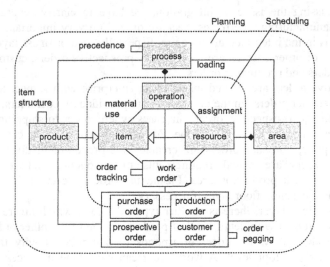

Figure 1: Primitive objects mapped on planning layer and scheduling layer

3.3 Collaboration schema

Using the objects in the model, our general framework of planning and scheduling integration is provided in Figure 2. This figure shows two modules of planning and scheduling and some relative objects to execute the modules as input or output. All types of orders indicated in Figure 1 can be shown in this chart. In addition to them, some transactional data such as capacity, inventory, pegging and tracking are addressed in the chart.

Collaboration between planning and scheduling is established through the feature that production orders and purchase orders can be made both by production planning and production scheduling. When a new production order is created in a shop floor, corresponding business processes will be carried out until the next cycle of production planning. In addition, inventory and capacity information are circulated as a bridge between planning and scheduling, while they have different granularities each of which corresponding to the different levels of planning and scheduling.

In conventional production management, production planning and detailed production scheduling perform on a sequential procedure. A production plan is decided first followed by a decision of scheduling that fulfills the production orders. Comparing to this, our model allows schedulers to make their own schedule even if it does not meet the given plan. They can create or modify production orders or purchase orders for themselves.

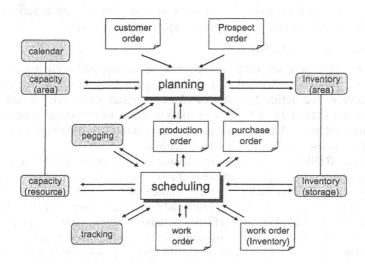

Figure 2: Integration framework of planning and scheduling

3.4 Object hierarchy across the interface

In order to establish two ways communication between production planning and production scheduling, there are object hierarchies across the interface. First of all, the hierarchies between item and product, resource and area, and operation and process are primarily important to make connections. All the scheduling level objects such as item, resource and operation should be represented as a child of the planning level objects. This is a kind of static relations across the interface.

In execution of production planning and production scheduling, order information is the key to coordinate the plan and the schedule. Consequently, work orders producing an item are defined in conjunction with a production order or customer order of a product. In this case, work order and production order refer to operation and process respectively. This dynamic relationship is represented by pegging objects that are generated and maintained by the planning and scheduling modules collaboratively.

The final and most meaningful relations are the self referring hierarchy in capacity objects and inventory objects. There are at least two levels in capacity objects: an area level capacity and a resource level capacity. Usually area capacity corresponds to capacity of a bottleneck resource in the area. On the other hand, inventory also has two levels: area level inventory and storage inventory. Area level inventory represents the volume of particular products or materials regardless of their status whither they have been made or not. In other wards, the number of inventory increases when the substance comes inside the area and decreases when it goes out from the area. One layer's information of inventory or capacity is effective as a constraint of the other, and vice versa.

4. IMPLEMENTATION ISSUES

In order to apply the model to real industries, system development usually needs a RDB schema, which can be also represented by object models. However, the detail of the schema is different from the general model because it should have technology specific aspects such as efficiency of table search. This section describes how to translate the general model to appropriate objects of RDB schema.

4.1 Object models and RDB

Considering RDB schema, other objects can be represented with respect to table information, property information and data type information. Considering RDB implementation, we define four object classes: master table, relation table, order table and other data table. All tables in RDB used for planning and scheduling will be implemented as a table of those four, having information defined in one of the general objects described in the previous sections.

Regarding RDB features, all master tables have property information about id, description, price, cost, capacity and registration. For the primary objects, we define sub classes of the master including party master, area master, storage master, item master, process master, operation master and resource master. Relation tables that relate one object to another are defined including source table, assignment table, loading table, material use table, item structure table, and precedence table.

Table for orders can be defined using prospective order table, customer order table, purchase order table, production order table, and work order table. This classification is the same as the object model. Note that these order tables can represent not only orders but also the results or progress of the orders.

Finally, the transactional data generated during the planning and scheduling decisions are created as the other data classes. They include MPS table, MRP table, lot tracking table, order pegging table, inventory table, capacity table, and calendar table. The lot tracking table and the order pegging table have links to two orders, connecting one to the other. The inventory table and the capacity table perform in both a planning and a scheduling.

4.2 Table generation procedure

The object model modified for RDB implementation does not always meet particular cases in the real manufacturing. Therefore, we provided a typical model as a template, which can vary depending upon each environment. In order to adjust the model, we propose the following procedure.

Step 1: Create and delete primitive objects for the application

Many kinds of objects may be added to the template for practical industrial applications. Some of them come from a schema in legacy systems. Furthermore, it is necessary to delete some template objects that are not used in the application. Sometimes, separation of an existing object is required. Sometimes, two objects are merged.

Step 2: Modify relational information among the objects

Relations among primitive objects can be changed. Depending on the cardinality of relationship between two objects, some relational objects are deleted where the target number can be reduced to single. Some new relational objects are created where one-to-many relations are additionally required.

Step 3: Modify properties for each object

In order to modify detail information of a table object, you can attach and/or detach any property candidates to the table object. New property candidates can be created if necessary. The content of each property of tables can be modified by choosing its attributes.

Step 4: Move or duplicate attribute through the links

With respect to efficiency of the RDB, some attributes of properties need to move or duplicate from one table object to other through the directed link between them. For example, consider that an item table has a link to a storage table, and then the storage table has a property of quantity. In this case, the property of quantity can move to the item table, so that the item table directory shows the inventory amount.

Step 5: Define data type object for each attribute

The attributes of each property have a particular data type such as string, integer, float, Boolean, and so on. The final specification of the data type completely depends on RDB management systems. Since the data type in object model represents in a conceptual level, the final data types of RDB systems should be determined.

5. CONCLUDING REMARKS

As industrial case studies, we investigated three application systems for different kinds of manufacturers: a 2nd tier automotive manufacture, which repetitively makes pipes for gasoline, oil and air supply; a manufacture, which provides one-of-a-kind laser cutting fabrication services for high precision parts; and a pump manufacturing enterprise, which provides their products in make-to-order basis. After the investigation of the detail of the current RDBs, the whole part of the data in the three enterprises can be represented by our object model with some appropriate extensions. Their object models also regenerated the corresponding RDB schema according to the procedure.

This experimental study showed that the proposed object model is practical enough for real industries. Furthermore, we found out that representations using the model are much easy to understand while we discuss the additional modules for the enterprises. Unfortunately, those three enterprises did not have a scheduling module as a computer system. They made daily schedules on white boards or on paper charts. However, in this case study, we successfully made preliminary design of database for those scheduling systems according to the object model.

From the experiment, this paper argues that the approach using the object model for planning and scheduling integration is more successful than the past. In traditional approach, in which a planning system and a scheduling system are developed independently, interfaces are so simple that any special advantages can not be provided from the integration. Moreover, in terms of step-by-step development, the representation of the object model is more useful because of independence between the computer-specific implementation aspects and the object model in manufacturing enterprises. The three enterprises of the case studies keep improving the system development according to this framework.

This paper proposed an object model for planning and scheduling integration in practical fields of discrete manufacturing. Different from the traditional system of production management, the integrated model was defined in order to adjust the plan and schedule collaboratively to the frequent changes of the market. With respect to system development for real industries, the object model can be translated to another object model for RDB schema, which has special features of tables, properties and data types. Since the RDB dependent model was provided as a template for any modification according to each case of manufacturing enterprises, we also proposed a modification procedure to meet each requirement in the industries.

Acknowledgments

This study is partially depending on the valuable discussions in the project of PSLX consortium. Author appreciates all the contribution of PSLX technical committees and the other members of the project.

REFERENCES

ESPRIT Consortium AMICE (Eds.) (1993), CIMOSA: Open System Architecture for CIM, Springer-Verlag

Fox, M.S., Gruninger, M., (1998), Enterprise Modelling, AI Magazine, AAAI Press, Fall 1998, pp. 109-121

IFIP–IFAC Task Force on Architectures for Enterprise Integration (1999), GERAM: Generalised Enterprise Reference Architecture and Methodology, [http://www.cit.gu.edu.au/~bernus/taskforce/geram/versions/geram1-6-3/v1.6.3.html]

ISA (2000), Enterprise-Control System Integration Part 1: Models and Terminology, ISA-95.00.01, The Instrumentation, Systems, and Automation Society

Nishioka, Y. (2004), Collaborative agents for production planning and scheduling (CAPPS): a charange to develop a new software system architecture for manufacturing management in Japan, International Journal of Production Research, (to appear).

PSLX consortium (Ed.) (2003), PSLX Engineering specification, PSLX consortium, [http://www.pslx.org]

Williams, T. J. (Eds.) (1989), A reference model for computer integrated manufacturing (CIM), Instrument Society of America

19. B2B Applications, BPEL4WS, Web Services and .NET in the Context of MDA

Jean Bézivin[1], Slimane Hammoudi[2],
Denivaldo Lopes[1,2] and Frédéric Jouault[1,3]

1 University of Nantes Email: Jean.Bezivin@lina.univ-nantes.fr
2 ESEO Email: {shammoudi, dlopes}@eseo.fr
3 TNI-Valiosys Email: Frederic.Jouault@lina.univ-nantes.fr

Recently, Model-Driven Architecture (MDA) has been proposed to take into account the development of large software systems, such as B2B applications on the Internet. However, before this becomes a reality, some issues need solutions, such as the definition of various Domain-Specific Languages (DSL) and also automatic transformation between these domain languages representing business concerns and those offering platform executability. In this paper, we provide some insights into transformation between some specific DSL particularly relevant to Business-to-Business (B2B) applications.

1. INTRODUCTION

Business-to-Business (B2B) applications existed before the Web. On the one hand, the emergence of the Internet and its services (such as the Web) has diffused the B2B applications. On the other hand, B2B applications on the Internet are often large and complex software systems.

Model Driven Architecture (MDA[TM])[71] (OMG, 2001) has been proposed for supporting the development of large software systems, such as B2B applications. However, before this becomes a reality some issues need to be resolved such as the model transformation. The objective of this paper is to provide some insights into the creation of meta-models, mappings and model transformation rules. We proceed here using UML[72] to create Platform-Independent Models (PIM) and transforming them into a Platform-Specific Models (PSM) based on Business Process Execution Language for Web Services (BPEL4WS) (Tony Andrews, 2003), Web Service (W3C, 2004) and .NET (Alex Ferrata, 2002). The model transformation is described using the Atlas Transformation Language (ATL) (Jean Bézivin1, 2003).

This paper is organized as follows. Section 2 is an overview of B2B applications in the context of MDA. Section 3 presents meta-models and mappings. Section 4 discusses some problems found during our research. The last section is the conclusion of our research.

---

[71] MDA[TM] is a trademark of the Object Management Group.
[72] Unified Modeling Language (UML) version 1.4.

2. OVERVIEW

One of the first domains in which computers were applied was B2B applications. In fact, B2B applications have profited from and financed advances in computer science.

Recently, the computer science community was confronted to new problems, such as the fast evolution of B2B applications, and the integration between different technologies. In order to make face to this new context, a change of paradigm was necessary. On the one hand, the object paradigm has given all that it could and does not seem in a position for giving much more. On the other hand, the service and the model paradigm have been developed and applied for meeting these new requirements.

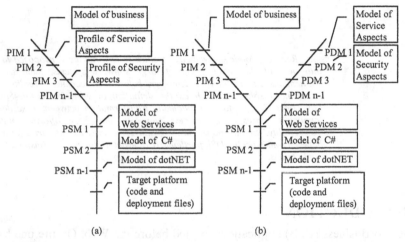

(a)   (b)

* This work had been done in the context of the INTEROP European Network of Excellence.

PIM –Platform-Independent Model PSM – Platform-Specific Model
PDM – Platform-Description Model
Figure 1: B2B Application Design and MDA

Figure 1 presents the B2B application design in the context of MDA. According to this figure, two techniques are possible: (a) marking and (b) weaving. Marking is based on UML Profiles for decorating a Platform-Independent Model (PIM) with aspects such as services and security. Weaving is based on the idea of making a texture between a PIM and a Platform-Description Model (PDM) (Jean Bézivin, 2002) before generating a Platform-Specific Model (PSM). In our research, we have privileged the weaving technique, but we will not describe this technique here. In this paper, we are more concerned by the model transformation that is one of the main challenges of the MDA Approach. According to figure 1, a PIM (e.g. business model) is transformed into a PSM (e.g. based on Web Service and BPEL4WS), refined in other PSMs (e.g. based on C# and .NET), until exported as code, deployment files, and config files. So, many levels of PIM and PSM are possible.

3. FROM UML INTO BPEL4WS, WEB SERVICE AND .NET

Before applying a model transformation for generating a target model from a source model, we need to obtain two things: a meta-model of each participant (i.e. BPEL4WS, Web Services and .NET) and a mapping from one into another meta-model. For this purpose, this section presents a meta-model for each platform and mappings.

3.1 Business Process and BPEL4WS Meta-model

Business process languages can support the definition and execution of a business process. Some business process languages were created for defining generic business processes (Assaf Arkin, 2002). Other business process languages consider a process as a composite Web Service (Frank Leymann, 2001). Other process languages were created for defining processes using Workflow (WfMC, 2002). So, a business process can be created in different ways, either as generic or a specialized one.

BPEL4WS (Tony Andrews, 2003) defines a model and a grammar for describing the behavior of a business process. BPEL4WS is the result of the merging of WSFL and XLANG (Satish Thatte, 2001). In fact, it has some concepts of WSFL and XLANG, such as directed graphs and block-structured language, respectively. It depends on the WSDL (W3C, 2001b), i.e. a BPEL4WS Process makes references to portTypes of the services involved (see section 3.3). In the right side of Figure 2 is presented a BPEL4WS meta-model (fragment) with the following main elements:

- Process – composed of activities, partners, correlation sets, fault handlers and compensation handlers. It has some attributes such as abstractProcess specifying whether the process is defined as abstract or as executable.
- PartnerLinks - defines the different parties that interact within a business process in execution. It is characterized by a PartnerLink that specifies the conversational relationship between two services through the declaration of their roles. Each role specifies only one WSDL portType that a partner needs to implement.
- Partners - defined as a subset of PartnerLink (i.e. references), it introduces a constraint on the functionality that a business partner provides.
- Variables - defines the data variables used by a process, and allows processes to maintain state data and process history based on messages exchanged. A variable can be defined in terms of WSDL message types (i.e. messageType) or XML Schema simple types (i.e. type) or XML Schema elements (i.e. element).
- Activity - structured in some parts such as control flows (e.g. Switch, While), message flow (e.g. Invoke, Receive, Reply), data flow (data is transferred between activities using Assign), transaction flow (Long-Running Transaction is only supported within a single business process) and extensibility (name-space codified on URI can be used in some BPEL4WS elements).

3.2 Mapping from UML into BPEL4WS

In Figure 2, we use a graphical notation for illustrating a mapping from UML (activity diagram) into BPEL4WS. The UML meta-model is presented on the left side, the mapping in the center, and the BPEL4WS meta-model on the right side. The mapping part is formed by four main graphical elements: connection (source and target), association, transformation rule and composition.

A connection links one or more meta-model element(s) to a transformation rule. The association shows a relationship between rules. The composition shows a tight relationship between rules. The transformation rule takes a source element and generates a suitable target element. We have used ATL transformation language for defining transformation rules (Jean Bézivin, 2003).

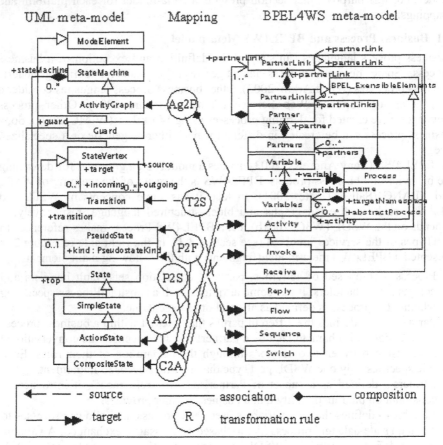

Figure 2 : Mapping from UML (activity diagram) into BPEL4WS

According to Figure 2, we have the following mappings (we will present only a few ATL transformation rules). (A transformation rule takes an element from a source meta-model and transforms it into an equivalent element in a target meta-model. Elements are equivalents, if they have equivalent semantics and structures. The rule P2S take a UML PseudoState of type 'Choice' and transform it into BPEL Switch.)

The UML PseudoState (Choice) is mapped into BPEL4WS Switch through the rule P2S:

-- Helper for P2S

helper context UML!PseudoState def:

helperGetCondition(): Collection(BPEL!BooleanExpr) =

  -- (body omitted due to space limitations)

helper context UML!PseudoState def:
helperGetActivity():Collection(Activity) =
 -- (body omitted due to space limitations)

helper context UML!PseudoState def:
helperGetOtherwise():Collection(Activity) =
 -- (body omitted due to space limitations)
rule P2S{
 from ps : UML!PseudoState(ps.kind=#pk_choice)
 to sw: BPEL!Switch
 mapsTo ps(
  name ← ps.name,
  case ← cases,
  otherwise ← otherw   ),
 cases : BPEL:Case(
  condition ← helperGetCondition(ps),
  ref ← helperGetActivity(ps)   ),
 otherw : BPEL:Otherwise(
  ref ← helperGetOtherwise(ps)   ) }

3.3 Web Service Meta-model

A service is an abstraction of programs, business process, and other artifacts of software defined in terms of what it does. Services can be organized in a Service Oriented Architecture (SOA) (W3C, 2004). SOA is a form of distributed system architecture based on the concept of services that is characterized by the following properties (W3C, 2004):

- Logical view - a service is a logical view of a system.
- Message oriented - the communication between an agent provider and an agent requester is defined in terms of messages exchanged.
- Description orientation - a service is described using meta-data.
- Granularity - services communicate using a small number of large and complex messages.
- Platform neutral - services communicate using messages codified in one platform-independent representation.

Figure 3 presents a simplified SOA model. An agent provider has services. These services are described through a meta-data representation, i.e. ServiceDescription. Afterwards, the agent provider stores information of its services in a Registry. An agent requester searches in the Registry for a specific service following a determined criterion. The Registry returns information of a desired service. The agent requester finds the meta-data of this service and uses it to exchange messages with the service. According to this figure and Web Service Architecture (W3C, 2004), we have the following similarities: Universal Description, Discovery, and Integration (UDDI) (UDDI.ORG, 2002) is the registry; Web Service Description Language (WSDL) (W3C, 2001b) is the service description; the service uses Simple Object Access Protocol (SOAP) (W3C1, 2001a) as a communication protocol for exchanging messages.

In this paper, we will only present a meta-model for WSDL. In the right side of Figure 4, a WSDL meta-model is presented. It is formed by:

- Definition - the main element of this meta-model which has a set of imports, types, messages, portTypes[73], bindings and services.
- ImportType - allows the association of a namespace with a document location.
- TypesType - employed for defining a simple or complex type defined by XML Schema.
- MessageType - describes the abstract format of a particular message that a Web Service sends or receives. It has parts (i.e. PartTypes) which describe each part of a message.
- portTypeType - defines the interface of a service. It has a set of messages that a service sends and/or receives.
- BindingType - describes a concrete binding of interface components, i.e., describes how to call up a service.
- ServiceType - describes a service, its interface (i.e. PortTypeType) and its endpoints.

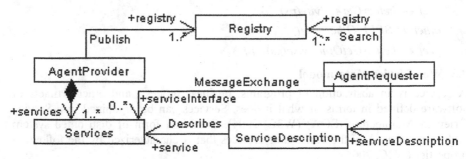

Figure 3: Service-Oriented Architecture

3.4 Mapping from UML into Web Service

Figure 4 presents a mapping from UML (class diagram) into WSDL.

According to Figure 4, we have the following mappings (we will present only a few ATL transformation rules). (UML Operation is equivalent to WSDL OperationType. An Operation Type is composed of Paramtype,..., Message.)

The UML Operation is mapped into WSDL OperationType through the rule O2O:

rule O2O{
   from op : UML!Operation
   to wsdlop : WSDL!OperationType
   mapsTo op(name ← op.name, parameterOrder ← op.getOrder(),
              input ← inp, output ← out),

---

[73] PortType was renamed to Interface in WSDL 1.2

inp : WSDL!ParamType (message ← inp_m),

out : WSDL!Paramtype(message← outp_m),

inp_m : WSDL!Message(name ← op.owner.name + '_' + op.name),

outp_m: WSDL!Message(name ← op.owner.name + '_' +
*   op.name + 'Response'),*

wsdlob: WSDL!BindingOperationType (name ← op.name,
*   input ← inputb,   output ← outputb)*

inputb : WSDL!StartWithExtensionType(
* encodingStyle ← http://' + 'schemas.xmlsoap.org/soap/encoding/',*
* use ← 'encoded',*
* namespace ← 'urn://'+ op.feature.owner.name + '.wsdl'),*

outputb : WSDL!StartWithExtensionType(
* encodingStyle ← 'http://' + 'schemas.xmlsoap.org/soap/encoding/'*
* use ← 'encoded',*
* namespace ← 'urn://'+*
* op.feature.owner.name + '.wsdl') }*

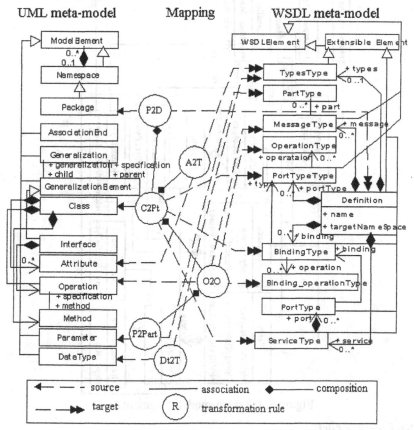

Figure 4 : Mapping from UML (class diagram) into WSDL

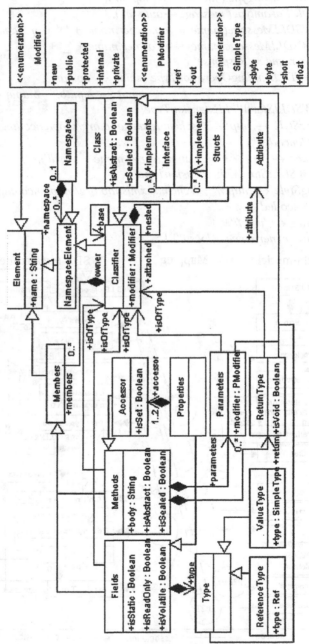

Figure 5: C# meta-model (fragment)

3.4 .NET meta-model

In our research, we have used .NET as target platform. The .NET platform for Web Services can be formed by C#, .NET Framework and Internet Information System (ISS) (Alex Ferrata, 2002). Figure 5 depicts a C# meta-model (fragment).

This C# meta-model is formed by:

- Namespace - a container for all other elements.
- Classifier - a generalization for members, classes, interfaces and structures. It has a set of members that can be fields, methods, and properties.
- Class - a specialization for Classifier and implements interfaces.
- Interface - another specialization for Classifier (only with methods' signatures).
- Structs - similar to Class, but it does not have heritage.
- Attribute – a declarative information that can be attached to programs' entities (such as Class and Methods) and retrieved at runtime. All attribute classes derive from the System.Attribute base class provided by the .NET Framework. Although Attribute belongs to C# API (i.e. model or M1 layer), we have used it as part of the C# metamodel, in order to manipulate it in the meta-model layer (i.e. M2 layer). C# Attribute does not have the same meaning of UML Attribute. In UML, an attribute is a feature within a classifier that describes a range of values whose type is a classifier.
- Fields - a composition of one type that can be a ValueType or a ReferenceType.
- Methods - has the signature of operations (return type, identifier and parameters).
- The creation of a Web Service in .NET using C# is made using the classes:
- WebService - base class for all Web Services.
- WebServiceAttribute - an attribute that can be associated with a class implementing a service.
- WebMethodAttribute - an attribute associated with the methods of a class implementing a service.

Figure 6 presents a simplified template and .NET meta-model, detailing the main elements for creating a Web Service. The template indicates that a Web Service is implemented using a class that extends the WebService class. This class is attached to the attribute named WebServiceAttribute. The methods accessed as services are attached to the attribute named WebMethodAttribute. The .NET template uses the Application Programming Interface (API) from .NET Framework to define Web Services, thus this template belongs to the model layer (or M1 layer). The .NET meta-model presents the deployment files web.config and disco needed to deploy a Web Service.

The deployment files Web.config and Disco are necessary for deploying Web Services using IIS and .NET framework. Web.config has specific information, which enables IIS for running a Web Service. Disco is employed for advertising a Web Service publicly and it has the location of the description of the service, i.e. WSDL.

3.5 Mapping from UML into .NET platform

Figure 7 presents a mapping from a UML (class diagram) into C# (fragment). According to Figure 7, we have the following mappings (we will present only a few ATL transformation rules). The UML Package is mapped into C# Namespace through the rule P2N:

```
rule P2N{
  from pck : UML!Package
  to cn : Csharp!Namespace
  mapsTo pck(
    name ← pck.name ) }
```

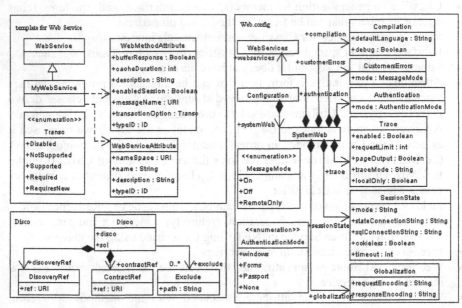

Figure 6 : .NET template and meta-model

The UML AssociationEnd is mapped into C# Field through the rule Ae2F:

```
-- Helper for getOtherEnd()
helper context UML!AssociationEnd def: getOtherEnd() :
UML!AssociationEnd =
            self.association.connection-> select(e|e <> self)->first();
rule Ae2F{
 from ae : UML!AssociationEnd
 to cf : Csharp!Field
 mapsTo ae(
 name ← ae.name,
 modifier ← if ae.visibility = #vk_public then  #public else #private endif,
 isVolatile ← false,
 isStatic ← false,
 isReadOnly ← false,
 isOfType ← ae.participant,
 owner ← ae.getOtherEnd().participant ) }
```

Afterwards, a model generated in this step is refined using the template and the .NET meta-model.

4. DISCUSSION

During our research, we have been confronted with real and unexpected problems, which merit a brief discussion, hence:

- Is it relevant to take into account APIs in the MDA context? - Someone could doubt the importance to take into account APIs in the model driven approach. However, Web Services, EJB, CORBA and other platforms are implemented using APIs. Moreover, one of the characteristics of MDA is to consider everything as model, thus it should include APIs too.
- Is it relevant to take into account Attribute-Oriented Programming in the context of MDA? - C# uses attributes for attaching information to program's entities such as Class and Methods. This information can be used by tools for generating complementary code or deployment files, or it can be retrieved at runtime. For developing Web Services in .NET, we need to use attributes. Thus, in Figure 5 we presented a C# meta-model where Attribute is one of its elements. However, the importance of taking this into account in the meta-model layer must be examined more carefully.

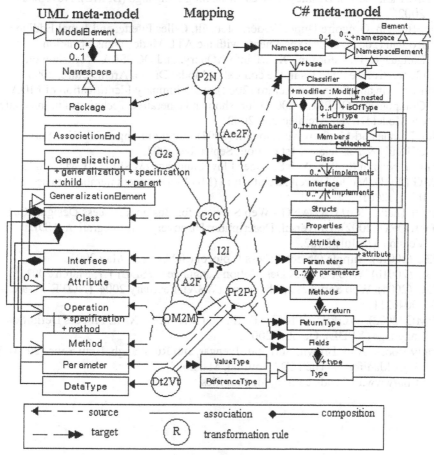

Figure 7 : Mapping from UML (class diagram) into C#

5. CONCLUSION

This paper emphasized solutions for some issues, mainly creation of meta-models and mapping specification for the development of B2B applications in Web Service platforms using an MDA approach. However, the integration of security and availability in B2B models from the start is still an open issue. We started presenting a BPEL4WS, a Web Service, and a .NET meta-model, afterwards we discussed some possible mappings between the UML meta-model and these meta-models. The material presented in this paper is a good illustration of the current trend in model-driven engineering with such recent proposal as IBM's MDA manifesto (Grady Booch 2004).

REFERENCES

Tony Andrews, Francisco Curbera, Hitesh Dholakia, and *et al*. (2003) Business Process Execution Language for Web Services (BPEL4WS) version 1.1, May 2003.

Assaf Arkin (2002). Business Process Modeling Language (BPML), November 2002.

Jean Bézivin, Grégoire Dupé, Frédéric Jouault, Gilles Pitette, and Jamal Eddine Rougui (2003). First Experiments with the ATL Model Transformation Language: Transforming XSLT into XQuery. 2nd OOPSLA Workshop on Generative Techniques in the context of Model Driven Architecture, 2003.

Jean Bézivin and Sébastian Gérard (2002). A Preliminary Identification of MDA Components. OOPSLA 2002 Workshop on Generative Techniques in the context of Model Driven Architecture, 2002.

Alex Ferrata and Matthew MacDonald (2002). Programming .NET Web Services. O'Reilly & Associates, 1st edition, September 2002.

Frank Leymann (2001). Web Services Flow Language (WSFL 1.0), May 2001.

OMG (2001). Model Driven Architecture (MDA)- document number ormsc/2001-07-01, 2001.

Satish Thatte (2001). XLANG - Web Services for Business Process Design, 2001.

UDDI.ORG (2002). Universal, Description, Discovery and Integration (UDDI) Version 3.0, July 2002.

W3C (2001a). Simple Object Access Protocol (SOAP) 1.1, May 2001.

W3C (2001b). Web Services Description Language (WSDL) 1.1, March 2001.

W3C (2004). Web Services Architecture (WSA), February 2004. (NOTE-ws-arch-20040211)

WfMC (2002). Workflow Process Definition Interface – XML Process Definition Language (XPDL), October 2002.

Grady Booch, Alan Brown, Shridhar Iyengar, Jim Rumbaugh, Bran Selic (2004) An MDA Manifesto, The MDA Journal, Mai 2004, http://www.bptrends.com/publicationfiles

20.   A Research Framework for Operationalizing Measures of Enterprise Integration

Ronald E. Giachetti,[1] Paula Hernandez,[1] Alba Nunez[1] and Duane P Truex[2]
1 Industrial & Systems Engineering, Florida International University,
Miami, FL Email:giachetr@fiu.edu
2 Computer and Information Systems, Georgia State University, Atlanta, GA

This paper develops a research framework to investigate measures of enterprise integration. In our view the term enterprise integration is an umbrella term that incorporates what we term integration types. The integration types are connectivity, information sharing, interoperability, coordination, and alignment. To determine which technology and/or enterprise integration method is best in a given situation we believe measures of integration are needed and must be grounded in empirical findings.

1. INTRODUCTION

This paper proposes a research framework to help answer two questions that arise when enterprises attempt to integrate technical and social processes and systems: First, what type of integration is needed? and second, what social and technical systems should be integrated for a particular business problem?

To answer these questions we develop a research framework to operationalize five integration types. The operationalization involves a definition of the constructs, relationships between those constructs, and how to measure them. The way in which we state the research questions, i.e. *what type of integration* indicates our presumptions concerning the solution. We assume a contingency-based view of the enterprise integration problem; i.e. we believe the best integration type is dependent on the particular business problem being addressed.

1.1 Background

We define an enterprise as an organization composed of interdependent resources (people, technology, infrastructure and machines) which must coordinate their functions and share information in order to achieve common enterprise goals. We refer to this as the enterprise integration problem rather than as simply coordination for two reasons. First, integration is a broad term that includes many integration types of which coordination is a single type. Second, enterprise integration conveys that the integration problem is not just a technical problem to be solved with IT, but also a social or organizational problem. Crucial issues facing enterprise systems managers and integrators are how much to integrate, what to integrate and how to achieve this coordination. Historically, organizational work systems were designed, built, and optimized to solve the local needs. There is little regard for how the local system would fit into the entire enterprise. These local systems utilize various data representation formats, have different data semantics, are built using different programming languages, employ different work process models and are launched on various hardware platforms. Management and information theorists have long understood the need for greater inter and intra organizational interaction such that the problem of how to integrate these heterogeneous systems has been a significant research agenda for more than twenty-five years (Petrie 1992; Patankar and Adiga 1995; Vernadat 1996; Vernadat 2002).

It has been established and is generally accepted that integration leads to improved enterprise performance (Armistead and Mapes 1993; Frohlich and Westbrook 2001; Brunnermeier and Martin 2002). Many researchers take this relationship as the starting point to develop and specify solutions to the integration problem. However, in a review of over 150 integration studies we find differing definitions of integration (Giachetti 2004). In a study commissioned by NIST, Brunnermeier and Martin (2002) estimate that poor interoperability between systems in the US automotive supply chain cost one billion dollar annually. The study was limited to one aspect of integration, interoperability between applications and did not consider other types of integration. Others have focused solely on information sharing (Lee and Whang 1998) or coordination of decisions (Malone and Crowston 1994). If there are many types of integration then the question remains, what is the most appropriate type of integration for a particular business situation? We have been unable to find an answer in the literature to this question.

One solution widely suggested is to install a single monolithic system, i.e. an enterprise resource planning (ERP) system, as a solution to all integration problems. Today, almost every Fortune 500 company has implemented an ERP system. ERP is a single vendor solution and thus interoperability problems are in theory avoided. In practice, while ERP replaces the many independent information systems companies operated (e.g. accounting, billing, order entry, and so forth); these same companies have found they still must maintain other applications, which must be integrated with the ERP system (Themistocleous *et al.* 2001). Moreover, the complexity of ERP implementations means that many companies fail to realize the promised benefits of integration (Kumar and Van Hillegersberg 2000). One reading of the literature suggests that using ERP systems to solve the integration problem is not a silver bullet. An alternative strategy is to have decentralized and highly distributed systems. These distributed systems are integrated via middleware or enterprise application integration (EAI) (Linthicum 2000).

In summary, enterprise integration (EI) has been shown to contribute to higher levels of performance of enterprise systems. However, studies show that EI is poorly understood and poorly applied in industry. EI research is needed to lead to a better understanding of integration and how it can be achieved. It seems advances can be made by considering the many different types of integration. Not all integration types are appropriate for every business situation, and research is needed to understand when and how to use each integration type.

2. ENTERPRISE INTEGRATION

There is a significant body of literature on enterprise modelling and enterprise integration. A prevalent research approach is the development of enterprise reference architectures that describe the enterprise from many different viewpoints in order to deal with the complexity of the enterprise system. Reference architectures that have been developed include CIMOSA (AMICE 1993), GRAI (Doumeingts *et al.* 1987), PERA (Williams 1994), and GERAM (Bernus 2001). The reference architectures embody knowledge of what enterprise engineers should analyze and how they should analyze it. The reference architectures decompose the enterprise into different viewpoints and levels of genericity. For example, CIMOSA has four complementary views of function, information, organization, and resources. Whenever you decompose a system the problem is how to integrate or relate the analysis and design done of the subsystems. The reference architectures provide guidance on how to accomplish this. The enterprise modelling research has matured to the point such that we have available validated modelling constructs, a convergence in reference architectures, and ontologies and other developments to formalize the collective enterprise knowledge gained. We believe to move the field forward there is a pressing need for the identification and quantification of the enterprise integration parameters. In other words, identifying constructs that have high impact on enterprise integration and defining measures for those constructs. We believe that measurement is a necessary component to further establishing a science base for enterprise integration.

3. RESEARCH FRAMEWORK

The research framework is limited to intra-enterprise integration. Thus, we avoid the issues specific to inter-enterprise integration such as studied in supply chain management. In order to conduct the study the unit of analysis needs to be defined. In an organization integration could be studied at different levels. For example, department level, person to person level, or between systems. We choose the business process as the level of analysis. This choice is in accordance with the underlying concept governing CIMOSA (Vernadat 1996).

The research framework shown in Figure 1 articulates a contingency perspective of enterprise integration. In this model the enterprise will realize positive performance impacts when the enterprise matches or fits the right integration type with the enterprise characteristics. Consequently, the best integration type is contingent on the characteristics of the enterprise to be integrated. In the following subsections each of the constructs are described and then relationships between constructs are proposed.

3.1 Enterprise Integration Types

As an initial conceptual model of the enterprise we see a layered framework of related independent activities sharing common goals that, taken as a whole, describe essential aspects of an organization. Some of the layers might be seen as 'technical' elements and others as 'social elements'. Taking a dualistic view separating the technical and social as if they were wholly independent of one another would be a mistake. This is because technologies contain embedded assumption about work practices, cultural values and norms and because social units act in reference to technologies. The social and the technical are intricately interrelated. For instance a data model may be seen as a model of business rules and practices. Similarly, software applications embed values, norms and work practices. Taken independently each of these layers can be seen as a view of the enterprise. We say this because each layer models and supports core assumptions about its fit in the hierarchy and a particular type of integration goal.

For the purposes of this analysis we consider five broadly defined subsystems each with its own specific integration issue to be addressed. The levels are termed: network, information, application, work processes, and organizational levels (Giachetti 2004). The enterprise integration types are shown in Figure 2 and each level is described next.

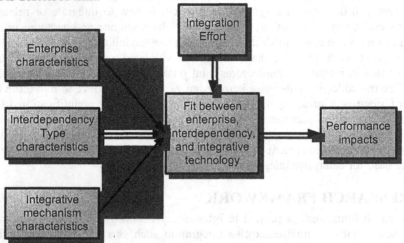

Figure 1. Enterprise integration research framework

Connectivity

At the network level, the integration issue is the physical heterogeneity of the hardware, machines, devices, and their operating systems found in a physical network. The integration goal at the network level is *connectivity* defined as the linkages between systems, applications, and modules.

Data Sharing

Data sharing is the ability of one organizational subunit to understand and use the data originating from another subunit. There are two components of this definition. First, the subunits must exchange data. Second, the data exchanged must be understood by the receiver. This second requirement is harder to satisfy then the first, because semantic differences among units and subunits are still prevalent in many companies.

Interoperability

The application level, describes the systems used by the business. The integration goal is interoperability, which is the ability of one software application to access/use data generated by another software system. Interoperability of software applications is usually achieved by developing interfaces to a system such as through an application protocol interface (API), with middleware, or with other enterprise application integration (EAI) technologies (Ruh *et al*. 2000).

Coordination

The work process level describes the tasks and the manner and order in which the tasks are conducted in order to produce an output. The problem of task dependencies occurs at this level and the integration issue is called coordination. Coordination has been defined as the "management of the dependencies that arise between business tasks" (Malone and Crowston, 1994). Coordination is achieved by integrating decisions. Mintzberg (1979) defined six broad categories of coordination mechanisms that organizations can use to coordinate their tasks. These are 1) standardization of norms, ideology, and culture,2) standardization of skills, 3)standardization of outputs, 4) standardization of work processes, 5) direct supervision, and 6) mutual adjustment.

Goal Alignment

The organizational level addresses the way that the three key elements of business strategy, organizational design strategy and information systems strategy must all be aligned with one another. A change in any of these elements requires an adjustment in the others. Thus alignment is the integration task at this level of analysis (Venkatraman and Henderson 1993; Joshi 2003).

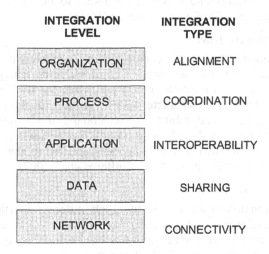

Figure 2. A framework to assess levels of enterprise integration (adapted from Giachetti 2004)

The enterprise integration level assessment framework is intended to unite different perspectives of enterprise integration as can be found in the literature review. For example, middleware approaches focus on interoperability at the application level, database approaches on the data level, and cross-functional teams at the process level. Each integration type will allow for constructions of measures of enterprise integration. Enterprise integration within a single company implies alignment within and between the different levels into a cohesive enterprise system. Inter-enterprise integration can occur at any level.

3.2 Interdependency Characteristics

In general an interdependency exists when there is any type of flow between people, organizational units, or applications within the enterprise. Flow types are material, information, decision, economic, and resources. Managing interdependencies between tasks and organizational units is viewed as critical to the smooth operation of a business (Crowston 1997; Camarinha-Matos and Pantoja-Lima 2001; Albino *et al*. 2002).

There are many different types of dependencies and authors have developed taxonomies to categorize and classify them (Malone and Crowston 1994; Whang 1995; Crowston 1997; Kim 2001). Identifying the existence of these interdependencies can be achieved through modelling activities such as data flow diagrams or IDEF0 to capture information flows or the GRAI methodology to capture decision control flows reveals the structural nature of the interdependency. For example, by showing a sequential dependency in which the output of one task is a required input to another task we have identified the existence of a particular type of dependency. What is also needed is a characterization of the strength of that interdependency. Strength of the interdependency has been modelled by the frequency of the communication (Christensen *et al*. 1996). Another approach is to use psychometric measures and survey participants to obtain a measure of the perceived interdependency (Wybo and Goodhue 1995).

3.3 Enterprise Characteristics

The enterprise characteristics may become obstacles to certain integrative type solutions or may facilitate other solutions. The literature on enterprise integration was examined to reveal what important enterprise characteristics impact integration. In this study we study three characteristics: 1) functional differentiation, 2) uncertainty, and 3) task analyzability to understand the nature of enterprise integration. These factors are based on an information processing theoretical view of the organization (Galbraith 1977) and have been utilized in many similar studies (Daft and Lengel 1986; Albino *et al*. 2002; Koufteros *et al*. 2002).

Functional Differentiation

Functional Differentiation is the degree to which different functional units (e.g. design, accounting, finance, and so forth) have different cultural norms, goals, methods, and vocabularies (Lawrence and Lorsch 1967; Daft and Lengel 1986). When functional differentiation is high it can become an obstacle to enterprise integration. Moreover, when functional differentiation is high, some forms of integration, for example data integration may reduce the flexibility individual sub-units need to deal with their environment and thus could have a negative impact on performance (Wybo and Goodhue 1995).

A method to measure goal incongruence is by a card sorting method (Christensen *et al*. 1996). In this approach all the possible goals are individually written on a card. Then each actor sorts the cards according to goal priority. The difference between any two actors selection is a measure of goal incongruence.

Task Uncertainty

Galbraith (1977) defines uncertainty as the gap between the amount of information required to perform a task and the information already possessed by the organization. Galbraith identified factors that contribute to uncertainty at the organization level such as goal diversity and labor diversity. What contributes to the uncertainty is likely industry specific. For example, Flynn and Flynn (1999) identified several additional factors that contribute to uncertainty in the manufacturing enterprises.

To cope with task uncertainty Galbraith identified two general strategies. The first is to reduce the information processing needs of the organization. The second is to increase the information processing capability of the organization. Clearly, integration is one means to implement the second strategy. To increase the information processing capability you can share data (data sharing integration type).

Task uncertainty is usually measured through survey methods (Daft and Lengel 1986; Victor and Blackburn 1987; Rosenzweig *et al.* 2003). An analytical approach would be to define the information requirements for a task and then to identify what proportion of the information requirements are available locally for the organizational unit to perform the task and the proportion that must be retrieved from other sources.

Task Analyzability

Task analyzability is whether the task can be managed by a defined set of procedures (Perrow 1967). Tasks that are routine and can be addressed by well-defined procedures are termed analyzable. When tasks are unanalyzable employees use judgement to make decisions. Task analyzability is measured through survey methods (Van de Ven and Delbecq 1974; Rice 1992).

3.4 Performance Impacts

Enterprise integration has been found to lead to improved enterprise performance. In the context of the research model performance impact means the integrative type when used in the presence of the enterprise environment and interdependencies will improve some unit level performance measure. For example, improved efficiency, improved effectiveness, improved quality, or other performance measures are possible.

Objective measures are frequently difficult to find at this level of analysis and also to isolate the impact from the integrative type is difficult. The approach taken is to use user evaluations of performance impact. The model of (Goodhue 1996) for measuring user evaluation can be used for this purpose.

3.5 Integration Effort

Integration effort is the difficulty level of achieving integration and is measured in terms of cost, time, and amount of resources that must be used in order to achieve the desired integration. Similar to performance, measuring effort is not always straightforward in an organization. A user evaluation of the effort required for integration is potentially the best measurement approach.

3.6 Relationships

The model shown in Figure 1 propositions several relationships between the constructs and enterprise performance. The first relation is between functional differentiate and various integration types.

Proposition #1:   When functional differentiation is high the integration effort required will be high.

Functional differentiation increases when organizational units must specialize to complete their tasks. Highly specialized units develop their own vocabulary which would make data integration more difficult since there would be greater semantic differences between the organizational units. Moreover, applications for highly differentiated organizational units tend to be optimized for local needs. Therefore, it is unlikely their software applications were designed for interoperability. Finally, highly differentiated units will have different goals. Collectively these characteristics would impose the need for greater effort in order to achieve the integration. Additionally, high functional differentiation would suggest the following integrative types would be more effective.

Proposition #2:   In enterprise environments with high functional differentiation integrative strategies that include goal alignment and data sharing would have the greatest performance impacts.

The justification for this proposition is that goal alignment and data sharing are meant to overcome the difficulties associated with functional differentiation. Whereas, achieving interoperability or coordination through the mechanisms identified by Mintzberg (1979) will have less impact in the presence of continued goal and semantic differences.

Proposition #3:   In enterprise environments with high uncertainty integrative strategies that include data sharing and interoperability will have greatest performance impacts.

High uncertainty defined as the absence of information is addressed by increasing information processing capacity. Frequently, the information to complete a task is needed from other organizational units. Consequently, it is expected that integrative technologies that increase information processing capacity such as data sharing and interoperability will have positive impacts on performance.

Proposition #4:   In enterprise environments with low analyzability integrative strategies that include data sharing and interoperability will have minimal performance impacts.

Low analyzability describes tasks that even with additional information the performance of the task is not improved because the decision making relies on judgement and not an analysis of the information available.

Proposition #5:   Interdependency and performance impacts are proportional such that the greater the interdependency the greater the performance impact.

The justification for the last proposition is that when organizational units are interdependent then integrating them leads to increases in efficiency and effectiveness for each unit in performing their tasks. When organizational units how low interdependence then integrating them will have little or no impact on their performance.

4. CONCLUSION AND FURTHER WORK

The primary contribution of this paper was to identify important environmental factors that are hypothesized to impact integration technology choice and performance.   Our model is more specific about the constructs and explicit concerning the relationships than previous research. Approaches to measuring the constructs were identified. The majority of the measures rely on previously validated measuring instruments.

While significant advances have been made in enterprise modelling and understanding enterprise integration there is a lack of measurement. Defining constructs and how to measurement them is at the foundation of developing theory in a field (Wacker 1998). In this paper we presented a contingency-based model of enterprise integration.   We identified three classes of constructs: enterprise, interdependency, and integration type. Propositions were formulated to determine the fit between each of these constructs and the overall impact on performance.

The future work is to test the model by collecting data and performing statistical analysis to validate each of the proposed relationships. Work is going forward on using the research framework to understand enterprise integration in the South Florida cruise industry. If strong evidence is found to support both the construct measures and the relationships then we would have achieved a strong theoretical basis for thinking about the impact of various integration types on enterprise performance. It would be possible to build a software tool for computer aided enterprise integration to aid analysts in studying business situations and designing integrative solutions.

Acknowledgments

The work was partially funded by the National Science Foundation under grant DUE-0220667 and by NASA AMES Research Center under grant NAG2-1612.

REFERENCES

Albino, V., P. Pontrandolfo, *et al.* (2002). "Analysis of information flows to enhance the coordination of production processes." International Journal of Production Economics 75: 7-19.

AMICE (1993). CIMOSA: CM Open Systems Architecture. Berlin, Springer-Verlag.

Armistead, C. and J. Mapes (1993). "The impact of supply chain integration on operating performance." Logistics Information Management 6(4): 9-15.

Bernus, P. (2001). "Some thoughts on enterprise modelling." Production Planning & Control 12(2): 110-118.

Brunnermeier, S. B. and S. A. Martin (2002). "Interoperability costs in the US automotive supply chain." Supply Chain Management: An International Journal 7(2): 71-82.

Camarinha-Matos, L. M. and C. Pantoja-Lima (2001). "Cooperation coordination in virtual enterprises." Journal of Intelligent Manufacturing 12: 133-150.

Christensen, L. C., T. R. Christiansen, *et al.* (1996). "Modeling and simulation in enterprise integration -- a framework and an application in the offshore oil industry." Concurrent Engineering Research and Applications Journal 4(3): 247-259.

Crowston, K. (1997). "A coordination theory approach to organizational process design." Organizational Science 8(2): 157-175.

Daft, R. L. and R. H. Lengel (1986). "Organizational information requirements, media richness, and structural design." Management Science 32(5): 554-571.

Doumeingts, G., B. Vallespir, *et al.* (1987). "Design methodology for advanced manufacturing systems." Computers in Industry 9: 271-296.

Flynn and Flynn (1999). "Information-processing alternatives for coping with manufacturing environmental complexity." Decision Sciences 30(4): 1021-1052.

Frohlich, M. T. and R. Westbrook (2001). "Arcs of integration: an international study of supply chain strategies." Journal of Operations Management 19: 185-200.

Galbraith, J. R. (1977). Organizational Design. Reading, MA, Addison-Wesley.

Giachetti, R. (2004). "Enterprise Integration: An information integration perspective." International Journal of Production Research 42(6): 1147-1166.

Goodhue, D. (1996). "Understanding user evaluations of information systems." Management Science.

Joshi, K., Porth (2003). "Alignment of strategic priorities and performance: an integration of operations and strategic management perspectives." Journal of Operations Management 21(3): 353-369.

Kim, H. W. (2001). "Modeling inter- and intra-organizational coordination in electronic commerce deployments." Information Technology and Management 2: 335-354.

Koufteros, X. A., M. A. Vonderembse, *et al.* (2002). "Integrated product development practices and competitive capabilities: the effects of uncertainty, equivocality, and platform strategy." Journal of Operations Management 20: 331-355.

Kumar, K. and J. Van Hillesgersberg (2000). "ERP: Experiences and evolution." Communications of the ACM 43(4): 23-26.

Lawrence, P. R. and J. W. Lorsch (1967). Organization and Environment. Cambridge, MA, Harvard Graduate School of Business.

Lee, H., L. and S. Whang (1998). Information sharing in a supply chain, Graduate School of Business, Stanford University: 1-22.

Linthicum, D. S. (2000). Enterprise Application Integration. Reading, MA, Addison-Wesley.

Malone, T. W. and K. Crowston (1994). "The interdisciplinary study of coordination." ACM Computing Surveys 26(1): 87-119.

Mintzberg, H. (1979). The structuring of organizations. Englewood Cliffs, NJ, Prentice-Hall.

Patankar, A. K. and S. Adiga (1995). "Enterprise integration modeling: a review of theory and practice." Computer Integrated Manufacturing Systems 8(1): 21-34.

Perrow, C. (1967). "A framework for the comparative analysis of organizations." American Sociological Review 32(2): 194-208.

Petrie, C. J. (1992). Enterprise integration modeling. Proceedings of the first international conference on enterprise integration, Cambridge, MA, MIT Press.

Rice, R. (1992). "Task analyzability, use of new media, and effectiveness: a multi-site exploration of media richness." Organizational Science 3(4): 475-500.

Rosenzweig, E. D., A. V. Roth, *et al.* (2003). "The influence of an integration strategy on competitive capabilities and business performance." Journal of Operations Management.

Ruh, W. A., F. X. Maginnis, *et al.* (2000). Enterprise Application Integration: A Wiley Tech Brief. New York, NY, John Wiley & Sons Inc.

Themistocleous, M., Z. Irani, *et al.* (2001). "ERP and application integration: exploratory survey." Business Process Management Journal 7(3): 195-204.

Van de Ven, A. and A. Delbecq (1974). "A task contingent model of work-unit structure." Administrative Science Quarterly 19: 183-197.

Venkatraman, N. C. and J. C. Henderson (1993). "Continuous strategic alignment: exploiting information technology capabilities for competitive success." European Management Journal 11(2): 139-149.

Vernadat, F. (2002). "Enterprise Modeling and Integration (EMI): Current Status and Research Perspectives." Annual Reviews in Control 26: 15-25.

Vernadat, F. D. (1996). Enterprise modeling and integration. London, UK, Chapman and Hall.

Victor, B. and R. S. Blackburn (1987). "Determinants and consequences of task uncertainty: a laboratory and field investigation." Journal of Management Studies 24(4): 387-403.

Wacker, J. (1998). "A definition of theory: research guidelines for different theory-building research methods in operations research." Journal of Operations Management 16: 361-385.

Whang, S. (1995). "Coordination in operations: a taxonomy." Journal of Operations Management 12: 413-422.

Williams, T. J. (1994). "The Purdue reference architecture." Computers in Industry 24(2-3): 141-158.

Wybo, M. and D. Goodhue (1995). "Using interdependence as a predictor of data standards: theoretical and measurement issues." Information & Management 29: 317-329.

21. A Vision of Enterprise Integration Considerations

A holistic perspective as shown by the Purdue Enterprise Reference Architecture

Hong Li[1] and Theodore J. Williams[2]

1. Water-Logic, Inc., 1933 E. Dublin-Granville Rd., # 151, Columbus, OH 43229 USA,
Phone: 614-296-7644
Fax: 866-206-9568, Email: hong@water-logic.com
2. Institute for Interdisciplinary Engineering Studies, Purdue University, 1293 Potter Center,
West Lafayette, IN 47907-1293 USA, Phone: 765-494-7434, Fax: 765-494-2351, Email:
tjwil@ecn.purdue.edu

Enterprise Integration (EI) is a key concept of Enterprise Engineering (EE) programs. This paper modifies the definition of Enterprise Integration through a broad vision of the field. Typical approaches are studied and reclassified based on recent results from the use of the Purdue Enterprise Reference Architecture. Theories of descriptiveness and prescriptiveness are proposed to support the newly established concept of Approach 2 Architectures as well as their general requirements.

1. INTRODUCTION

One of the authors wrote years ago that it was unfortunate that the subject of enterprise integration had generally (to date) been presented with a strong technology view only (Williams, 1996a). Although efforts have been made to improve both research and practice in the field of enterprise integration (EI) since then, most solutions offered by scientists and technologists have still (to date) been heavily technically oriented as before, and as so defined, are often not in alignment with management goals in business. Without the justified link with business strategies, even the most advanced information technology could be easily devalued as some pure technical proposals that did not matter much with diminished strategic importance (Carr, 2003).

In this article, the authors will review the key concepts of EI and address some of the open issues in this field. The authors further categorize the current efforts into two different approaches. Then they will emphasize the key characteristics of a holistic approach to EI, which is missing from the purely technical orientation. Under this holistic view, they will also present the focal points of the Purdue Enterprise Reference Architecture (PERA)*, including the recent developments in PERA in the hope of calling attention from both academicians and practitioners to reevaluate the present solutions as well as their limits, most of which are still based primarily on electronic connectivity of information only. The business aspects of solutions to EI that have been so far overlooked have actually offered us solid help to think outside the existing technical box and find the right direction to unify

---

* The authors assume that the readers are familiar with PERA and other Type 2 Enterprise Reference Architectures, mainly CIMOSA and GRAI.

technical solutions with business strategies, which indicates why and how EI should be considered more than a pure technical endeavor.

2. ISSUES IN THE BASIC CONCEPT OF EI

Many authors have noticed the importance of the business aspects of EI, which are all inclusive in nature. (Petrie, 1992) defined EI as an issue of improvement in enterprise performance. He stated that EI was not "simply a matter of improving connectivity among computer systems". Instead, "EI occurs when there is an improvement in the task-level interactions among people, departments, services, and companies".

Goranson (Goranson, 1992) explained the philosophical aspects of EI in terms of epistemology. He pointed out the epistemology of EI defined the nature of the domain. "This philosophy is, by definition not primarily driven by technical concerns. Rather, business and sociological constraints of information interaction prevail".

Vernadat (Vernadat, 1996) commented on a common misconception about EI:

Integration of enterprise activities has long been confused with information system integration under the influence of computer science developments. In fact, activity integration should drive information integration. In other words integration needs must be defined by business users, not by computer scientists!

Vernadat later in the same book gave his definition of EI as follows:

Enterprise integration is concerned with facilitating information, control, and material flows across organization boundaries by connecting all the necessary functions and heterogeneous functional entities (information system, devices, applications, and people) in order to improve communication, cooperation, and coordination within this enterprise so that the enterprise behaves as an integrated whole, therefore enhancing its overall productivity, flexibility, and capacity for management of change (or reactivity).

Mische (Mische, 2002) also noticed the all-inclusive nature of EI engagements. He too believed that the goal of EI was on the performance improvement side through harmonization and unification among the information processing environment, technologies, human performance, and business processes.

Even with above broad vision of EI, however, definitions of EI explicitly based on connectivity are still common. For example, after Kosanke (Kosanke, 1998) gave a definition of EI based on performance improvement, he soon stated, "The prime goal of enterprise integration is to use information technology for integration of the enterprise operations." (Kosanke, 1999) gave another definition of EI as below:

Enterprise integration: provide the right information at the right place and at the right time and thereby enable communication between people, machines and computers and their efficient cooperation and coordination.

Another typical connectivity-based definition of EI was from the Next-Generation Manufacturing (NGM) Project in US. It (Bloom, 1997) defined EI as

A system that connects and combines people, processes, systems, and technologies to ensure that the right people and the right processes have the right information and the right resources at the right time.

Given the needs for information integration in the field of EI, the above physically connectivity-based definitions are still correct but in a narrow sense, because technically the common links between components of an enterprise, or between enterprises, are indeed informational in nature. Nevertheless, in reality, connected electronically by the means of information technology does not always mean integrated properly for business at all. The correctness of the physical connectivity is still subject to interpretation. In other words, information integration alone may not necessarily lead to performance improvement (Mische, 2002; Vernadat, 1996; Williams, 1996a; 1996b, and 1999). Information technology as pointed out by many professionals in the field can never be more than an enabler of change (Mische, 2002).

The needs for physical connectivity in the area of EI have grown into a much bigger and richer perspective, Enterprise Interoperability, which is about both information and functionality sharing between concerned parties (Vernadat, 2003). In the above technically oriented definitions of EI, however, the criteria for the *right* connectivity remain largely unaddressed.

During enterprise development processes, the information description will need to present flows such as decision and control, material, energy, etc. Together they work as the "mortar" for the enterprise in question to hold together its elements or components as the "bricks," such as people, organizations, technologies and related equipment, and business processes, etc., which substantiate the business model and fulfill the business mission.

This same metaphor can be applied to the connections between enterprises. For the tasks of EI, Inter- or Intra-, the focus is on fulfillment of the "building" design through the coordination between all elements, or components. The ingredients of the "mortar" are carefully selected because they have to satisfy the binding requirements, which depend on the business relations in consideration. The "building" architecture must be responsible for managing all relationships including the information architecture.

Associated with the flows discussed above, the existence of many "non-programmed" tasks, as defined by Simon years ago (Simon, 1977), presents one of the key issues of information presentation. Note that complexity involved in the enterprise development is often more than the issues of pure formalization because of unstructured and intangible problems that "are sometimes difficult to describe, measure or standardize." (Uppington, 1998)

Even the performance-based definitions of EI may not sufficiently describe the most important functions of EI. Although EI will bring up enterprise-wise performance improvement, localized performance improvement itself may not be sufficiently justifiable as the ultimate driver for EI. In PERA's term (Williams, 1994; 1996b; and 1999) the driver of EI has to be the business mission of the enterprise(s) involved.

The concept of performance improvement may also imply that the enterprise performance could be improved by implementing EI projects after the enterprise systems in question were established, instead of being considered as part of the establishment in the first place. That is, EI could be treated only as an effort after the event. However, that approach has been proved to be less desirable since it may not be always cost-effective. In the viewpoint of PERA, without global considerations, this approach at best may only offer partial results (Rathwell, 1996;

Williams, 1994; 1996a, 1996b; and 1999). For the same reason, the renovation or disposal of legacy systems should not be considered blindly as localized phenomena either.

A holistic approach will still be desirable even for a performance improvement project. A common mistake in that situation is to concentrate on a localized improvement right from the beginning without a proper investigation on its connections with other systems or subsystems that interact with it, technically and non-technically. In order to prevent the project from failing into a partial result, a global view of the impact from the planned improvement and consequently an alignment between the local improvement and the global objectives from the outset will always be necessary.

Although claims based on physical connectivity may promise readily available and even unlimited connections at the beginning, the inefficiency and ineffectiveness, or even the infeasibility of such initiatives in a real-world setting usually signal an unsustainable solution in the end. In other words, although the narrow focus shared between connectivity-based technical proposals and localized performance improvement projects often makes it possible for the two to go hand in hand, neither of them may well be sufficient to generate the desired global impact once finished.

3. MODIFICATION OF THE DEFINITION OF EI

As discussed above, EI should play an integral part at the center of the stage for major enterprise development programs. In terms of business, EI must be dedicated by the top leadership of integration program management where business relations are defined and organized. In terms of technologies, EI must be responsible and accountable for coordination of the multidisciplinary efforts, not depending on any technical implementation paradigm. The basic concepts of EI should prepare both the user community and the suppliers of EI with a viable definition that helps both sides set up a shared vision at the right level. Therefore, the authors feel obligated to restate here our previous definition of enterprise integration (Williams, 1999) as shown below:

Definition of Enterprise Integration:

Enterprise Integration is the coordination of all elements including business, processes, people, and technology of the enterprise(s) working together in order to achieve the optimal fulfillment of the mission of that enterprise(s) as defined by enterprise management.

This definition does not confine the concept of EI within the area of either operational or any physical connectivity in any predefined manner. The goal of integrating all elements is set specifically for a holistic view of the fulfillment of the business mission based upon decisions made by the management of the enterprise(s). The needed connectivity is not emphasized for the sake of connectivity, but expected as a result of subsequent actions driven by the business requirements.

This definition demonstrates a firm conviction of the authors that the concern for implementation technology is only one of the elements in the integration effort. Clearly with this definition, performance improvement should only be considered a means of mission fulfillment of the enterprise(s) in question. Even EI itself is not an end but a means of the mission.

In the early stages of setting up organized efforts to study the issues of EI, one of the major approaches represented by ICEIMT (the International Conference on Enterprise Integration Modeling Technology) was oriented to study only the technical issues or possibilities of computerized technology in EI (Büscher, 1997; Goranson, 1992; 1997; Kosanke, 1998; 1999; and Petrie, 1992). It was understandable that the initial assumption at that time was that computer-based technology, particularly computer-based enterprise modeling, was considered "the key to EI" (Goranson, 1997).

However, after the organized efforts that have been made for more than ten years, the discovery is that the breadth and depth of EI are far beyond mere implementation technology or electronic connectivity. The findings should have been impressive enough to remind the leaders and sponsors of all organizations involved to reconsider the limitations of the initial definitions and orientations. While technical studies of electronic connectivity still remain necessary, open issues beyond technical implementation in the field of EI are already demanding more attention.

Wortmann (Wortmann, 1997) noticed a flaw in the claim of descriptiveness in CIMOSA's approach of computer-based enterprise modeling. People may claim that powerful modern information technology is able to connect or "describe" virtually everything physical with computerized means. However, Wortmann pointed out, the values of such a description would be questionable if it did not prescribe any effective constraints to enforce feasible solutions meaningful in the real world.

4. REVIEW OF TWO TYPES OF ENTERPRISE REFERENCE ARCHITECTURES

The International Task Force on Architectures for Enterprise Integration was another major international endeavor exploring approaches to enterprise integration, which started about the same time as ICEIMT. In order to answer the challenge of these open issues, technical and so-called non-technical included, it identified two types of Enterprise Reference Architectures (Williams, 1996c; 1997; and 1999), Type 1 and Type 2.

In terms of the connectivity concern discussed previously, the Type 1 Architectures describe the architecture or physical structure of some component or part of the integrated system such as the computer system or the communications system. They are those that are responsible for carrying out the physical or electronic connections needed by EI. However, Type 1 Architectures themselves must be substantiated through a project or projects that will ensure that the technical implementations will be established properly through a quality program to meet the business needs.

Therefore, the Task Force identified a second type, the Type 2 Architectures that describe the structure of the development and implementation programs themselves. In other words, the Type 2 Architectures are those that are capable of describing the implementation processes of the Type 1 Architectures. In terms of this descriptive capability, they may also be called "Enterprise Models," especially because they are about the enterprise entities where the deliverables of Type 1 architectures are produced. A fundamental assumption here is that the process described in this enterprise model will help the enterprises involved develop and then implement the Type 1 Architectures. (Williams, 1996c)

Please note that the differentiation between the two types of architectures was significant because many issues in EI, other than the physical technologies studied by Type 1 Architectures, proved to be more significant as key success factors to the efforts in EI. The definition of the Type 2 Architectures allowed the Task Force to study those key issues arising in the development processes, which are identified as the life cycles of the enterprise development programs or processes.

The Task Force reached a consensus on the importance of the Type 2 Architectures, which later became the foundation of GERAM (Generalized Enterprise Reference Architecture and Methodology) proposal and its requirements as produced by the Task Force. (Bernus, 1996a and Williams, 1997)

Table 1 summarizes the concepts of the two types of Enterprise Reference Architectures. Because the Type 2 Architectures must by definition include technical solutions in all types of the physical forms, the Type 1 Architectures are actually a subset of the Type 2 Architectures as conveniently shown in the form of Set theory in the table.

Table 1. Categorization of enterprise reference architectures

Category	Definition	Purpose	Content Relationship of the Two Types
Type 1 (T1)	Those which describe an architecture or physical structure of some component or part of the integrated system such as the computer system or the communications system	Direct the development of technical solutions of EI and their implementation rules	T1 ⊂ T2
Type 2 (T2)	Those which present an architecture or structure of the project which develops the physical integration, i.e., those that illustrate the life cycle of the project developing the integrated enterprise	Direct the development process for both technical and non-technical solutions of EI and the rules of the implementation process	

However, while it was necessary and mutually beneficial, the consensus within the Task Force, which was mainly aimed at the completeness of GERAM, came at a price. The "complete" GERAM, as a big "container" for the three contributing candidate Type 2 Architectures, CIMOSA, GRAI (Doumeingts, 1992a; 1992b; and 2000), and PERA, concealed some fundamental and strategic difference among the candidates. The authors believe that the important difference was rooted in their different visions of EI.

5. TWO APPROACHES TO EI

Among the three candidate architectures, CIMOSA and PERA most typically represented the two different approaches to EI. The former has taken an approach of construct-based computerized modeling; the latter an approach of holistic enterprise integration management. The two different approaches to EI can be expressed below:

- Enterprise integration with construct-based enterprise models and tools represented by CIMOSA, developed by the AMICE Consortium under the ESPRIT Program of the European Community (AMICE, 1993; Kosanke, 1995; 1998; 1999; and Vernadat, 1996; 2003). We will term this as *Approach 1* Architecture or Approach 1 hereafter.
- Enterprise integration with holistic enterprise integration management represented by PERA, the Purdue Enterprise Reference Architectures and Methodology (Li, 1994; Rathwell, 1996; and Williams, 1992; 1994; 1996b; 1997; 1998; 1999). We will term this as *Approach 2* Architecture or Approach 2 hereafter.

CIMOSA pioneered a novel approach to construct-based enterprise modeling and enterprise integration. As a Type 2 Architecture, CIMOSA developed its Modeling Framework, Integrating Infrastructure, and System Life Cycle in order to guide and support the process of creating a set of computer executable models of a subject enterprise. The intended deliverables from CIMOSA will be computer models that are capable of analyzing, monitoring, and operating the subject enterprise. Given the degree of complexity in EI, the computing power offered by modern science and its achievements should make this approach a very attractive candidate to provide EI.

The CIMOSA computerized modeling approach philosophically has however left itself in a somehow vulnerable position to certain open issues. Wortmann's question on the value of the descriptiveness of CIMOSA models mentioned above (Wortmann, 1997) did reveal a loophole in the enterprise models that CIMOSA offered. Virtually, modern computer technology may present or describe everything. But which of the presentations will be fully capable of becoming reality is most probably beyond what a computer can answer. A computer-executable enterprise model in the virtual world does not necessarily mean a "good enough" operatable enterprise in the real world, which must depend upon physical platforms beyond the discretion of the "virtual reality" presented by computers.

Bernus (Bernus, 1996c and 2002) came to the conclusion that the completeness and consistency of formal machine processed models were insufficient for pragmatic purposes. This conclusion might well indicate again the inherent limit of CIMOSA enterprise models, which would probably never allow the models to take over completely the operations of the subject enterprises as originally promised by the CIMOSA initiatives.

Nell (Nell, 1997) saw that the global impact of enterprise integration was larger than any hardware or software. Therefore, he felt that EI standards should be somewhat platform-independent. Similarly, the International Standard ISO 15704 (ISO 15704, 2000) also acknowledged a broad vision of EI as well as Enterprise Engineering (EE).

Inevitably, as researchers proceed in the field of EI, one has to decide if his efforts should be limited within computerized tools and technology only. Both the power and the limits of computer technologies have to be recognized. The inherent conflict between seemingly omni-descriptiveness and platform-dependent prescriptiveness represents the fundamental limit of CIMOSA's modeling approach.

As an automated aid, computer models will sure help one way or another. But real-world applications always need to have many requirements defined including those "non-programmed", which may well be beyond the descriptive power of

CIMOSA's construct-based approach. The introduction of the Approach 2 Architecture above signifies to the readers that comprehensive enterprise integration should not always be identified with the construct-based approach.

6. IDENTIFY A CANDIDATE ARCHITECTURE FOR APPROACH 2

In order to become a candidate for Approach 2 Architecture, a generic enterprise reference architecture has to satisfy the following two necessary and sufficient conditions:

(1) Descriptiveness of Approach 2 Architecture

Type 2 Enterprise Reference Architecture that describes the full and complete life cycle of an EE implementation program should be based on the descriptive capacity of the semantics of its architectural formalism. It should not be compromised by any formal syntax of physical or digital machines.

(2) Prescriptiveness of Approach 2 Architecture

Type 2 Enterprise Reference Architecture that defines a methodology for the full life cycle management should prescribe the generic process paradigm of the life cycle. It should not be compromised by any specific process paradigm of physical or digital implementation.

Each of the two conditions is individually necessary. The first condition defines the requirement for the descriptive capacity of the Approach 2 Architecture. The second condition clarifies the purpose and scope of the descriptive power. They both further state their genericity that is technically independent, which can be consequently summarized in the following theorem:

Theorem 1. (Technical Independence of Approach 2 Architecture)

The architectural formalism of Approach 2 Enterprise Reference Architecture is independent of implementation technologies.

Since the machine syntax alone may never sufficiently fulfill the descriptive requirements for those unstructured and intangible aspects of an enterprise development program, the restraint of machine excutability must be removed from the definition of an Approach 2 Architecture before it is able to describe complex subjects such as strategic decision making processes or the Human and Organizational Architecture as PERA does.

Note that the first necessary condition does not exclude the formal syntax of machines from the architectural formalism. The machine syntax should be best suitable for the descriptions of those "programmed" (Simon, 1977) procedures. However, whenever the effective and efficient human communications need to be specified, machine languages may never replace the functions of human languages (Dress, 1999 and Rosen, 2000). While it is possible to train humans to understand machine languages, the true challenge to the study of electronic communications does not lie in the direction of letting humans think in terms of machines, but it is the other way around. That is the study of programming the machines so that they could think and communicate like humans.

As discussed previously, it was the important findings of non-technical factors that have led to the identification and the initiative of Approach 2 Architecture. The principles of generic enterprise architecture and methodology of Approach 2 are turned out fundamentally different from those of implementation concerns with Approach 1. Therefore, they cannot be replaced or interfered by those of Approach 1.

A typical example can be seen through the first two Phases of PERA, Identification of Enterprise Business Entity and Concept (Williams, 1996b; 1997; 1998; and 1999). During the two phases, business strategies that are free from implementation concerns are developed. The forming process of functional considerations will not start until the third phase, which is Definition as shown underlined in Figure 1. In the first two phases, Identification and Concept, there is therefore no room for so-called Function View, Resource View, or Organization View as defined by CIMOSA's Modeling Framework.

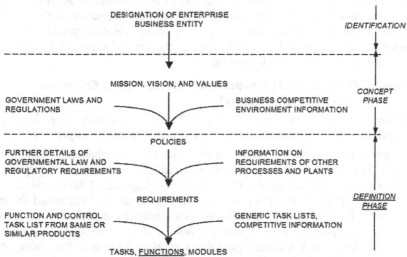

Figure 1 The earliest appearance of Functions in PERA is during the Definition Phase

The approach of CIMOSA's Four Views will be justifiable once the first two phases of PERA deliver the business requirements of "What" to the Definition phase, particularly if the subject enterprise in question is implemented as an IT program. The premature attempt to include the Four Views into the first two phases of PERA is a serious methodological error, which will lead to only two possible consequences. The efforts to fill in those empty Views will either be in vain, or even worse, will violate the principle of the PERA methodology by imposing premature implementation decisions well before the business strategies are formulated. Neither of them could represent the acceptable practice of the PERA Methodology.

The reconfirmation of the technical independence in Theorem 1 above represents the cornerstone of the conceptual integrity of Approach 2 Architecture. The definition of EI presented previously by the authors demands that EI be a unifying endeavor between business and technical professionals, between different implementation domains, and between generations of changes, both technical and non-technical included.   Only then may a technically independent architectural formalism offer the needed all-embracing capacity.

Business strategies are an inseparable part of the life cycle of enterprise development programs. As Goranson pointed out (Goranson, 2003), the business world is strategically different from the operational world. As shown by PERA methodology, nothing but a technically independent formalism will provide a neutral ground for the communication and cooperation between the two worlds. A similar rationale applies to the communication and cooperation between different technical domains as well.

Requirements from change management represent another important strategic reason for the technical independence of the architectural formalism. Because of the broadness, depth, and dynamics of enterprise development programs, whether an enterprise reference architecture is economically sustainable largely depends on whether it will be able to maintain the stability of its presentation as all kinds of the related implementation paradigms change. Therefore the formalism of the Approach 2 must be able to stay neutral through any future changes in these paradigms.

As a result of the above discussions, another theorem of Approach 2 Architecture can be readily expressed as the following:

**Theorem 2. (Organizational Independence of Approach 2 Enterprise
      Reference Architecture)**

The architectural formalism of Approach 2 Enterprise Reference Architecture is independent of the organizational paradigm of the subject enterprise(s).

Please note that the independency defined in the two theorems may not be reversible. The technical and organizational paradigms of the subject enterprise will be dependent upon the descriptions of the particular Approach 2 Architecture.

The PERA study of enterprise development programs has continued for more than 15 years, because both EI and EE are practically too important to be ignored. Changes in management theories and technology development have become the norm of the day. New business paradigms such as Extended Enterprise, Agile Manufacturing, Virtual Enterprise, etc. emerged in the past ten years one after another (Browne, 1998). The speed of development of new applied technologies in industries seemed to be ever faster. However, "At the end all have to do with enterprise engineering and are contributing to enterprise integration." (Kosanke, 1998)

To be open or interoperatable at the level of enterprise reference architectures with their architectural formalism as defined by PERA (Li, 1994), the broad descriptiveness of an Approach 2 Architecture will make it fully adaptable to those changes. This also enables the methodology so defined to be generic enough to guide EE development programs in any specific industry, process or discrete, and in any specific organization, hierarchical or flat.

7. PRESCRIPTIVE NATURE OF TYPE 2 ARCHITECTURES

However, this descriptiveness of Approach 2 Architecture is by no means unlimited. As a Type 2 Architecture, its focus is placed upon the Process Paradigm of the implementation process, or the Process Paradigm of the Life Cycle of the EE project in question.   Therefore, a theorem deducted from the definitions of Type 2 Architectures and its two Approaches defined in this article can be expressed below:

Theorem 3 (Prescriptiveness of Type 2 Architectures)

A Type 2 Architecture prescribes its process paradigm of enterprise life cycles, which imposes a constraint on the architectural descriptiveness of the Architecture.

Because the paradigm of the development process will demand the orientation of development organization, we may have a Corollary of organizational paradigm below:

Corollary of Theorem 3 (Organizational Paradigm of Development)

A Type 2 Architecture prescribes its organizational paradigm of enterprise life cycles based on its process paradigm, which imposes a constraint on the architectural descriptiveness of the Architecture.

Now that we have established the two theorems of Type 2 Architectures, we may have a better understanding of the inherent limit of Approach 1 Architectures since they mainly present the development process in the space of software engineering or IT to develop computer-based models. Given the limitation of the architectural formalism of Approach 1 Architecture, it may at most prove the integrity of a computerized model strictly within its own formal boundary.

Table 2 Categorization of Two Approaches to EI with Type 2 Architectures

Category	Definition	Purpose	Content Relationship of the Two Types
Approach 1 (A1)	Those which present an architecture or structure that illustrate the life cycle of computer-based enterprise modeling process	Direct the development of computerized Enterprise Models and their implementation rules	
Approach 2 (A2)	Those which present an architecture or structure that illustrate the life cycle of the implementation of enterprise engineering programs developing the integrated enterprise	Direct the holistic enterprise integration development process for both technical and non-technical solutions of EI and the rules of the implementation process	$A1 \subset A2$

Table 2 summarizes the concepts of the two different Approaches to EI. Because the Approach 1 Architectures are a special type of integration efforts, the Approach 1 Architectures are considered a subset of the Approach 2 Architectures as shown in the form of Set theory in the table.

By comparing the set relations in Table 1**Error! Reference source not found.** and Table 2, the following relationships of architectural contents can be immediately obtained:

$$T1 \subset A1 \subset A2 \tag{1}$$

Where, $T1 \subset T2$; $A1 \subset T2$; and $A2 \subseteq T2$.

As shown in Equation (1), Type 1 Architectures are highly dependent of technical implementations. On the other side of the relationship, Approach 2 Architectures are independent of technical implementations, which allow them to manage the relationships of all types of implementation paradigms for the subject enterprises in question. Equation (1) also indicates that Approach 2 Architectures will accept those implementation paradigms with their detailed concerns under the condition that they should commit those details to the global missions of the subject enterprise(s).

As an Approach 2 Architecture, PERA has a long history of following the path of systems engineering (Li, 1994 and Williams, 1999). It never self-imposes any constraint from a specific engineering domain. In order to manage the limited space for regular academic publication, the authors have to restrain themselves from more detailed and formal presentation of PERA modeling theories. In the next section, they will however give brief highlights to demonstrate how PERA is able to pass the test of Approach 2 Architecture.

Differentiation between the two Approaches to EI by no means spells an end for what computer models may provide. As a matter of fact, the Approach 1 Architectures represented by CIMOSA have paid their major attention to model the behavior of the subject enterprises to be integrated. Less attention has been given to the behavior of the EE development processes in general, which should not be centered mainly on software engineering processes. It seems that recent researches have regenerated more interest in such studies of the development processes (Cieminski, 2002; Levi, 2002; Nell, 2002a; Webb, 2002; and Weston, 2002).

Initiatives on educational programs of EI and EE have been discussed in the research community (Nell, 2002b and Vernadat, 1996). The candidates for hosting the educational programs proposed include Industrial Engineering, Business Management, and Computer Science, etc. Given the broad definition of the Approach 2 Architectures, the authors believe that a Department of Industrial Engineering with modern IE orientation is probably the best place to experiment with these programs since the basic concepts and practice needed for practicing EI or EE are very close to those of the original orientation of a modern IE department (Rouse, 2004 and Turner, 1993).

8. HIGHLIGHTS OF PERA AS AN APPROACH 2 ARCHITECTURE

Being an Approach 2 Architecture, PERA has demonstrated the following key features, which should satisfy the qualifications for the Approach 2 Architecture:

(i) PERA presents a full life cycle of implementing a general enterprise development program.

Among all three Type 2 Architectures with the Task Force, PERA was considered the most complete one with greatest details (Bernus, 1996b). PERA life cycle was used within the Task Force as the tool to identify enterprise requirements as completely as possible. However, the holistic life cycle model embraced by PERA is much more important than just a checking list for requirements identification for the following reasons:

- PERA's full life cycle model, with its firm embrace for enterprise Mission, Vision and Values (Williams, 1992 and 1996b), signifies the importance of the decision-making level where the decisions on EE projects should be made. If the broad definition of EI given by the authors previously is acceptable, the scope and depth of its impacts may only be properly appreciated and evaluated at the highest possible position within the organization involved. Decisions at this level are in a "different category" (Nell, 1997) from technical only. Approach 2 Architecture should prepare its users for communications suitable for this level.
- PERA's full life cycle model, with its end-to-end presentation at a proper level of abstraction, demonstrates a strong affinity with the pragmatic patterns of decision-making in business circles. It can also be readily combined with different modern management theories on business life cycles, such as product life cycles, project life cycles, change management, and organizational ecology, etc (Li, 2003). Approach 2 Architecture should provide a firm support to justify the integration strategy along with help from other business strategies.
- PERA's full life cycle model, with its detailed step-by-step methodology, also outlines the necessity of closely linking high-level business strategies with implementation solutions. Keeping the technical implementations in alignment with business decisions during EE projects is probably one of the most important missions of a multidisciplinary team of EI. PERA has demonstrated how Approach 2 Architecture should be able to help users turn business strategies into concrete action plans in the context of the project life cycle.

(ii) PERA presents a program methodology that will manage all components of an enterprise, Business, Processes, People and Organization, and Technology.

The all-inclusive nature of EI demands comprehensive enterprise architectures. In order to manage the embedded complexity, PERA identifies the priorities and dependencies among the components and their integration requirements to ensure seamless integration at the enterprise level as follows:

- PERA acknowledges the general objectiveness of every type of business, and defines the business of the subject enterprise in the form of Mission, Vision, and Values (MVV) to prepare its overall business model in a disciplined manner during the master planning process.
- PERA defines business processes of the subject enterprise directly driven by the defined business MVV to fulfill the necessary business functions and other related business constraints.
- PERA defines a "people architecture," human and organization, in the first place in the process of developing business solutions, by following a fundamental philosophy of automation that is the values of the human being are always above the values of the machine.
- PERA recognizes the needed physical technologies in implementation architectures to develop technical solutions. Technical development however must always respect human and organizational development. This respect in return simplifies the design effort of the implementation architectures.

(iii) PERA presents a descriptive architectural formalism for the EE programs.

The descriptiveness of the architectural definition offered by PERA is not about the specifications of a machine-executable language, but fundamentally more about the semantic capability of complexity management in general. This is required by the Approach 2 Architecture because of the following reasons:

- Such a "platform-independent" feature allows the Enterprise Architects involved to produce innovative business solutions as much as possible without constraints unnecessarily placed by any existing management paradigm or implementation technology.
- Such a "platform-independent" feature also allows effective and efficient program communications in the multidisciplinary approach required by EI.
- Such a descriptive definition will simplify the change management. Any changes incurred either by a management paradigm shift or by new technology will have a minimum impact on the architectural definitions that are independent of any management paradigm or implementation technology.
- Such descriptiveness will also simplify the program management of the life cycle because it provides a stable foundation for the program decomposition or partition, which does not have to change as another change happens to either the management paradigm or the implementation technology.

The reason for PERA to become such a descriptive architecture is not by an intentional design but is an inevitable result after years of pursuing an approach of systems-engineering style (Checkland, 1999; Klir, 2001; Sage, 1992; 2000; Thomé, 1993; and Williams, 1961) in the field of EI.

The goal of the PERA research has never been to look for a replacement of the current professions. Instead it has been to identify the gaps between the new challenges from the field of EI and the available capabilities of established majors, management and engineering alike. Particularly, being a result of PERA research itself, the principle of "people-first" of PERA (Williams, 1998) commands that PERA entertain the general audience in need in the field of EI.

In order to help implementation of the real-world enterprise development programs, the gap identified was a missing step-by-step guide for Master Planning (Williams, 1992 and 1996b). The best possible process of PERA Master Planning then identified should meet the following requirements:

- The Master Plan should focus on finishing the contents of "What" for the designs of the program involved, which are architecture designs of the subject enterprise without detailed organizational or technical consideration. These "What" designs play the key role of bridging between business requirements and technical specifications.
- After the "What" designs are finished, the Master Plan should then complete the initial assignments of the needed functional carriers, either human or machine, so that more detailed technical or organizational designs will be able to get started.
- The Master Plan should also complete a program management plan for the execution of the Master Plan, including the necessary program decomposition and partition.
- The Master Plan then should be maintained and revisited through program management practice during its execution process.
- Domain-dependent design details, which are about organizational or technical "How," are largely known to professionals within those domains or majors, and therefore should be considered outside the scope of the Master Plan. As a result, the main structure of PERA Master Planning process is kept descriptive.

As an overarching framework across many different domains or professions, such a domain-independent descriptiveness ensures the integrating capability across those domains under Approach 2 Architecture.

9. SUMMARY

The authors present their concerns and conclusions about the major issues of EI and EE. In order to clarify the clouded concepts by pure technical orientation, they restate the definition of EI to demonstrate their holistic view in this field. They also revisit the core concepts of two Types of Enterprise Reference Architectures to further differentiate two different approaches among so-call Type 2 Architectures. The authors introduce a new category, Approach 2 Architecture, to define a broad research area for EI and EE with supporting theories supplied. General requirements including architectural formalism for the Approach 2 Architectures are also discussed to guide further efforts. During the discussions of verifying PERA as an Approach 2 Architecture, the focus is placed upon the business and engineering practicalities of PERA methodology.

REFERENCES

AMICE (1993) *CIMOSA: Open System Architecture for CIM*, second revised and extended edition, Springer-Verlag.

Bernus, P. (2002) "Quality of Virtual Enterprise Reference Models" In *Enterprise Inter- and Intra-Organizational Integration, Building international consensus, IFIP TC5/WG5.12 International Conference on Enterprise Integration and Modeling Technology (ICEIMT'02)*, April 24-26, 2002, Valencia, Spain, Kosanke, K., Jochem, R., Nell, J. G., and Ortiz Bas, A., Eds. Kluwer Academic Publishers, Boston, Dordrecht, London, 2003, pp. 135 – 146.

Bernus, P., and Nemes, L. (1996a) "A Framework to Define a Generic Enterprise Reference Architecture and Methodology" *Computer Integrated Manufacturing Systems*, Vol. 9, No. 3, pp. 179 – 191.

Bernus, P, Nemes, L., and Williams, T. J., Eds. (1996b) *Architectures for Enterprise Integration*, Chapman and Hall.

Bernus, P., Nemes, L., and Morris, B. (1996c) "The Meaning of an Enterprise Model" In *Modeling and Methodologies for Enterprise Integration*, Bernus, P., and Nemes, L., Eds. Chapman and Hall, London, pp. 182 – 200.

Bloom, H. M., (1997) "Enterprise Integration – A United States View" In *Enterprise Engineering and Integration: Building international consensus, Proceedings of ICEIMT '97*, International Conference on Enterprise Integration and Modeling Technology, Torino, Italy, October 28-30, 1997, K. Kosanke, and J. G. Nell, Eds. Springer, pp. 6 – 19.

Browne, J., Goranson, H. T., and Katzy, B., *et al.* (1998) "Part 1. New Paradigms for the Manufacturing Enterprise" In *Handbook of Life Cycle Engineering, Concepts, models, and technologies*, Molina, A., Kusiaka A., and Sanchez, J. Eds. Kluwer Academic Publishers, Dordrecht, Boston, London.

Büscher, R. (1997) "Enterprise Integration and Standardization – A European View" In *Enterprise Engineering and Integration: Building international consensus, Proceedings of ICEIMT '97*, International Conference on Enterprise Integration and Modeling Technology, Torino, Italy, October 28-30, 1997, Kosanke, K. and Nell, J. G. Eds., Springer, pp. 3 – 5.

Carr, N. G. (2003) "IT doesn't matter" *Harvard Business Review*, Vol. 81, No. 5, pp. 41 – 49.

Checkland, P. (1999) *Soft Systems Methodology: a 30-year retrospective*, John Wiley and Sons, LTD. Chichester, New York.

Cieminski, G. Macchi, M., and Garettti, M., *et al* (2002) "Proposal of a Reference Framework for Manufacturing Systems Engineering" In *Enterprise Inter- and Intra-Organizational Integration, Building international consensus,* IFIP TC5/WG5.12 International Conference on Enterprise Integration and Modeling Technology (ICEIMT'02), April 24-26, 2002, Valencia, Spain, Kosanke, K., Jochem, R., Nell, J. G., and Ortiz Bas, A., Eds. Kluwer Academic Publishers, Boston, Dordrecht, London, 2003, pp. 167 – 176.

Doumeingts, G., Vallespir, B., Ducq, Y., and Kleinhans, S. (2000) "Production management and enterprise modelling", *Computers in Industry*, Vol. 42, Nos. 2-3, pp. 245 – 263.

Doumeingts, G., Vallespir, B., Zanettin, M., and Chen, D. (1992a) *GIM, GRAI Integrated Methodology, A Methodology for Designing CIM Systems*, Version 1.0, Unnumbered Report, LAP/GRAI, University Bordeaux 1, Bordeaux, France.

Doumeingts, G., Chen, D., and Marcotte, F. (1992b) "Concepts, models and methods for the design of production management systems", *Computers in Industry*, Vol. 19, No. 1, pp. 89 – 111.

Dress, W. B. (1999) "Epistemology and Rosen's modeling relation", Plenary presentation on the 43rd Meeting of the International Society for the Systems Sciences, June 27 - July 2, 1999, Pacific Grove, California, available at http://hypernews.ngdc.noaa.gov/hnxtra/Dress_paper.html.

Goranson, H. T. (2003) "Architectural support for the advanced virtual enterprise", *Computers in Industry*, Vol. 51, No. 2, pp. 113 – 125.

Goranson, H. T. (1997) "ICEIMT in Perspective – 92 to 97" In *Enterprise Engineering and Integration: Building international consensus, Proceedings of ICEIMT '97*, International Conference on Enterprise Integration and Modeling Technology, Torino, Italy, October 28-30, 1997, Kosanke, K. and Nell, J. G. Eds. Springer, pp. 167 – 174.

Goranson, H. T. (1992) "The Suppliers' Working Group, Enterprise Integration Reference Taxonomy" In *Enterprise Integration Modeling, Proceedings of the First International Conference*, Petrie, C. J., Ed. The MIT Press, Cambridge, Massachusetts, London, England, pp. 114 – 130.

ISO 15704 (2000) Industrial automation systems - Requirements for enterprise-reference architectures and methodologies, International Standards Organization, Geneva.

Klir, G. J. (2001) *Facets of Systems Science*, 2nd Edition, Kluwer Academic/Plenum Publishers, New York.

Kosanke, K., Vernadat, F., Zelm, M. (1999) "CIMOSA: Enterprise Engineering and Integration," *Computers in Industry*, Vol. 40, No. 2-3, pp. 83 – 97.

Kosanke, K. and Vernadat, F. B. (1998) "CIMOSA – Life Cycle Based Enterprise Integration" In *Handbook of Life Cycle Engineering, Concepts, models, and technologies*, Molina, A., Kusiaka A., and Sanchez, J. Eds. Kluwer Academic Publishers, Dordrecht, Boston, London.

Kosanke, K. (1995) "CIMOSA – Overview and Status," *Computers in Industry*, Vol. 27, No. 2, pp. 107 – 109.

Levi, M. H. (2002) "The Business Process (Quiet) Revolution, Transformation to Process Organization" In *Enterprise Inter- and Intra-Organizational Integration, Building international consensus*, IFIP TC5/WG5.12 International Conference on Enterprise Integration and Modeling Technology (ICEIMT'02), April 24-26, 2002, Valencia, Spain, Kosanke, K., Jochem, R., Nell, J. G., and Ortiz Bas, A., Eds. Kluwer Academic Publishers, Boston, Dordrecht, London, 2003, pp. 147 – 158.

Li, H. (2003) *Manage Consulting Business with Purdue Enterprise Reference Architecture*, Workshop on Applying Enterprise Reference Architecture in Consulting Business, Tsinghua University, Beijing, PR China, Jan. 3–8, 2003.

Li, H. (1994) A Formalization and Extension of the Purdue Enterprise Reference Architecture and the Purdue Methodology, Ph.D. Thesis, Purdue University, West Lafayette, Indiana. Also published as Li, H. and Williams, T. J., A Formalization and Extension of the Purdue Enterprise Reference Architecture and the Purdue Methodology, Technical Report 158, available at http://iies.www.ecn.purdue.edu/IIES/PLAIC/PERA/Publications, Purdue Laboratory for Applied Industrial Control, Purdue University, West Lafayette, IN 47907, April, 1995.

Mische, M. A. (2002) "Defining Systems Integration" In *Enterprise Systems Integration*, 2nd Edition, Myerson, J. M. Ed. Auerbach Publications, Boca Raton, London, New York, Washington, D. C., pp. 3 – 10.

Nell, J. G., delaHostria, E, and Engwall, R. L., *et al.* (2002a) "System Requirements: Products, Processes and Models" In *Enterprise Inter- and Intra-Organizational Integration, Building international consensus*, IFIP TC5/WG5.12 International Conference on Enterprise Integration and Modeling Technology (ICEIMT'02), April 24-26, 2002, Valencia, Spain, Kosanke, K., Jochem, R., Nell, J. G., and Ortiz Bas, A., Eds. Kluwer Academic Publishers, Boston, Dordrecht, London, 2003, pp. 245 – 252.

Nell, J. G., and Goranson, H. T. (2002b) "Accomplishments of the ICEIMT'02" In *Enterprise Inter- and Intra-Organizational Integration, Building international consensus*, IFIP TC5/WG5.12 International Conference on Enterprise Integration and Modeling Technology (ICEIMT'02), April 24-26, 2002, Valencia, Spain, Kosanke, K., Jochem, R., Nell, J. G., and Ortiz Bas, A., Eds. Kluwer Academic Publishers, Boston, Dordrecht, London, 2003, pp. 15 – 23.

Nell, J. G. (1997) "A Standardization Strategy that Matches Enterprise Operation" In *Enterprise Engineering and Integration: Building international consensus, Proceedings of ICEIMT '97*, International Conference on Enterprise Integration and Modeling Technology, Torino, Italy, October 28-30, 1997, Kosanke, K. and Nell, J. G. Eds., Springer, pp. 54 – 63.

Petrie, C. J. (1992) "Introduction" In *Enterprise Integration Modeling, Proceedings of the First International Conference*, Petrie, C. J., Ed., The MIT Press, Cambridge, Massachusetts, London, England, pp. 1 – 14.

Rathwell, G. A. and Williams, T. J. (1996) "Use of the Purdue Enterprise Reference Architecture and Methodology in Industry (the Fluor Daniel example)" In *Modeling and Methodologies for Enterprise Integration*, Bernus, P., and Nemes, L., Eds. Chapman and Hall, London, pp. 12 – 44.

Rosen, R. (2000) *Essays on Life Itself*, Columbia University Press, New York, 2000.

Rouse, W. B. (2004) "Embracing the Enterprise", *Industrial Engineer Magazine*, Vol. 36, No. 3, pp. 31 – 35.

Sage, A. P., and Armstrong Jr., J. E. (2000) *Introduction to Systems Engineering*, John Wiley and Sons, Inc., New York, NY.

Sage, A. P. (1992) *Systems Engineering*. John Wiley and Sons, Inc., New York, NY.

Simon, H. A. (1977) *The New Science of Management Decision*, Revised Edition, Prentice-Hall, Inc., Englewood Cliffs, N.J.

Thomé, B., Ed. (1993) Systems Engineering, Principles and Practice of Computer-Based Systems Engineering, John Wiley & Sons, Inc., New York, NY.

Turner, W. C., Mize, J. H., Case, K. E., and Nasemetz, J. W. (1993) *Introduction to Industrial and Systems Engineering*, Third Edition, Prentice Hall, Englewood Cliffs, NJ.

Uppington, G. J. (1998) Identifying the Road Ahead: Enhancing and extending the Enterprise Integration Identification Phase, Ph.D. Thesis, Griffith University, Brisbane, Australia.

Vernadat, F. B. (2003) "Enterprise Modeling and Integration, From Fact Modeling to Enterprise Interoperability" In *Enterprise Inter- and Intra-Organizational Integration, Building international consensus,* IFIP TC5/WG5.12 International Conference on Enterprise Integration and Modeling Technology (ICEIMT'02), April 24-26, 2002, Valencia, Spain, Kosanke, K., Jochem, R., Nell, J. G., and Ortiz Bas, A., Eds. Kluwer Academic Publishers, Boston, Dordrecht, London, pp. 25 – 33.

Vernadat, F. B. (1996) *Enterprise Modeling and Integration: principles and applications*, Chapman and Hall, London, Winheim, New York, Tokyo, Melbourne, Madras.

Webb, P. (2002) "Enterprise Architecture and Systems Engineering" In *Enterprise Inter- and Intra-Organizational Integration, Building international consensus,* IFIP TC5/WG5.12 International Conference on Enterprise Integration and Modeling Technology (ICEIMT'02), April 24-26, 2002, Valencia, Spain, Kosanke, K., Jochem, R., Nell, J. G., and Ortiz Bas, A., Eds. Kluwer Academic Publishers, Boston, Dordrecht, London, 2003, pp. 159 – 166.

Weston, R. H., Ang, C. L., and Bernus, P., *et al.* (2002) "Virtual Enterprise Planning Methods and Concepts" In *Enterprise Inter- and Intra-Organizational Integration, Building international consensus,* IFIP TC5/WG5.12 International Conference on Enterprise Integration and Modeling Technology (ICEIMT'02), April 24-26, 2002, Valencia, Spain, Kosanke, K., Jochem, R., Nell, J. G., and Ortiz Bas, A., Eds. Kluwer Academic Publishers, Boston, Dordrecht, London, 2003, pp. 127 – 134.

Williams, T. J. and Li, H. (1999) "PERA and GERAM–Enterprise Reference Architectures in Enterprise Integration" In *Information Infrastructure Systems for Manufacturing*, Mills, J. J., and Kimura, F., Eds. Kluwer Academic Publishers, Norwell, MA, pp. 3–30.

Williams, T. J. (1998) "Characterization of the Place of the Human in Enterprise Integration" In *Handbook of Life Cycle Engineering, Concepts, models, and technologies*, Molina, A., Kusiaka A., and Sanchez, J. Eds. Kluwer Academic Publishers, Dordrecht, Boston, London, 1998.

Williams, T. J. and Li, H. (1997) "The Task Force Specification for GERAM and its Fulfillment by PERA," *A. Rev. Control*, Vol., 21, pp. 137 – 147.

Williams, T. J. (1996a) "The needs of the field of integration" In *Architectures for Enterprise Integration*, Bernus, P, Nemes, L., and Williams, T. J., Eds. Chapman and Hall, pp. 21 – 31.

Williams, T. J., Rathwell, G. A., and Li, H., Eds. (1996b) A Handbook on Master Planning Implementation for Enterprise Integration Programs Based on the Purdue Enterprise Reference Architecture and the Purdue Methodology, available at http://iies.www.ecn.purdue.edu/IIES/PLAIC/PERA/Publications, Purdue University, West Lafayette, Indiana, US.

Williams, T. J., Bernus, P., and Nemes, L. (1996c) "The Concept of Enterprise Integration" In *Architectures for Enterprise Integration*, Bernus, P., Nemes, L., and Williams, T. J., Eds., Chapman and Hall, pp. 9 – 20.

Williams, T. J. (1994) "The Purdue Enterprise Reference Architecture," *Computers in Industry*, Vol. 24, Nos 2 – 3, pp. 141 – 158.

Williams, T. J. (1992) *The Purdue Enterprise Reference Architecture*, Instrument Society of America.

Williams, T. J. (1961) *Systems Engineering for the Process Industries,* McGraw-Hill Book Company, New York, NY.

Wortmann, J. C. (1997) "Enterprise Reference Architectures – A Research Portfolio" In *Enterprise Engineering and Integration: Building international consensus, Proceedings of ICEIMT '97*, International Conference on Enterprise Integration and Modeling Technology, Torino, Italy, October 28-30, 1997, Kosanke, K. and Nell, J. G. Eds. Springer, pp. 20 – 26

22.   Enterprise Integration Engineering as an Enabler for Business Process Management

Arturo Molina[1], Jorge Garza[2], Guillermo Jiménez[2]

1 CSIM-ITESM armolina@itesm.mx
2 CII-ITESM garza.jorge@itesm.mx,guillermo.jimenez@itesm.mx

This paper describes how Business Process Management has been implemented based on a Reference Framework defined based on Enterprise Integration Engineering concepts. The Reference Framework includes the following components: strategy definition (competitive, supply chain, operational), performance evaluation system, process design/re-design, and enabling technologies. It describes how all these issues have to be considered in an integrated way to align the company strategy with process improvement projects in order to achieve excellent performance. One case study is reviewed to describe how the reference model has been used in a OEM (Original Equipment Manufacturer) to achieve change management and best manufacturing practices implementation.

1. INTRODUCTION

Emerging economies, social and political transitions, and new ways of doing business are changing the world dramatically. These trends suggest that the competitive environment for manufacturing enterprises in 2020 will be significantly different than it is today. To be successful in this competitive climate, manufacturing enterprises of 2020 will require significantly improved technological and organizational capabilities. The acquisition of these capabilities represents the challenge facing manufacturing. Two important concepts have emerged to support companies in this new challenging scenario (Vernadat 2002, Bernus *et al.* 2003, Grigoria *et al.* 2004):

Enterprise Integration Engineering (EIE) is the collection of modeling principles, methodologies and tools that allow to engineer different entities' life cycles in an enterprise (e.g. enterprise, project, product, processes). The foundation relies on the creation of models of the structure, function and behavior of the different entities. EIE allows a detailed description of all the key elements of an entity (activities, data/information/knowledge, organizational aspects, human and technological resources). In an enterprise model, this description provides the means to connect and communicate all the functional areas of an organization to improve synergy within the enterprise, and to achieve its mission and vision in an effective and efficient manner.

Business Process Management (BPM) is the set of theories, techniques, methods, tools and applications that support the design and development of Business Process Management Systems (BPMS) which are software platforms that support the definition, execution, and tracking of business processes. Proper analysis of BPMS execution logs can yield important knowledge and help organizations improve the quality of their business processes, including the production of goods and services to business partners, as well as the enterprises' own management activities. This analysis is known as Business Process Intelligence.

Among all these issues, business process management, integration and coordination remain challenging because of its knowledge intensive nature. Therefore there is a need for systematic methodological- and technology-supported approach to develop and sustain a successful company.

This paper describes a Framework for Enterprise Integration Engineering that has been defined and developed to support Business Process Management in Mexican companies. An example based on a case study of an OEM company is presented to demonstrate the usage of the reference model.

2. ENTERPRISE INTEGRATION ENGINEERING (EIE) REFERENCE FRAMEWORK

2.1 Components of the EIE Reference Framework

The components of the reference model are depicted in Figure 1. Each of the different components provides guidelines, methodologies and tools to engineer business process changes. The components are:

- A strategy realization process and performance evaluation systems support the definition of three types of strategies in the company, namely: Competitive-, Value Chain- and Production/Service Strategy. All these strategies are associated with performance measures to evaluate the impact of the strategy pursued in the organization.
- Reference Models for Enterprise Modeling supports the visualization of enterprise knowledge, processes and associated performance measures in order to identify areas of opportunities for improvement.
- Decision making and simulation models support the evaluation of different strategies and implementation of best manufacturing practices using different simulation tools such as: dynamic systems and discrete event simulation. Best practices are defined in terms of logic program models to describe its impacts on business performance.
- Business Process Management Systems and Business Process Intelligence tools support the execution and analysis of process using business and IT perspectives. Business Process Management Systems allow process design, execution and tracking based on workflow technology. The Business Process Intelligence analysis supports decision making for predicting and optimizing processes.

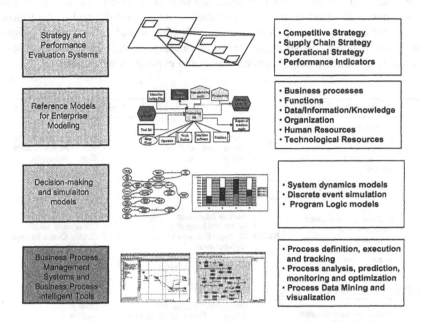

Figure 1 Components of the Enterprise Integration Engineering Reference
Framework for Business Process Management

2.2 Strategy and performance evaluation systems

Analysis tools and guidelines are provided to define three propositions to achieve competitive advantage: product innovation, operational excellence and customer focus (Hope and Hope 1997).

The competitive strategy should be translated into a set of decisions of how the organizations can deliver value to the customer. Value Chain strategy is about making decisions of how a company will establish an organizational model (external and internal) that will exploit the different possibilities to build an effective and efficient value chain. Different decisions can be conceived in value chain strategy: Vertical Integration, Structuring into Strategic Business Units, Horizontal Integration and Establishment of a Collaborative Organizational Structure.

The last strategy defines how the company will produce or deliver its products or services. The production/services strategy is based on the following factors: product description, characterization of customers and suppliers, and process definition. All these factors are defined by order-qualification and order-winning criteria (Hill 1989). The criteria are: price, volume, quality, lead-time, delivery speed and reliability, flexibility, product innovation and design, and life cycle status. Based on all these performance measures the following production strategies may be defined (Molina and Medina 2003):

- Production Strategy: Make to Stock (MTS), Make to Order (MTO), Assemble to Order (ATO), Configure to Order (CTO), Build to Order (BTO) and Engineer to Order (ETO).
- Service Strategy: Services on Catalogues (SoC), Configuration of Services (CoS) and Design of Services (DoS)

The impact of these strategies in a company should be able to be measured using a performance evaluation system. Performance measures are defined in the following dimensions: Quality, Time, Cost, Volume, Flexibility and Environment. Figure 2 depicts the process of strategic decision making, using different analyses for strategic decisions and performance measures to evaluate their impacts.

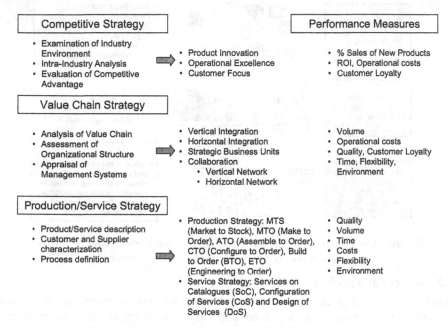

Figure 2 Guidelines for strategy definition process

2.3 Reference Models and Enterprise Modelling

The reference model used in this component, it is based on the Extended Enterprise concept (Browne, et. al. 1999; Vernadat, 2002) and the ENAPS Reference Models (Rolstadås 1998). It comprises 8 business processes to describe a generic structure of an ideal intra and inter integrated-extended enterprise. Below is a brief description of the business processes of the Integrated Extended Enterprise Reference Model: Co-Engineering, Customer Driven Design, Supplier Relationship Management, Customer Relationship Management, New Product Development, Obtaining Customer Commitment, Order Fulfillment/Supply Chain Management, and Customer Service

The reference model can be particularized to any enterprise and its core processes are chosen for modeling and simulation, in order to evaluate process improvement through Business Process Management. Extended Event-Process-Chain (eEPC) diagrams are used to model at different levels of detail the core processes (Figure 3). The detail level is defined according to the specification level of the activities included. The first level considers only general process functions; the second level considers specific activities of each function from the first level; and in the third level a deeper specification of activities is achieved for the specific functions from the second level, furthermore, material and information flows can also be included. In order to guarantee an effective global analysis, it is necessary to develop models covering the function and control views (Scheer, 1999).

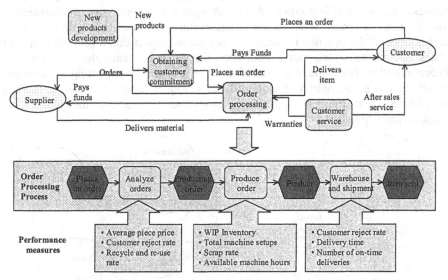

Figure 3 Extended Enterprise Reference (represented as eEPC diagrams)

2.4 Decision making and simulation models

Simulation allows the dynamic visualization of systems, and the interaction among their components in order to have a more realistic picture of the process or processes selected, and to understand process behavior. In this research, system dynamics simulation, discrete event simulations and program logic models were used:

System Dynamics simulation: The applied theory of system dynamics and dynamic systems modeling method come primarily from the work of Jay Forrester (Forrester 1980). The models are built based on feedback loops of key performance measures, cause-and-effect models, feedback influences and impacts of effects. Therefore enterprise models of behavior have been developed to demonstrate the effects and impacts of best practices implementation on performance measures (Molina and Medina 2003). An example of a dynamic model of a company including key manufacturing performance measures is presented in Figure 4.

Discrete event simulation: simulation is the most common method used to evaluate (predict) performance. The reason for this is that a quite complex (and realistic) simulation model can be constructed using actors, attributes, events and statistics accumulation. Business processes simulation can be performed, for example, in order to evaluate resource usage and to predict performance measures such as delivery time and cost, capacity usage, etc.

Program Logic Models: A Logic Model can be seen as a conceptual map that supports the evaluation of the possible impact in the implementation of a manufacturing practice. A logic model states short and long term impacts and what resources and methods are to be used ((Coffman, 1999; Alter, *et al.*, 1997). The manufacturing practices are described and organized as program logic models (results, effects, impacts and benefits) allowing evaluation and planning of changes in the business process. For example, SMED (Single Minute Exchage of Dies) requires people to be trained, design a new set-up process, and implement the new procedure. The results of each of these activities are people trained and set up process designed and implemented. The changes that are required to implement the SMED practice are: flow of activities, abilities of operators and new set up instructions. If the practice is successful one might expect that a reduction of set up times will be achieved, more production time will be available, WIP (Work in Process) and costs will be reduced and the company will be expecting to increase its profit. This is a description of system dynamic model, where a cause-effect impact of different performance measures is described to evaluate the impact of a manufacturing practice.

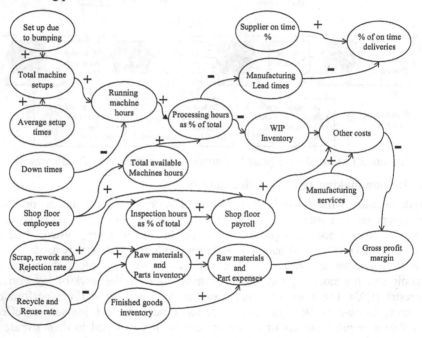

Figure 4 Key manufacturing performance measures described in a system dynamic model.

From several research and consulting projects developed at our research group, a Database of Best Manufacturing Practices has been collected. Results of this investigation are organized using Program Logic Models. The best manufacturing practices database has been developed and organized according to a logic model structure, which describes the benefits expected, and performance measures that a practice might influence, in order to evaluate their feasibility and effectiveness of implementation to optimize critical performance measures (Table 1).

Once the enterprise's (manufacturing and service) process is understood using the simulation tools, it is usually necessary to propose changes in order to improve the opportunity areas identified, e.g. by implementing SMED (a Best Manufacturing Practices). In the database, manufacturing practices are organized into activities, outputs, changes/effects, impacts and benefits – allowing the evaluation of the impact of implementing a best practice. The combination of System Dynamic models with Logic Models allows a systemic understanding of the impact of the implementation of best manufacturing practices.

Table 1. Description of best manufacturing practice using Program Logic Models

Activities	Outputs	Changes/Effects	Impacts	Benefits
All necessary activities to implement a best manufacturing practice: Train Design Implement Evaluate	Immediate results of activities: People trained Process designed Process executed	Changes in business processes: Flow of activities and information Availability of data, Information, and knowledge Human capital: knowledge, skills and abilities Technological capital: capacity, capabilities, and usage. Organization: practices, procedures, methods, and tools.	Impact on performance measures: Quality Volume Time Cost Flexibility Environment	Operational: value added per strategy, process and resource Economics: Profit / ROI Strategic: innovation, excellence, customer focus.

2.5 Business Process Intelligence tools and Business Process Management Systems

Business Process Management (BPM) is the set of theories, techniques, methods, tools and applications that support the design and development of Business Process Management Systems (BPMS). BPMSs are software platforms that support the definition, execution, and tracking of business processes. Proper analysis of BPMS execution logs can yield important knowledge and help organizations improve the quality of their business processes and services to their business partners. BPMSs allow the execution of company processes based on workflow technology. In addition, Business Process Intelligence (BPI) allows users to analyze completed process executions from both a business- and an IT perspective. IT analysts will be interested in viewing detailed, low-level information such as average execution time per process or the length of the work queues of human or technological resources. Business users will instead be interested in higher-level information, such as the number of 'successful' process executions, or the characteristics of processes that did not meet the customer's expectations. The analysis capabilities of BPI can also be applied to analyze the design of a process model – in particular for identifying techniques to improve an existing process definition and/or the use of Information Technology. Therefore the utilization of Business Process Management Systems, together with BPI analysis capabilities, allow companies to support change using a technology driven approach.

3. EXPERIENCIES IN APPLYING THE EIE FRAMEWORK

An OEM (Original Equipment Manufacturer) has been working on the improvement of its Product Delivery System (PDS) through an integrated flow based on a Business Process Management System in order to satisfy their customer needs. A

Product Delivery System is divided in the following cycles: customer order cycle (customer-retailer), replenishment cycle (retailer-distributor), manufacturing cycle (distributor-manufacturer), and procurement cycle (manufacturer-supplier). These cycles are co-ordinated and aligned in order to decrease the Total Cycle Time (receipt of order, planning, supply, manufacturing, warehousing and delivery). The EIE reference model was used to guide the design and implementation of the PDS in the following manner:

Competitive strategies were defined in order to achieve Operational Excellence. The following strategies were selected to achieve this objective: integration of the product delivery system flow, competitive excellence tools deployment (5S, TPM, Setup, Mistake Proofing, Root Cause Analysis), cost savings, commercial, operational and financial key initiatives and people cultural change.

Value Chain Strategies were defined in order to support these competitive strategies. These included: collaboration approach for the domestic market with customers and suppliers, aiming to materialize the concept of the 'virtual factory'. Horizontal integration was achieved by sharing commercialization resources with Business Units in the Northern American Operations and consolidating Asian Suppliers.

Production Strategies were set to satisfy customers' demands of different nature. These were defined as follows: Make to Stock (MTS) for the domestic market and Make to Order (MTO) for the exports market.

The core process defined was Order Processing (Product Delivery Process), which was supported by the PDS. The process was divided into different cycles, which could be addressed in turn in order to achieve internal goals to reduce weaknesses, always having in mind that these goals had to lead to the drivers of the company and to specific results

Performance indicators were defined to provide feedback about the company's progress toward achieving its strategic objectives:

- Competitive Strategy: cost reductions and time reductions
- Value Chain Strategy: cost reduction and increased flexibility by using local suppliers. Cost reduction by sharing resources for commercialization and purchasing.
- Production Strategy: reduced inventory level for exportation market (MTO), and setting of optimal inventory levels for national market (MTS).
- Process measures:
  - Reduction of Customer Order Cycle and Replenishment Cycle
  - Manufacturing Cycle: minimize setup times and increase mix model production
  - Procurement Cycle: reduction of suppliers' lead-time negotiation, cost savings.
- Strategic decisions to evaluate:
  - Customer Order Cycle: 80% Sales by Web, Forecast planning by Web, Customer Orders Status and Shipment using Web; and Warranty Online
  - Replenishment Cycle: outsource logistic operator and automated receiving and warehousing process.

- Manufacturing Cycle: facility's re-layout based in material flow concept and line flexibility, implement SMED (Single Minute Exchange of Dies), and redesign allocation algorithm for mix model production
- Procurement Cycle: 90% suppliers online (automated purchase orders), redesign supplier's negotiation process and 50% of part numbers in Kanban Online.

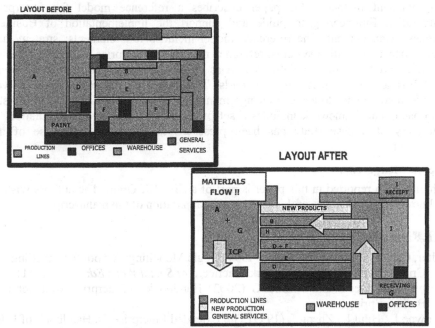

Figure 5 Layout redesign to allow an improved Product Delivery Process

The strategies were evaluated and the decision to redesign the Product Delivery Process and implement a Business Process Management system to support the process execution was made. The BPM system allows customer and supplier to use the Web for different operations, and the system is connected to the ERP (Enterprise Resource Planning) and MES (Manufacturing Execution System) on the shop-floor. The layout of the plant was redefined and the best manufacturing practice of SMED (Single Minute Exchange of Dies) was implemented (Figure 5). The new process was monitored and analyzed in order to evaluate the impact on the different performance measures.

4. CONCLUSION AND FURTHER WORK

A few years ago, manufacturing industry, especially in Mexico, was characterized as a labor intensive sector; however, the trend is changing, since this industry is not distinguished any more because of the low wages it used to pay. Today manufacturing in Mexico is evolving to a more knowledge-based industry, and it is hoped that this will continue in the future. Since the characteristic component of Enterprise Integration Engineering and Business Process Management is knowledge, this research emphasizes the point where the future of manufacturing lies.

There is a need for better practices of knowledge acquisition,   visualization and use in manufacturing companies. Therefore it is important to develop new strategies, methodologies and tools that allow enterprise to document, evaluate and apply changes in its business process, using formal enterprise reference models. These models can be used to analyze, monitor and determine positive or negative impacts of best practice implementation using a low risk and systematic process improvement method. This paper describes a reference model for Enterprise Integration Engineering to guide and support the implementation of Business Process Management. The reference model includes four elements: strategy and performance evaluation systems, reference models for enterprise modeling, decision making and simulation models, as well as business Process Management Systems and Business Process Intelligence tools. This reference framework has allowed Mexican companies to achieve change management using a systematic and holistic approach. The framework includes a set of tools for modeling and simulation. A summary of a case study has been presented to demonstrate the use of the framework.

Acknowledgments

The research reported in this paper is part of a CEMEX Grant. The authors wish to acknowledge the support of this grant in the preparation of the manuscript.

REFERENCES

Alter, Catherine and Egan, Marcia (1997) Logic Modelling: A Tool for Teaching Critical Thinking in Social Work Practice. *J of Social Work Education* 33(1)

Bernus P., Nemes L. and Schmidt G. (2003) Handbook on Enterprise Architecture. Berlin : Springer.

Browne J., Hunt I., Zhang J. (1999) The Extended Enterprise. In Handbook of Life Cycle Engineering: Concept, Methods and Tools. Molina A., Kusiak, A. and Sanchez (Eds). Kluwer Academic Publishers. 3-29

Coffman L. (1999) Learning From Logic Models: An Example of a Family/School Partnership Program. Harvard Family Research Project, http://www.gse.harvard.edu/hfrp/pubs.html.

Forrester, J.W. (1968). Principles of systems : text and workbook. Wrigh-Allen Press, Cambridge, MA.

Grigoria D., Casatib F., Castellanos M., Dayalb U., Sayalb M., Shanb M-C. (2004) Business Process Intelligence, *Computers in Industry 53*(3) 321–343.

Hill, T. (1989) Manufacturing Strategy. Homewood, IL: Richard D. Irwin.

Hope J., and Hope T. (1997) Competing in the Third Wave. Boston: Harvard Business School Press

Molina A. and Medina V. (2003) Application of Enterprise Models and Simulation Tools for the evaluation of the impact of best manufacturing practices implementation. *Annual Reviews in Control,* 27(2) 221-228.

Rolstadås, A. (1998) Enterprise performance measurement. *International Journal of Operations & Production Management.* 18(9) 989-999.

Scheer, A-W. (1999) ARIS – Business Process Modelling. New York : Springer

Vernadat F.B. (2002) Enterprise Modeling and Integration (EMI): Current Status and Research Per-spectives. *Annual Reviews in Control* 26. 15-25.

23. Deriving Enterprise Engineering and Integration Frameworks from Supply Chain Management Practices[74]

Angel Ortiz[1], Víctor Anaya[1] and Darío Franco[1]

1 Research Center on Production Management and Engineering, Polytechnic University of Valencia, Spain Email:{aortiz,vanaya,dfranco}@cigip.upv.es

Enterprise Engineering and Enterprise Integration have been leveraged as key topics in Enterprise Management. Since the 80s multiple approaches, methodologies, languages and, frameworks have been proposed. Despite the numerous results currently existing, new trends and solutions are continuously emerging. This paper provides a landscape of the current problems on Enterprise Engineering and Integration, the strategies, solutions and our vision about future trends.

1. INTRODUCTION

During the past 20 years, firms have faced continues changes in managerial and technological solutions in order to cope with new market objectives and challenges. Companies have moved from individual strategies, where each enterprise did their own work without considering the collaboration with other enterprises, to collaborating strategies, where sharing and exchanging information is necessary to give complete solutions that users demand. On the other hand, enterprises have evolved technologically from an all manual activities situation, to a situation with intensive support by Information Technology (IT).

Analysing multiple solutions, approaches and proposals that historically have tried to improve the management of business entities, we can differentiate two complementary fields of research, *Enterprise Engineering* and *Enterprise Integration*. Although complementary, they impact each other in some way, mainly Enterprise Engineering over Enterprise Integration, because to integrate something it is needed to know things that are going to be integrated.

Enterprise Engineering is the art of understanding, defining, specifying, analysing, and implementing business processes for the entire enterprise life-cycle, so that the enterprise can achieve its objectives, be cost-effective, and be more competitive in its market environment (Vernadat, 1995).

Enterprise Integration consists in breaking down organizational barriers to improve synergy within the enterprise so that business goals are achieved in a more productive and efficient way (Vernadat, 2002).

---

[74] This paper was developed in the framework of the INPREX Project (DPI2004-02594). This Project is partially funded by the CICYT of the Spanish Government.

The paper summarises the main topics and future trends that the authors envision in a near future of the area. This vision is stated from the authors' research background and experiences in European research projects, projects with companies and national projects.

- European projects: In ECOSELL (GRD1-2001-40692) and V-CHAIN (DPI2002-11149-E), tile, furniture, automotive and motorcycle, enterprises were analysed and where a lack of integration was detected among and within participating enterprises' processes, taking into account the management of their supply chains. UEML (IST-2001-34229) and INTEROP (IST-1-508011) are concerned with the (mainly inter-)integration of enterprise models and the alignment of these models with the information systems that support them, also including the analysis of enterprise architectures and ontologies and their impact when interoperating.
- *Projects with Firms*: mainly with SMEs, where solutions to process management and the development of their whole life-cycle were tackled. A main problem has been in these projects that there is a lack of tools that are customizable, accessible at a low cost, and easy to use and update..
- *National Projects*: where enterprise integration methodologies were developed, problems were identified, classified and possible solutions proposed.

2. LANDSCAPE

After analysing the state of the art of Enterprise Engineering and Enterprise Integration the authors have concluded that existing solutions can be classified according to two frameworks (see Figure 1). Each framework is composed of three components at least: Methodologies, Tools and Languages (cf ISO 15704:2000). Methodologies are a set of steps grouped in processes and phases that describe the actions that must be carried out to build up a business, from the conceptual idea to the operation of the enterprise. Methodologies may propose the use of different languages (modelling or implementation languages), according to the application domain, view, and phase within the life-cycle (Petit, M. *et al*, 2002), although it is not mandatory for a methodology to do so. In the same way, methodologies may refer to tools that could be used to carry out different phases (for example, tools for modelling business processes, or tools to implement the information systems that will execute the processes envisioned in the design phase). On the other hand, there is a need for enterprise engineering tools (cf ISO 15704:2000) that support different phases of the enterprise's construction and support different modelling languages, that permit integrating solutions at different levels of a life-cycle or that permit different integration approaches depending on the languages used.

2.1 Enterprise Engineering Framework

Our research center has long been working in this context. In 1999, (Ortiz, 1999a) proposed the IE-GIP framework embracing tools, methodologies and languages. That proposal defined a methodology aiming to cover the entire life-cycle of business entities. The methodology is based on the PERA proposal, and from the architectural point of view, the CIMOSA proposal was adopted whenever possible. On one hand, the life cycle concept of the PERA proposal and several aspects related with human teams, strategic approaches and master planning issues, have been adapted to the business process perspective of IE-GIP (for a description of the phases of the life-cycle see (Ortiz, 1999a)). On the other hand, CIMOSA plays a key role in the lower level phases from the Requirements Definition phase to the Implementation Description phase.(see Figure 2).

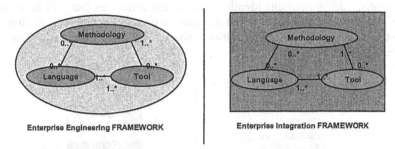

Enterprise Engineering FRAMEWORK   **Enterprise Integration FRAMEWORK**

Figure 1. Enterprise Engineering and Integration Frameworks

Finally, IE-GIP defined a computer tool called GIPMODEL (Modelling and Management of Integrated Processes, acronym in Spanish) aiming to give a computer-assisted modelling support to the application of the proposal. Furthermore, CILT (CIMOSA Learning Tool) and VR-CILT (Virtual Reality-CIMOSA Learning Tool) tools were developed (Ortiz, 1999b) to cover the conceptual aspects related with the CIMOSA proposal.

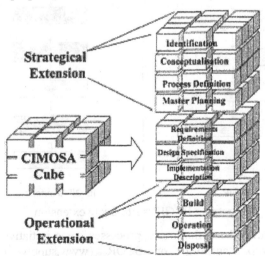

Figure 2. IE-GIP language extension

IE-GIP has been used in a series of projects with SMEs. From this expertise and from the new trends, approaches and technologies we extended IE-GIP's methodology in different ways. Now, it is emphasized the capabilities to automate the generation of software from the enterprise modelling, and to align the strategic, operational and IT levels to keep track, assure enterprise's objectives and make the company more agile against changes. IE-GIP's extension is being refined and tested in a national project called INPREX (Interoperability between Extended Enterprise

Processes, acronym in English) (see Figure 3 – dark boxes are new or refined phases).

Below, we provide only a further description of those IE-GIP phases that have been extended (darkest boxes in Figure 3):

- *Processes Identification:* Identification of the processes that will be considered as important to be analysed and improve in order to achieve the business goals defined at the conceptualization phase (for instance, customer orders management process).

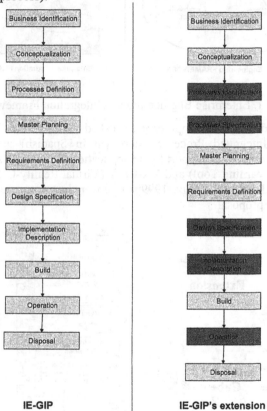

IE-GIP   IE-GIP's extension

Figure 3. IE-GIP's life-cycle extension

In order to standardise the definition of processes some initiatives have arisen as RosettaNet (www.rosettanet.org) or SCOR (www.supply-chain.org). These proposals also cover standard definition of business processes and specific activities that will be identified in the next phase (Process Specification Phase). The new technological proposals give a more appropriate support to this phase, as for example, the repositories and the enrichment of information with semantics. These repositories contain a formal description of the processes, with their associated semantics, and can be instantiated and parameterised. Examples of these repositories are the ebXML libraries (www.ebxml.org/specs). From an ontological point of view, processes can be described using formal languages such as Description Logic, Frame Logic, DAML or OWL; or semi-formal languages such as UML itself.

- *Processes Specification:* processes identified in the previous phases are analysed more deeply. Processes and activities are defined and some of the entity objects are identified. The diagrams are refined until it is enough knowledge in order to analyse the viability of the project at the Master Planning phase.

Enterprise Modelling is a way to express this processes specification. We encourage the use of BPMN (business process modelling notation) as the language to specify processes at an abstract level, although some principles defined in other existing standards (or standard proposals) must be also considered. The choice of BPMN is due to the capabilities to be mapped on BPML and after that, to be supported by some systems in order to run and simulate models.

Some of the previously stated standards are ISO TC184 SC5, WG1 (Business representation), ISO 14258 (concepts and rules for enterprise models) and ISO 15704 (methodologies requirements and enterprise reference architectures). Other standards are CEN TC310 WG1 for high level enterprise modelling and architectures, ENV 12204 for enterprise modelling constructs.

- *Processes Design*: The enterprise models are enriched and customized for facilitate their execution over platforms.. Processes are classified as executable directly on an IT platform (executable processes) or carried out by humans (manual processes).

Examples of tools able to execute processes in some sense are workflow management systems, systems able to execute in a distributed way business processes (e.g., Vitria®) or more recent business process execution systems such as n^3 from Intalio®. Business Process Management is living a great momentum, with the support of strong groups putting effort on it. For example, BPMI (Business Processes Management Initiative), WfMC (Workflow Management Coalition) and the OMG (Object Management Group.

- *Processes Implementation/Implantation*: The implementation of IT platform executable processes has been optimised in the IE-GIP extension.

We have taken profit of proposals such as MDA (model driven architecture) to achieve a tighter alignment and to generate (semi)automatically IT systems embedding the logic of business processes (see Figure 4). Further, in (Franco, D., 2003) we define derivation rules that allowing the automatic generation of OWL-S descriptions (semantic descriptions for web services) from enterprise models.

The use of MDA permits us to align Information System (IS) solutions with software requirements (Harmon, 2004), ensuring that software requirements are compliant with enterprise requirements.

The Model Driven Architecture (MDA, www.omg.org/mda) is a proposal of the Object Management Group (OMG) for the generation of software from models. The main idea of MDA is to deduce a model from other model until it is transformed into the code of the application, assuring the compliance between models at different abstraction levels.

MDA distinguishes among three kinds of models (Object Management Group, 2003): the computation-independent model (CIM), the platform independent model (PIM) and platform-specific model (PSM). CIM, also called a domain model, shows the environment in which the system will operate. PIM depicts the information system without considering specific platform/technology. PSM represents the model of the IS considering specific platform details. Finally, at the lowest level considers the code of the application in a specific platform.

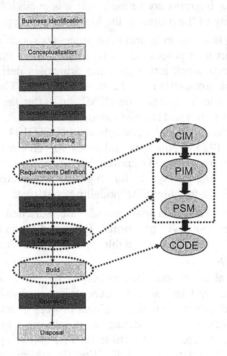

IE-GIP's extension

Figure 4. Phase of Implementation driven by the MDA

- *Operation*: The IT systems or formal procedures defined in the previous phase are executed. Thus, processes are transformed from a static state to a dynamic state and the execution of theses processes can be managed.

After running processes (automatically or manually), data can be gathered, mainly considering the key performance indicators associated with each process and that where defined in the processes definition phase. With this information, a deep analysis can be done (Business Performance Management) in order to improve processes cost or their execution time, etc; predictions can be made by means of data mining techniques. Thus, an analysis is necessary to check the achievement of current enterprise goals and strategies against the enterprise model/s and to propose new versions.

2.2 Enterprise Integration Framework

In order to cope with the global solutions demanded by customers, companies need to collaborate. Collaborative Networks Organisations (CNO) require an extreme exchange of flows (information & knowledge, material & services, and money (Ortiz A. *Et al*, 2003)), a strong support of information technologies and a big motivation of managerial staff in order to achieve integration solutions that provide the visibility and exchange of transactions necessary to do agile network of enterprises.

Interoperability is the action by means of which two or more active business items (applications, companies, departments within a company, etc) exchange events or flows of information (that is, control, data and decisional information flows, etc) to collaborate.

Enterprises can exchange entity objects at different levels (see Figure 5.). However, we have focused our attention to the exchange at business- and enterprise models level and the integration and collaboration of Enterprise Application.

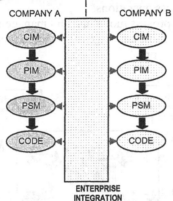

Figure 5. Multilevel Enterprise Integration

2.3 UEML

As we have previously mentioned, Enterprise Modelling is the art of externalizing the knowledge of an enterprise to be shareable. Thus, when companies collaborate on a network of enterprises (for instance, a supply chain), they need to exchange enterprise models with other firms. These models are represented in graphical or textual languages, and usually are represented in heterogeneous languages. This fact requires techniques combination and transformation of models across different languages and tools to achieve the required interoperability and integration. UEML (Berio *et al*, 2004) (Unified Enterprise Modelling Language) is an enterprise modelling language aiming to exchange enterprise models represented using different modelling languages. UEML v1.0 was developed at UEML project[75] funded by IST Programme of the European Commission 5th Framework.

UEML has been defined as an enterprise modelling language which constructs synthesize concepts appearing commonly in different enterprise modelling languages. In this way, it was found out that many of the building elements that enterprise modelling languages provide, although represented with different terms (syntax), they represent the same or a similar concept. Therefore, UEML appears as an intermediate language use to translate a models between different languages reducing the number of interfaces needed to exchange models in a network of enterprises compared with a peer-to-peer approach (see Figure 6).

Furthermore of the exchange capabilities offered by UEML, It also supports consistency of various model views, insofar as models representing different views of an enterprise (decisional, organizational, functional etc) using different languages

---

[75] UEML IST-2001-34229

can be put all together, and keeping links between these views by means of the intermediate relationship offered by UEML as a common connector (see Figure 6).

A list of current and real business problems in which a UEML can play a central role is (Jochem, 2003):

- Lack of Integration of information systems encoding fragmented non sharable enterprise knowledge.
- Shortage of coordination of business processes.
- Multiple views of business operations.
- Poor interoperability of process modelling and management tools.
- Insufficient coverage by most languages of required modelling views.
- Diverse visual representations.

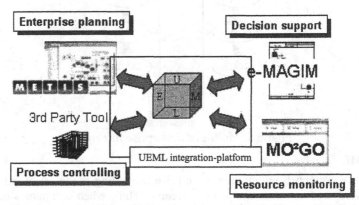

Figure 6. Translations of EMs by using a UEML (Berio, 2004)

2.4 Enterprise Application Integration

In a more technological level, there are multiple solutions to achieve a tool-to-tool integration (EAI, B2Bi or eHubs), however these solutions are very specific of the tools that interoperate and they are not very reusable. XML has become a standard to structure messages that enterprise applications exchange in order to communicate with one another.

Now, with the Service Oriented Approach, enterprise integration will be more affordable as far as a better encapsulation is available, and messaging a distribution of applications is well defined.

ATHENA is a European project trying to provide solutions at this level (Chen, 2003).

2.5 The General Picture

Figure 7 shows the research paths followed by the authors of this article. Only the main items are shown (those where most effort has been spent), with some smaller efforts omitted from the figure.

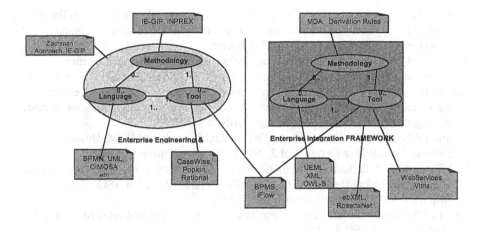

Figure 7. Research Summary

3. CONCLUSIONS

Despite many results generated in the enterprise integration / enterprise engineering field, problems still remain in the area. Some of these are listed below:

- Improper or poor use of methodologies and tools, when practicing enterprise engineering, frequently the models generated are of poor quality and become obsolete very fast.
- Low visibility of end-to-end processes and associated information within network of enterprises.
- Lack of alignment among strategic, operational and technological aspects of an enterprise, so enterprise can not manage properly changes necessaries to cover customer or market evolving requirements/needs.

The authors envisage that solutions to these problems (most of them commented at the previous ICEIMT (Kosanke *et al*, 2003)):

- Developing and disseminating easy-to-use and user-friendly enterprise modelling languages (mainly addressing problems of SMEs) as a means to exchange information between enterprises, but also within the enterprise.
- Following clear and intuitive methodologies that assure the quality of business documents and enterprise models. In this way, our extended methodology can be customized to specific needs and follows proposals and languages that permit the semi-automatic implementation of applications from enterprise models. Further, a better aligned of strategic, operational and IT levels is achieve. Thus, we can evaluate the IT solutions developed are contributing to the achievement of the business entity goals.
- Enforcing industry standards whenever possible.

REFERENCES

Berio,G, Anaya,V, Ortiz,A (2004) Supporting Enterprise Integration through a Unified Enterprise Modeling Language. In INTEROP-EMOI Proceeding. A Caise2004 Workshop. Vol 3. pp. 165-176. ISBN:9984-9767-3-4. Riga, Latvia.

Chen,D, Doumeingts,G (2003) European initiatives to develop interoperability of enterprise applications—basic concepts, framework and roadmap. Annual Reviews in Control 27, pp153–162. Elsevier.

ENV 12204 : Advanced Manufacturing Technology - Systems Architecture - Constructs for Enterprise Modelling, CEN TC 310/WG1, 1996

Franco,D, Anaya,V, Ortiz,A (2003) Automatic Derivation of DAML-S Service Specifications from UML Business Models. Lecture Notes in Computer Science. Springer-Verlang Berlin Heidelberg. ISBN: 0302-9743.

Harmon,P (2004) The OMG's Model Driven Architecture and BPM. Business Process Trends Newsletter. Vol 2, N° 5 May. www.bptrends.com

ISO/TC 184/SC 5. Industrial automation systems – concepts and rules for enterprise models. Technical Report ISO 14258, Web version WG1 N4, ISO, 1999. http://www.mel.nist.gov/sc5wg1/std-dft.htm.

ISO 14258:1998, Industrial automation systems -- Concepts and rules for enterprise models, TC 184/SC 5, 1998

ISO 15704:2000, Industrial automation systems -- Requirements for enterprise-reference architectures and methodologies, TC 184/SC 5, 2000.

Jochem,R (2003) Common Representation through UEML-Requirements and Approach. International Conference on Enterprise Integration and Modelling Technology. Enterprise Inter- and Intra-Organizational Integration. Kluwer, Valencia, Spain.

Kosanke, K., Jochem R., Nell, J., Ortiz, A. (2003) Enterprise Inter- and Intra-Organizational Integration. Building International Consensus. Kluwer Academic Publishers, ISBN: 1-4020-7277-5.

Ortiz,A, Franco,R.D, Alba,M (2003) V-CHAIN: Migrating from Extended to Virtual Enterprise within an Automotive Supply Chain. PROVE'03 Proceedings. Processes and Foundations for Virtual Organizations.

Ortiz,A, Lario,F, Ros,L (1999a) Enterprise Integration—Business Processes Integrated Management: a proposal for a methodology to develop Enterprise Integration Programs. Computer in Industry, 40, pp.155-171, Elsevier.

Ortiz, A., Lario, F., Ros, L. and Hawa, M. (1999b) Building a Production Planning Process using an Approach based on CIMOSA and Workflow Management Systems Computers in Industry. Vol. 40. pp. 207-219. Elsevier

Object Management Group, "MDA guide version 1.0", OMG, 2003.

Petit, M, *et al* (2002) D1.1: State of the Art in Enterprise Modelling, UEML-IST–2001-34229, www.ueml.org.

Vernadat, F.B (1996) Enterprise Modelling and Integration: principles and applications. Chapman & Hall, London.

Vernadat,F.B (2002) Enterprise Modelling and Integration (EMI): Current Status and Research Perspectives.

24.   How to Model Business Processes with GPN

Günter Schmidt and Oliver Braun

Department of Information and Technology
Management, Saarland University, PO Box 15 11 50
D-66041 Saarbrücken, Germany [gs\ob]@itm.uni-sb.de

Organizations today face increasing pressure to reduce time to market, i.e. to improve the design and the operations of business processes in terms of lead time and meeting due dates. Formal analysis using a mathematical graph-based approach can help to achieve this kind of improvement.

We will apply business graphs to scheduling workflows in terms of time-based optimization. We will concentrate on performance measures like completion time, flow time and tardiness. From a business process network we derive two types of directed graphs, one representing the task net (task graph) and the other one representing the resource net (resource graph). In the task graph a node is representing a task and its duration and arcs are representing different kinds of precedence constraints between tasks. The resource graph is similar to a Petri net and represents resource constraints and flows of jobs.

In order to compute optimal or near-optimal workflow schedules the algorithms have to relate to the structure of the business graphs. We will show that a variety of data structures commonly assumed in modern scheduling theory can be represented within the framework of business graphs. Based on these data structures specific scheduling algorithms to optimize time-based performance measures can be applied with the objective to reduce time to market.

1. INTRODUCTION

Modelling languages are required for building models in various application areas. We shall focus on the management of business processes which require the modelling of time-based activities for planning and scheduling purposes. A business process relates to a stepwise procedure for transforming some input into a desired output while consuming or otherwise utilising resources. Some general examples for business processes are: 'Product Development', 'Procurement', or 'Customer Order Fulfilment'; some more special examples would be 'Claims Processing' in insurance companies or 'Loan Processing' in banks. The output of a business process should always be some kind of achievement (goods or services) which is required by some customer. The customer might be either inside or outside the organisation where the process is carried out (Schmidt, 2002).

Two major aspects of business process management are planning and scheduling. Planning is concerned with determining the structure of a process before it is carried out the first time. Scheduling in turn is concerned with assigning resources over time to competing processes. Both planning and scheduling focus on dependencies among transformations within one process or between different processes. Malone and Crowston (Malone and Crowston, 1994) formulated the need to merge the paradigms of business process planning and business process scheduling concerning the management of dependencies among transformations. The reason is not only to increase the potential of applying results from planning and scheduling theory to the management of business processes but also to consider the relevance of problems arising from business process management for a theoretical analysis within these research areas.

Planning and scheduling require a specialised model of the business process. To build the required process model we base our analysis on Generalised Process Networks (GPN) (Schmidt, 1996), a graphical language related to CPM type of networks (Slowinski and Weglarz, 1998). We will show that GPN are expressive enough to formulate problems related to planning and scheduling of business processes within the same framework. Doing this we use a semi-formal presentation of the syntax and the semantics of GPN.

We start with a short discussion of business processes. Then we introduce a framework for systems modelling to define requirements for business process models. Based on this we describe the different graph models within GPN and discuss its application to business process planning and scheduling. Finally, we use an example to demonstrate the modelling capabilities of the approach.

2. WHAT IS A BUSINESS PROCESS?

A business process is a stepwise procedure for transforming some given input into some desired output. The transformation is time and resource consuming. A business process has some form of outcome, i.e. goods or services produced for one or more customers either outside or inside the enterprise. There are two usual meanings attached to the term 'business process'; a business process may mean a process *type* or a process *instance*.

The process type can be described by defining general rules and structure of a process; the process instance is a real process following the rules and structure of a given process type. A process type can be interpreted as a pattern; the behaviour of a corresponding instance matches with the pattern. A process type might be a pattern called 'Product Development', and the corresponding instance would be 'Development of Product X' carried out according to the pattern of 'Product Development'. In the sequel a process instance will also be referred to as a workflow or a job.

The process type is defined by its input and output, functions to be performed, and rules of synchronisation. The process *input* and *output* are related to tangible and intangible achievements. For example the major shop floor functions in production have as input different kinds of raw materials which are transformed into various types of output called processed material; office functions are mainly transforming data or information into new data or new information. In general input and output will consist of both material and information.

A *function* represents the transformation of some input into some output. Functions are related through *precedence relations* which constrain the possible ways a process can be executed. E.g. a precedence relation requires synchronisation if the output of a predecessor function is part of the input of the successor function. Before a function can be executed certain *pre-conditions* have to be fulfilled and after a function has been executed certain *post-conditions* should be fulfilled.

Starting and ending a function is caused by *events*. In general an event represents a point in time when certain *conditions* come about, i.e. the conditions hold from that time on until the next event occurs. Conditions related to events are described by values of attributes characterising the situation related to the occurrence of an event.

These event values are compared to pre-conditions and post-conditions of functions. Before carrying out some function its pre-conditions must match with the conditions related to its beginning event and after carrying out a function the conditions related to its ending event must match with the post-conditions of the function. *Synchronisation* means that there must be some order in which functions might be carried out over time; in its simplest form a predecessor-successor relationship has to be defined.

To fully determine a process type a number of variables related to the input and output of functions need to be fixed. The input variables define the *producer* who is responsible for the execution of a function, the required *resources*, and the *required data*; the output variables define the *product* generated by a function, the *customer* of the product, and the *data available* after a function is carried out.

Once a process type is defined its instances can be created. A process instance is performed according to the definition of the corresponding process type. The input, output, functions, and synchronisation of a process instance relate to some workflow or real job which has to be carried out in accordance with the regulations documented by the process type definition. The input must be available, the output must be required. Functions that make up a process type have to be instantiated. A function instance is called task. It is created at a point in time as a result of some event and is executed during a finite time interval.

To ensure task execution various scheduling decisions need to be taken considering the synchronisation and the resource allocation constraints as defined by the process type and resource availability. Scheduling process instances or workflows means to allocate all instances of different process types to the required resources over time. The process type represents constraints for the scheduling decision (Blazewicz *et al*, 1996). In terms of modern scheduling theory an instance of a business process is a job which consists of a set of precedence constrained tasks.

Additional attributes to tasks and jobs can be assigned (Schmidt, 1996b). Questions to be answered for process scheduling are: which task of which job should be executed by which resource and at what time? Typically, performance measures for business process instances are time-based and relate to flow time, tardiness or completion time of jobs; scheduling constraints are related to due dates or deadlines.

3. WHAT HAS TO BE MODELLED?

Modelling is a major component in planning and scheduling of business processes. A framework for systems modelling is given by an architecture. An architecture is based on the requirements for building models and defines the necessary views on a system. Many proposals of architectures have been developed and evaluated with the objective to find a generic enterprise reference architecture (Bernus *et al*, 2003).

An architecture which fits in such a framework is LISA (Schmidt, 1999). LISA differs between four views on models:

* the granularity of the model,

- the elements of the model and their relationships,
- the life cycle phase of modelling, and
- the purpose of the model.

According to granularity models for process types (planning) and for process instances (scheduling) have to be considered. Concerning the elements and their relationships models of business processes should represent all relevant inputs (data, resources) and outputs (data, products), the organisational environment (producer, customer), the functions to be carried out, and the synchronisation (events, conditions, dependencies). Referring to life cycle phases of systems different models are needed for analysis, design, and implementation. Finally, concerning the purpose of modelling we need models for the problem description and for the problem solution. The problem description states the objectives and constraints and the problem solution is a proposal how to meet them. Fig. 1 shows the different views to be represented by business process models in the framework of LISA.

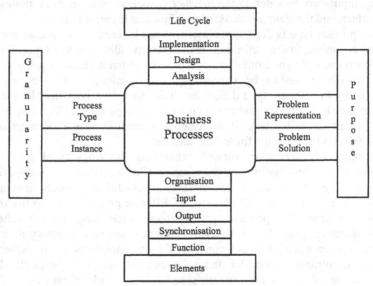

Figure 1. Views on business processes defined by LISA

We will concentrate here on the question how a model for the problem solution can be derived from a model for problem representation. The focus is scheduling of workflows. The modelling language is Generalized Process Networks.

4. GENERALISED PROCESS NETWORKS

There exist many modelling languages to describe business processes. Examples are Petri Nets (Petri, 1962), Data Flow Diagrams (DeMarco, 1978), Event Driven Process Chains (Keller *et al*, 1992), Workflow Nets (Aalst, 1998), Unified Modelling Language (Fowler and Scott, 1998), Dependency Graphs (Kumar and Zhao, 1999) and Metagraphs (Basu and Blanning, 2003). Most of these languages have been developed for planning purposes with a focus on the problem description. Models suited for scheduling purposes in particular for optimisation require a representation which is suited for combinatorial problem solving (Curtis *et al*, 1992). For this reason Generalized Process Networks (GPN) are developed.

The modelling language has to fulfil the following requirements:

- **Completeness and consistency:** all relevant views of a system must be covered and must be defined in a semantically consistent way,
- **Understandability:** the syntax and semantics must be easy to understand and easy to use by the target audience.

The relevant system views for business processes are modelled as defined in LISA. We shall differ between a model for a process type (used for planning) and a model for a process instance (used for scheduling). However, both models are build with the same language.

GPN = (E, F, A, O, I, L) is a directed And/Or graph with two sets of nodes E={e_1, e_2, ... , e_n} (events) and F={f_1, f_2, ... , f_m} (functions), a set of arcs A $\subseteq$ {E × F} $\cup$ {F × E}, sets of logical (AND, OR, XOR) output O $\subseteq$ {E × F} and input I $\subseteq$ {F × E} operators, and various sets of labels L assigned to events, functions, and arcs.

In a GPN graph events are represented by circles and functions are represented by boxes. Arcs are connecting events and functions or functions and events but never events and events or functions and functions. Events represent the dependencies in processing functions. We differ between six possible dependencies: three for beginning events and three for ending events.

- **begin-AND:** all functions triggered by this event have to be processed (default),
- **begin-OR:** at least one function triggered by this event has to be processed (arcs are connected by an ellipse),
- **begin-XOR:** one and only one function triggered by this event has to be processed (arcs are connected by an ellipse with a dot),
- **end-AND:** this event occurs only if all functions ending with this event have been processed (default),
- **end-OR:** this event occurs if at least one function ending with this event has been processed (arcs are connected by an ellipse),
- **end-XOR:** this event occurs if one and only one function ending with this event has been processed (arcs are connected by an ellipse with a dot).

In Fig. 2 a GPN graph is shown where seven events and seven functions are represented.

There are one **begin-XOR** related to event 1 and functions 12 and 14, one **end-XOR** related to functions 36 and 46 and event 6, and one **end-OR** related to functions 57 and 67 and event 7; all other logical input and output operators are of type **AND**.

The sets of labels L are related to six layers. The first layer defines the labels of the nodes and arcs (numbers), the second layer is dedicated to the functional specifications (name, time, cost, performance), the third to synchronisation aspects representing relationships between functions and events (value lists, pre- and post-conditions), the fourth to input and output data, the fifth to required resources and generated products, and the sixth layer describes the customer-producer relationship of a function. Labels related to the six layers are shown in Fig. 3.

Each function is described by the attributes function name, time and cost consumption which are related its execution, and certain performance measures under which the execution of the function is evaluated. Connected to each function is a list of pre-conditions and a list of post-conditions. The pre-conditions must be satisfied before the function can be carried out; post-conditions are satisfied as a result after performing the function. Additional labels may be assigned to the function:

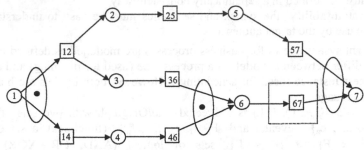

Figure 2. GPN example graph

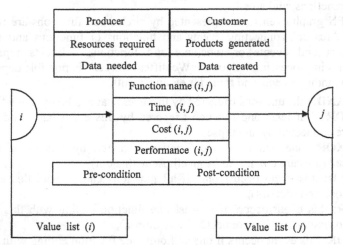

Producer	Customer
Resources required	Products generated
Data needed	Data created

Figure 3. Labels of GPN

- **Producer-Customer:** the producer is responsible for carrying out some function and the customer needs the results from this function. The inputs of the function are transformed under the responsibility of the producers, and the output of the function is consumed by the customers. A producer and a customer might be two distinct organisational units of the same enterprise.
- **Resource-Product:** resources are the physical inputs of the function, products are its physical outputs (resources required, products generated). Resources might be specific machines or employees with certain qualifications as well as material or incoming products to be processed. Products might be types of goods or services.
- **Data-Data:** input data represent the information required for performing a function and output data represent the information available after performing it (data needed, data created).

There are at least two events connected to each function; one represents its start and the other one its end. Events define constraints for the synchronisation of functions.

An event separating two functions represents the constraint that both functions can be executed only in a certain sequence. Functions which have no separating event can be performed in parallel. The occurrence of an event is a necessary condition to perform a function. Each event is described by a value list defining the environmental conditions represented by the event. The occurrence of an event is also a sufficient condition for performing a function if its value list meets the pre-conditions of the function adjacent to this event. An example of a GPN labeling schema is shown in Fig. 4.

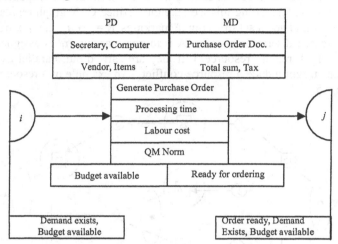

Figure 4. An example for GPN labeling

The function 'Generate Purchase Order' can be interpreted as an activity of a procurement process. Pre-conditions represent the assumption that there must be some 'Budget Available' for purchasing. The post-condition 'Ready for Ordering' which should be fulfilled after the function 'Generate Purchase Order' is processed. The meaning is that the purchase order is ready for sending out. Both conditions match with some values of the list of the beginning and the ending events. Data needed for preparing a purchase order are the 'Vendor' (address of vendor) and the 'Items' (list of items) to be purchased; data created are all purchase order related: 'Total sum' or 'Tax' (to be paid). Required resources might be a 'Secretary' and a 'Computer'; the product generated is a 'Purchase Order Document'. The manufacturing department 'MD' (the customer) asks the purchasing department 'PD' (the producer) to process the function 'Generate Purchase Order'. It can only be carried out if event i occured and not only demand for the items exist but also the budget is still available.

There are certain properties of a GPN related to a workflow:

- A workflow is *well structured, iff* all activities are represented by functions and events and there is only a single workflow instance for each situation.
- Redundancy measures the number of alternatives a workflow can follow.
- Each GPN representing a workflow has a critical path depending on the time consumption of the functions.

4.1 Resource Graph

In the process planning phase all required attributes of a business process are defined; their values are determined once an instance of a business process, i.e. a workflow is created. For example data for 'Vendor' or 'Items' might be 'Vendor ABC' and 'Item 123'. The emphasis of models for workflows is to find answers to scheduling questions, such as timing and resource allocation, taking into account competing process instances or workflows (jobs).

In order to model the resource setting we derive from the GPN graph a resource graph $RG = (GPN, R)$. Doing this we take the defined GPN graph representing the business process and add a set of resource markings $R = \{r_1, r_2, \dots r_k\}$ where each r_i represents a scarce discrete resource. A resource marking r_i is assigned to these functions f_j which require resource i if they are carried out. Parallel execution of these functions might induce a resource conflict. An example of a resource graph is shown in Fig. 5.

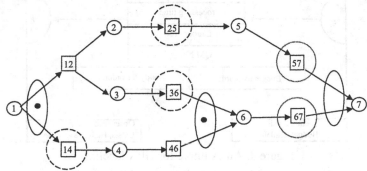

Figure 5. Resource graph example

There are two scarce resources which lead to markings of functions 14, 25 and 36 by r_1 and of functions 57 and 67 by r_2.

There are certain properties of a RG related to a workflow:

- A workflow is *resource restricted*, *iff* there is at least one resource marking in RG.
- A workflow is *resource constrained*, *iff* one resource can only be assigned to one function at any time.
- A workflow is *function constrained*, *iff* one function can only be assigned to one ressource at any time.

4.2 Task Graph

When it comes to operations there are many process instances or workflows (jobs) which have to be carried out in parallel. We will assume that these instances come from workflows which are not only resource restricted but also resource and function constrained.

To model these workflows we define a task graph $TG = (T, A_w, A_r, L)$. $T = \{T_{ij} \mid i=1,\dots, n; j=1, \dots, m\}$ is a set of tasks T_{ij} where index i relates to the function f_i of the workflow and index j relates to different workflows (jobs). $A_w \subseteq \{T_{kj}\} \times \{T_{lj}\}$ is a set of precedence arcs derived from the GPN (all chains of length two connecting predecessor-successor pairs). $A_r \subseteq \{T_{ik}\} \times \{T_{il}\}$ is a set of edges between tasks T_{ij} which cannot be executed in parallel due to resource constraints. L is a set of labels assigned to tasks and arcs.

In case two or more tasks cannot be processed simultaneously, a hyperedge is introduced between the corresponding tasks. Tasks associated with the same hyperedge create conflicts concerning the usage of resources. The scheduling decision has to resolve these conflicts such that a resource-feasible schedule can be generated (compare(Schmidt, 1989) and (Ecker *et al*, 1997)). To solve the problem all conflicting tasks have to be put in some sequence such that a resource-feasible schedule can be constructed. Algorithms to solve this kind of problem are given in (Ecker and Schmidt, 1993).

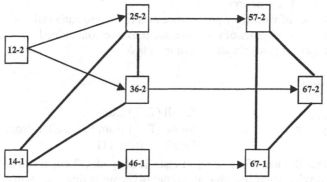

Figure 6. Task graph example

An example of a task graph with two workflow instances related to the GPN graph from Fig. 2 and the resource graph from Fig. 5 is shown in Fig. 6. Tasks in the task graph relate to functions in the GPN graph. The instance number is added to the function number. Resource markings r_1 and r_2 lead to two sets of edges $A_1 = \{(14\text{-}1,25\text{-}2), (25\text{-}2,36\text{-}2), (36\text{-}2,14\text{-}1)\}$ and $A_2 = \{(57\text{-}2,67\text{-}2), (67\text{-}2,67\text{-}1), (67\text{-}1,57\text{-}2)\}$.

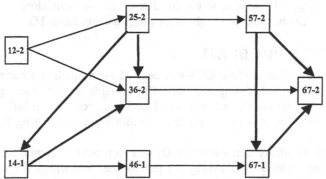

Figure 7. Scheduling solution related to task graph example

There are certain properties of a TG related to a workflow:

- TG is *acyclic, iff* each path build from A_w does contain each task node at most once.
- TG is *executable, iff* all paths build from A_w do contain each task node at most once.

4.3 Mathematical Programming

Once an acyclic task graph TG is set up the scheduling problem can be treated by means of mathematical programming. Although there is more than one alternative to formulate such a model for a scheduling problem we follow here the formulation given in (Adams *et al*, 1988).
Let

- the tasks from TG be numbered by $T_1, ..., T_{n+m}$
- p_i be the processing time of task T_i
- T_0 and T_{n+m+1} be two dummy tasks *start* and *end* with zero processing time
- P be the set of k resources
- A_w be the set of workflow precedence arcs related to each task pair (T_i, T_j)
- E_i be the set of all pairs of tasks that are resource constrained
- t_i be the earliest possible starting time of task T_i

minimize t_{n+m+1} (1)
    subject to

$t_j - t_i \geq p_i$	for all (T_i, T_j) from A_w	(2)
$t_j - t_i \geq p_i$ or $t_i - t_j \geq p_i$	for all (T_i, T_j) from E_i for all r_i from P	(3)
$t_i \geq 0$	for all T_i from TG	(4)

(2) ensures that the workflow order of tasks is obeyed; (3) ensures that there is only one task occupying each resource at a time; (4) ensures that each workflow instance is carried out.

The objective of the mathematical program (1)-(4) is related to the completion time of the last task of all workflow instances, i.e. the makespan. But also other time-based performance measures of workflow execution can be modelled. E.g., in case flow time is the objective we have to minimize $(t_{n+j} + p_{n+j} - t_0)$ with t_{n+j} as the earliest possible starting time of the last task n of workflow j and p_{n+j} as its processing time; if tardiness is a criterium we have to minimize $MAX\{0, t_{n+j} + p_{n+j} - d_j\}$ with d_j as the due date of the j -th workflow.

In Fig. 7 a feasible solution is given showing an executable TG.

5. EXAMPLE PROBLEM

We shall now demonstrate how GPN can be used for modelling a business process for planning and for scheduling purposes. The example is related to a procurement process. It deals with purchasing goods and paying corresponding bills. Let us start to explain how to build a model on the planning level considering the following setting.

If the manufacturing department (MD) of a company is running out of safety stock for some material it is asking the purchasing department (PD) to order an appropriate amount of items. PD fills in a purchase order and transmits it by E-mail or fax to the vendor; a copy of the confirmed purchase order is passed to the accounts payable department (APD). The vendor is sending the goods together with the receiving document to the ordering company; with separate mail the invoice is also sent.

Once the invoice arrives PD compares it with the purchase order and the goods sent via the receiving document. The documents are checked for completeness and for correctness. If the delivery is approved APD will pay the bill; if not PD complains to the vendor. Invoices for purchased goods come in regularly and have to be processed appropriately.

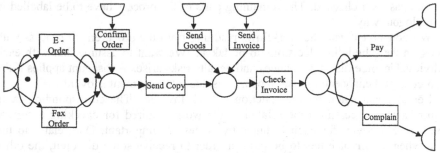

Figure 8. Procurement process

This process is shown in Fig. 8. Arcs leading from left to right represent the functions of the purchaser's process and arcs leading from the top to the bottom represent functions of the vendor's process. The purchasing order can be sent either by e-mail or by fax. This is represented by the two functions 'E-Order' and 'Fax Order'. Once the order is confirmed by the vendor a copy of the order is sent to APD represented by the function 'Send Copy'. If the ordered goods and the corresponding invoice have arrived the function 'Check Invoice' can be carried out. Depending on the outcome of the checking procedure the functions 'Pay' or 'Complain' are performed. In case there are complaints only about parts of the delivery both functions are carried out.

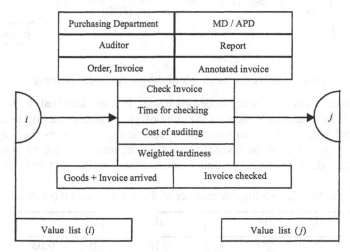

Figure 9. GPN representation of a selected function

In Fig. 8 the labels for most of the layers were omitted. In order to give a small example how labelling is done we concentrate on the function 'Check Invoice' using all six GPN layers. The result is shown in Fig. 9. We assume that PD is taking over the responsibility for this function and MD and APD need the results. The resource needed is an auditor who is generating a report. Data needed for the 'Check Invoice' function are the order and the invoice data; the function creates 'Annotated Invoice' data. Before the function can be carried out the ordered goods and the invoice should have arrived; after carrying out the function the condition holds that the invoice has been checked. The remaining parts of the process have to be labelled in an analogous way.

We now investigate the workflows resulting from the procurement process with focus on the scheduling decisions to be taken. We want to assume that with each individual invoice discount chances and penalty risks arise. A discount applies if the invoice is paid early enough and a penalty is due if the payment is overdue.

Let us focus again on the function 'Check Invoice'. The corresponding tasks require some processing time related to the work required for checking a current invoice. Moreover, for each instance two dates are important. One relates to the time when the invoice has to be paid in order to receive some discount, the other relates to the time after which some additional penalty has to be paid. For the ease of the discussion we assume that discount and penalty rates are the same. Let us furthermore assume that there is only one auditor available to perform these tasks and that the auditor is the only constrained resource of the procurement process. The resource graph is shown in Fig. 10.

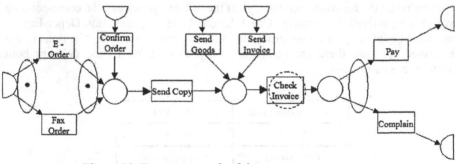

Figure 10. Resource graph of the procurement process

There are three invoices waiting to be processed. It is obvious that the sequence of processing is of major influence on the time of payment considering discount and penalty possibilities. Table I summarises the scheduling parameters showing invoice number (J_j), total sum of the invoice (w_j), time required to check an invoice (p_j), discount date (dd_j), penalty date (pd_j), and the rate for discount and penalty (r_j), respectively.

Table I   Scheduling parameters of the procurement workflow

J_j	w_j	p_j	dd_j	pd_j	r_j
J_1	200	5	10	20	0.05
J_2	400	6	10	20	0.05
J_3	400	5	10	15	0.05

In general there are n invoices with n! possibilities to process them using a single resource. The range of the results for the example data is from net savings of 30 units of cash discount up to paying additional 10 units of penalty depending on the sequence of processing. The task graph is shown in Fig. 11.

The data required for scheduling relate to the processing times p_j, the amount of the invoice w_j, the discount and penalty rates r_j, the discount dates dd_j, and penalty dates pd_j; the scheduling objective is assumed to be to maximise the sum of cash discount minus the penalty to be paid. The mathematical program (1)-(4) has to be revised in an analogous way. An optimal schedule is represented in Fig. 12.

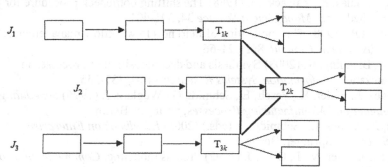

Figure 11. Task graph of the procurement process

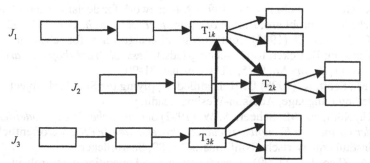

Figure 12. Scheduling solution of the procurement workflow

6. CONCLUSIONS

We have presented business graphs to describe business processes and to schedule corresponding workflows within a single model. Besides GPN resource graphs and task graphs are used as a formalism to create appropriate data structures for the application of mathematical programming formulations. The approach has the capabilities to structure problems from a descriptive point of view and to optimise workflows based on time-based criteria. It is easy to understand and easy to use, and it is especially suited for modelling sequence dependent decision problems having a combinatorial structure.

We have not presented algorithms to solve the arising scheduling problems. The scope of this contribution is to demonstrate that planning and scheduling problems can be modelled using a common and easy to use notational framework. We have illustrated this by an example. There are many business processes which can be analysed and optimised using the notational framework of business graphs.

There are promising areas for future research. One is to develop business graph-based workflow optimisation tools. Another one is a theoretical analysis of various workflows with respect to models investigated in modern scheduling theory. A third area might be the integration of business graphs in commercially available business process modelling tools.

REFERENCES

Adams, J., Balas, E., Zawack, D. (1988) The shifting bottleneck procedure for job shop scheduling, *Management Science* 34, 391-401

Aalst, W.M.P. (1998) The application of Petri nets to workflow management, *J. Circuits Systems Comput.* 8(1), 21-66

Basu, A., Blanning, R. (2003) Synthesis and decomposition of processes in organizations, *Information Systems Research* 14(4), 337-355

Blazewicz, J., Ecker, K., Pesch, E., Schmidt, G., Weglarz, J. (1996) *Scheduling Computer and Manufacturing Processes*, Springer, Berlin

Bernus, P., Nemes, L., Schmidt, G. (eds.) (2003) *Handbook on Enterprise Architecture*, Springer, Berlin

Curtis, B., Kellner, M. I., Over, J. (1992) Process modeling, *Communications of the ACM* 35(9), 75-90

DeMarco, T. (1978) Structured Analysis and System Specification, New York

Ecker, K., Gupta, J., Schmidt, G. (1997) A framework for decision support systems for scheduling problems, *Eur. J. of Operational Research*, 101, 452-462

Ecker, K., Schmidt, G. (1993) Conflict resolution algorithms for scheduling problems, in: K. Ecker, R. Hirschberg (eds.), *Lessach Workshop on Parallel Processing, Report No.* 93/5, TU Clausthal, 81-90

Fowler, M., Scott, K. (1998) UML Distilled: Applying the Standard Object Modelling Language, Addison-Wesley, Reading

Keller, G., Nüttgens, M., Scheer, A.-W. (1992) *Semantische Prozeßmodellierung auf der Grundlage Ereignisgesteuerter Prozeßketten (EPK)*, Veröffentlichungen des Instituts für Wirtschaftsinformatik Nr. 89, Saarbrücken

Kumar, A., Zhao, J. L. (1999) Dynamic routing and operational controls in workflow management systems, *Management Science* 45(2), 253-272

Malone, T. W., Crowston, K. (1994) The interdisciplinary study of coordination, *ACM Computing Surveys* 26(1), 87-119

Petri, C.A. (1962) Kommunikation mit Automaten, *Schriften des Instituts für Instrumentelle Mathematik* Nr. 3, Bonn

Schmidt, G. (1989) Constraint-satisfaction problems in project scheduling, in: Slowinski, R., and Weglarz, J. (1989), 135-150

Schmidt, G. (1996) Scheduling models for workflow management, in: B. Scholz-Reiter, E. Stickel (eds.), *Business Process Modelling*, Springer, 67-80

Schmidt, G. (1996b) Modelling production scheduling systems, *Int. J. Production Economics* 46-47, 109-118

Schmidt, G. (1999) Informationsmanagement - Modelle, Methoden, Techniken, Springer, Berlin

Schmidt, G. (2002) *Prozeßmanagement - Modelle und Methoden*, Springer, Berlin

Slowinski, R., Weglarz, J. (eds.) (1989) *Recent Advances in Project Scheduling*, Elsevier, Amsterdam

25. Enterprise Integration and Networking: Issues, Trends and Vision

Arturo Molina[1], David Chen[2], Hervé Panetto[3], Francois Vernadat[4] and
Larry Whitman[5]
IFAC TC 5.3 Enterprise Integration and Networking
1 CSIM-ITESM, armolina@itesm.mx
2 University Bordeaux 1, chen@lap.u-bordeaux1.fr
3 Université Henri Poincaré Nancy I, Herve.Panetto@cran.uhp-nancy.fr
4 Eurostat, European Commission, Francois.VERNADAT@cec.eu.int
5 Wichita State University, larry.whitman@wichita.edu

Enterprise Integration and Networking has been the topic of extensive research. Achievements deal with theoretical definition of reference models and architectures, modeling languages and tools, and development of relevant standards. The impact on today business has somehow been limited; therefore a revision of relevant issues and trends is required to establish a coherent vision for future research. This paper summarizes the underlying principles and challenges for enterprise modeling and integration, and its impact on enterprise networking.

1. ENTERPRISE INTEGRATION AND NETWORKING CONTEXT

The results of a study carried out in the United States for establishing the visionary manufacturing challenges for 2020 defined six grand challenges for manufacturers that represent gaps between current practices and the vision of manufacturing 2020 (NRC, 1998). These challenges are summarized in Table 1. In addition to these challenges, the Next Generation Manufacturing Systems (NGMS) will be more strongly time-oriented while still focusing on cost and quality. A NGMS should satisfy the following fundamental requirements:

- Enterprise integration and interoperability
- Distributed organization
- Model-based monitor and control
- Heterogeneous environments
- Open and dynamic structure
- Cooperation
- Integration of humans with software and hardware
- Agility, scalability and fault tolerance.

The requirements expressed above are related to the concept of the Networked Enterprise, be it an Extended, Fractal, Holonic or Virtual Enterprise (Camarinha-Matos et al., 1998) This is an emerging paradigm that results from the rapidly changing business environment forcing the complete supply chain from customers to suppliers to work in a more tightly-coupled mode. The Networked Enterprise relies to a large extent on Enterprise Integration and Enterprise Modeling techniques as defined in Vernadat (1996), Fox and Gruninger (1998) and Weston (1993).

Table 1. Manufacturing Challenges and Enterprise Integration proposals

CHALLENGES	BUSINESS	KNOWLEDGE	APPLICATIONS	COMMUNICATIONS (ICT)
Grand Challenge 1. Achieve concurrency in all operations	• Business and strategy models • Evaluation tools for decision making • Formalisms for modeling concurrency operations	• Knowledge about business processes and operations (functions, information, organization and resources) • Knowledge about core competencies (resources based view) • Knowledge based simulation	• Software to simulate operation to see parallelism and concurrency • Standards • Tools for monitoring and control of parallelism and concurrency	• Standards • Reliable communication networks
Challenge 2. Integrate human and technical resources to enhance workforce performance and satisfaction.	• Enterprise measurement systems (e.g. Balanced Score Card) • Enterprise trust systems • Compensation systems based on enterprise performance measures	• Description of Skills, Core Competencies, Organization roles and Knowledge assets • On line resources availability and capacity • Balanced automatic vs. manual tasks	• Integration of Enterprise Applications (ERP,MES, SCADA, Factory Automation Systems) • Workflow management systems (WfMS) • Computer Supported Cooperative Work (CSCM)	• Open platforms and architectures • Human Computer Interaction applications • Friendly User Interfaces
Challenge 3. "Instantaneously" transform information gathered from a vast array of diverse sources into useful knowledge for making effective decisions.	• Integration of business information • Networked enterprises • Ontologies • Consistent enterprisewide decision-making structure	• Interoperability of models • Standards (KIF, KQML) • Shared Ontologies • Explicit knowledge models • knowledge management system	• Standards • Interfaces • Interoperable databases • Data WarehouseMining • Decision support software	• Standards • Interfaces • Interoperability
Grand Challenge 4. Reduce production waste and product environmental impact to "near zero."	• Standards • International Regulations • Optimization Models • Sustainable development models • lean and clean production paradigms	• Reference models • Total quality control • Total quality maintenance • Design innovation • Use of new material & technology • Design for recycling	• Operations control software, Monitoring software • TQC support software • Preventive maintenance planning system • CAD software supporting innovation	• Operations control hardware • Benchmarking and Performance measures systems
Grand Challenge 5. Reconfigure manufacturing enterprises rapidly in response to changing needs and opportunities.	• Reference Models and architectures • New manufacturing paradigms (holonic manufacturing, intelligent agents, decentralized and autonomous production cells/unit)	• Reference models • Ontologies • Polyvalence of human operators • Model-based manufacturing and control • Modular design skills	• Component based enterprise applications • Modular and Reconfigurable systems • Components based software solutions (Plug in/Plug out) • Simulation software • Standard	• Interpretability • Standards
Grand Challenge 6. Develop innovative manufacturing processes and products with a focus on decreasing dimensional scale.	• Make/Buy Strategies, Reference Models (ATO, CTO, BTO) • Virtual enterprise paradigm • Decentralized decisionmaking and systems	• Integrated Product and Process Development • Product Life Cycle Management • Innovative design theories • CAD/CAM integration	• Product Data Management / Product Life Cycle Management • Advanced Planning Systems • Supply Chain Management • Logistic Management Systems • Knowledge Based Engineering	• Wireless local area network • Web-based applications • SmartCards applications • Nanotechnologyand MEMS

Enterprise Integration and Modeling (EIM) enable an enterprise to share key data/information/knowledge in order to achieve business process coordination and cooperative decision-making, and therefore Enterprise Integration (Chen *et al.*, 2001; Morel *et al.*, 2003). Thus, there is a need for better process management and for more integration within individual enterprises and among networks of enterprises

(Molina and Medina, 2003). The integration concept of providing quickly the right information at the right place at the right time under the right format throughout the enterprise is therefore evolving. Enterprise Integration now concerns (Vernadat, 2002; Whitman *et al.*, 2001):

- Efficient business process management, integration and coordination;
- Enterprise-wide consistent decision-making;
- Team collaboration supported by Computer supported collaborative work (CSCW) for concurrent design and engineering activities;
- Increased flexibility throughout the company;
- Product life cycle management throughout the existence of a product;
- Interoperability of IT solutions, systems and people to face environment variability in a cost-effective way.

Recent advances in information and communication technologies have allowed manufacturing enterprises to move from highly data-driven environments to a more cooperative information/knowledge-driven environment. Enterprise knowledge sharing (know-how), common best practices use, and open source/web based applications are enabling to achieve the concept of integrated enterprise and hence the implementation of networked enterprises.

2. NEEDS FOR ENTERPRISE INTEGRATION

The question to answer is how Enterprise Integration and Modeling can deal with the technological challenges that allow an enterprise to face global competition and fluctuating market conditions. Using a reference model for Enterprise Integration, the contributions of the research area of Enterprise Integration and Networking can be classified into: Business, Knowledge, Application and Communications. Table 1 summarizes how the different challenges can tackle the issues faced by next generation manufacturing systems.

2.1 Physical system integration, application integration, and business/knowledge integration

The literature has reported that different forms of integration have emerged over the last decades. These being (Chen and Vernadat, 2004):

- Physical system integration (ICT),
- Application integration, and
- Business/Knowledge integration.

Physical system integration (Information and Communication Technologies) essentially concerns systems communication, i.e. interconnection and data exchange by means of computer networks and communications protocols. Physical system integration dates back to the early 1970's and is still evolving. Work done has first concerned the 7-layer OSI/ISO standard definition, and then the development of specialized manufacturing and office automation protocols such as MAP, TOP, and field-buses. It now continues with developments on ATM, fast Ethernet, Internet and web services, SOAP (Simple Object Access Protocol), or RosettaNet. Message queueing systems (such as IBM's MQ Series) and message-oriented middleware (MOM) are important corporate components of the basic infrastructure at this level.

Application integration concerns interoperability of applications on heterogeneous platforms. This type of integration allows access to shared data by the various remote applications. Distributed processing environments, common services for the execution environment, application program interfaces (API's), and standard data exchange formats are necessary at this level to build cooperative systems. Application integration started in the mid 1980's and is still on-going with very active work concerning STEP, EDI, HTML, XML, or eb-XML for the exchange of common shared data, development of common services for open systems around the web (web-services), integration platforms for interoperable applications in distributed environments (e.g. OSF/DCE, OMG/CORBA, WSDL, and more recently J2EE or Java to Enterprise Edition environments and .NET). Other tools used at this level are workflow management systems (WfMS) and computer support to collaborative work (CSCW).

Business/Knowledge integration relates to the integration at the corporate level, i.e. business process coordination, collaboration, knowledge sharing, and consistent enterprise-wide decision-making. This mostly concerns enterprise interoperability and requires externalizing enterprise knowledge to precisely model business operating rules and behavior. Early work has only been pursued by major programs financed by governments such as the ICAM and IPAD programs. More recently, the CALS Initiative and the Enterprise Integration Program (EIP) in the United States, as well as CIMOSA by the ESPRIT Consortium AMICE, GRAI decisional approach by LAP/GRAI of University of Bordeaux, AIT Initiative or the IST program of EU in Europe plus the Globeman Project of the IMS program investigated the issue.

2.2 Business integration: towards the Networked Enterprise

Enterprise Integration can be approached from five different perspectives, or levels, as shown in Table 2. At the sub-enterprise level, the functionality of the integrated application or system is limited to a relatively homogeneous area, typically a single local site under a single ownership. For example, flexible manufacturing systems are at the integrated sub-enterprise level. Complete functional integration at the single-site enterprise level assures that business processes, manufacturing processes and product realization are united using a common architecture to fulfill a common goal. This is most likely for a single plant under single ownership, such as an automated factory.

The next three levels of EI – *multi-site, extended,* and *virtual* – occur over multiple geographic settings. Multi-Site enterprise integration is generally an issue faced by large enterprises (e.g., Boeing, IBM, General Motors, and EADS) in integrating heterogeneous systems throughout their facilities. An extended enterprise, which generally involves complex supply chains, concerns the integration of all members of the supplier and distribution chain to the common goal of market share capture through product realization. Virtual enterprises are very similar to extended enterprises, but they have the feature of being created and dissolved dynamically on a as-needed basis, and integration of member entities is largely electronic (Browne and Zhang 1999). All levels, to varying degrees, influence and are influenced by integrated product realization, integrated business systems, and tools enabling integration. While the objective is to support creation and operation of extremely efficient, flexible, and responsive extended manufacturing enterprises, the path to reach this will require capturing the wisdom achieved at each of the enterprise integration levels (Panetto *et al.*, 2004).

Table 2. Levels of Enterprise Integration

Level of Integration	Functionality	Geographic	Ownership	Homogeneity of Functional Systems	Stage of Maturity
Sub-Enterprise	Limited	Local	Single Owner	Homogeneous	State of Industry
Single-Site Enterprise	Complete	Local	Single Owner	Homogeneous	Leading Edge
Multi-Site Enterprise	Complete	Distributed	Single or Multi-Owner	Mixed	Leading Edge
Extended Enterprise	Static complete	Distributed	Multi-Owner	No, but may be mixed in some functions	Leading Edge
Virtual Extended Enterprise	Dynamic Complete	Global	Multi-Owner	No	Limited in 2004 More pervasive in 2015

3. ENTERPRISE MODELING AS THE MEANS TO ACHIEVE INTEGRATION

Collaboration and coordination between people, applications, and computer systems require models that are shared among all the actors in a cooperative environment. Therefore, enterprise models are a must in achieving enterprise integration (Chen *et al.*, 2002a; Whitman and Huff, 2001, Vernadat, 1996).

A core concept in enterprise modeling is the business process that encapsulates all the key elements of the enterprise, i.e. activities or functions, data/information/knowledge, human and technological resources. When a business model is created a representation of people interactions, roles and responsibilities, data/information exchange, resources required to execute certain activities, and procedures/instructions used to control functions are described in detail. The enterprise model is used as a semantic unification mechanism, or knowledge mapping mechanism, built by applying principles and tools of a given enterprise modeling method. Semantic concept definitions in the model can be expressed in the form of ontology, i.e. using a shared neutral knowledge representation format. The obtained enterprise model is also a means to represent shared concepts at a high level of abstraction and to capture stakeholders requirements (Panetto, 2001; Vernadat, 2002; Panetto *et al.*, 2003).

The aim of Enterprise Modeling is to provide:

- A visualization of enterprise knowledge to better understand how an enterprise is structured and how operates;

- Support change management using an enterprise engineering approach supported by structured analysis, rigorous design methods, simulation tools, and systematic decision-making; and
- A model used to control and monitor enterprise operations.

The main motivations for Enterprise Modeling are:

- Understanding how an enterprise is structured and behaves in order to manage system complexity,
- Capitalization of enterprise knowledge (know-what, know-how and know-why),
- Business process management based on enterprise engineering concepts,
- Improved change management in all types of processes,
- Achievement of enterprise integration and interoperability

The aims of Enterprise Integration are:

- To assist in the fulfillment of enterprise goals through the alignment of strategies, competencies and technologies in a company,
- To support the execution of collaborative, concurrently and distributed business processes,
- To enable business communication and coordination among various organizational entities of the extended enterprise,
- To facilitate knowledge sharing and information exchange between people and applications,
- To provide interoperability among heterogeneous, remotely located, independent vendor applications and Information and Communication Technologies (ICT).

4. TRENDS AND FUTURE VISION

4.1 Concerning Enterprise Modeling and Reference Models

New advances in Enterprise Engineering methods as well as a strong need to progress towards Enterprise Integration call for efficient enterprise modeling languages and advanced computer-based tools. Enterprise modeling is concerned with representation and analysis methods for design engineering and automation of enterprise operations at various levels of detail (e.g. coarse modeling, re-engineering, detailed design and analysis, performance evaluation, etc.).

The following methods or architectures for enterprise modeling are considered key in this evolution: IDEFX (Whitman *et al.*, 1997), GRAI-GIM (GRAI Integrated Methodology) (Doumeingts and Vallespir, 1995), CIMOSA (CIM Open System Architecture) (Berio and Vernadat, 1999), PERA (Purdue Enterprise Reference Architecture) (Williams, 1994), GERAM (Generalized Enterprise Reference Architecture and Methodology) (IFAC-IFIP Task Force, 1999; Bernus and Nemes, 1996).

Various methods and modeling techniques have been proposed over the last decade to cover different aspects of enterprise modeling, e.g., ARIS ToolSet, BONAPART, CimTool, FirstSTEP, IDEF methods, IEM, IBM's FlowMark, IMAGIM, METIS, PrimeObject and PROPLAN, to name a few.

The interoperability between these methodologies and tools is low. Most of the enterprises modeling tools are just graphical model editors. There are no model analysis functions and rules built in these tools. The proliferation of fancy and non inter-operable EM tools on the marketplace has created a prejudicial Tower of Babel situation. Unified languages are therefore proposed as consensus such as PSL (Process Specification Language) for manufacturing processes, supported by NIST in the US, or UEML (Unified Enterprise Modeling Language) for business processes, supported by EU (Panetto, 2002; Panetto, 2004; Vernadat, 2002). Also, while EM is widely used for well-structured processes, few EM tools deal with semi-structured or non-structured processes. Further developments are necessary in this area to better take into account human and organizational aspects. It is important to mention that Enterprise Modeling largely remains a concept or is even completely ignored by most SMEs (Small and Medium Enterprises). Various efforts are underway by standardization groups to propose standards (CEN/ISO WD 19440, ISO DIS 14258, ISO CD 15704, ODP Enterprise language and OMG ManTIS Task Force). Future trend in this area is to continue to develop UEML and to accelerate the standardization effort. For instance, the ISO TC184 SC5/WG1 is considering launching a new work item on process modeling language.

Concerning the reference architectures, literature survey shows that there are two types of architectures. Type 1 describes an architecture or physical structure of some component or part of the integrated system such as the computer system or the communications system. Type 2 presents an architecture or structure of the project which develops the physical integration, i.e., those that illustrate the life cycle of the project developing the integrated enterprise. Today, the architecture concept is not sufficiently exploited. One of the reasons is the lack of proper architecture representation formalism supporting significant characterization of features and properties of enterprise systems. Furthermore, existing architecture principles were not developed to a satisfactory level to allow bringing significant improvement to enterprise architecting. Further research is needed in this area.

Regarding reference models and in the area of standardization, some partial approaches can be mentioned. For examples, ISO 15531 MANDATE is a reference model focusing on information and resource views of manufacturing domain; the IEC 62264 series standard is a reference model on production management and control focusing on the information flow between the control domain and the rest of the enterprise. All these approaches are still on-going works and not mature. More recently, a European Technical Specification (CEN TS 14818: Decisional Reference Model) has been approved. It is based on the GRAI approach and shows a basic decision-making structure defined at high level abstraction.

4.2 Concerning Enterprise and Processes Models Interoperability

A significant initiative to develop interoperability between process models is ISO CD 18629 - Process Specification Language (PSL). In PSL a formal semantic approach (called PSL ontology) is used. However, important efforts are still needed to get effective implementation in industry. Another relevant initiative is the standard dealing with manufacturing software capability profiling (ISO 16100) carried out by ISO TC184/SC5/WG4.

The standard IEC/ISO 62264 (2002) defines models and establishes terminology (semantics) for defining the interfaces between an enterprise's business systems and its manufacturing control systems. It describes in a rather detailed way the relevant functions in the enterprise and the control domain and the objects normally exchanged between these domains. It is becoming the accepted model for B2M integration and interoperability.

To meet new industrial challenges, there is a shift from the paradigm of total integration to that of interoperation. Relevant standardization activity focusing on interoperability is just starting and most of work remains to be done in the future.

In order to reach a broad consensus for model information exchange between enterprise modeling tools, the UEML project (Panetto *et al.*, 2003a; Vernadat, 2002) has defined an initial set of generic constructs with the aim of achieving interoperability between them. In recent years, one of the most notable research efforts has been directed to improvement of interoperability (mainly software interoperability), a critical success factor for enterprises striving to become more flexible and to reduce the effort required to establish and sustain cooperation. Software interoperability has been especially addressed by specific software markets such as EAI and XML based solutions. However, these solutions mostly focus on compatibility of distinct formats without looking at the so-called modeling domain, i.e., the domain stating the rationale behind the software and providing reasons for building software. Information about the modeling domain, without taking into account any software issues, is essential to achieving greater interoperability. It is likely to be really difficult or even impossible to understand and recover this kind of information from software. As a consequence, this information should be associated with the software from the beginning and should be continuously maintained.

UEML could solve the issue of horizontal interoperability at the enterprise level. Thus, as information is controlled at the Automation level, it should need to be defined through a vertical interoperability approach from the product that produces it through the Manufacturing Execution System that consolidates it to the Enterprise Business Processes that use it. Standards such as the IEC/ISO 62264 together with the IEC 61499 function block draft standard for distributed industrial-process measurement and control systems could partially solve the vertical interoperability problem from the Business to the Manufacturing levels.

Consequently, as a prerequisite to building such a vertical information system dealing with physical process constraints, the TC5.3 UEML working group is aiming at defining and formalizing a practical and pragmatic language that should serve as a pivotal language ensuring a common understanding of the product information along its whole life cycle (Panetto *et al.*, 2003b). At the European level, the INTEROP network of excellence will further develop UEML v1.0 and deliver a extended UEML specification v2.0. ATHENA Integrated Project and in particular project A1 will use UEML 1.0 as a baseline to develop a set of modeling constructs for collaborative enterprise. Applying AUTO-ID (Morel *et al.*, 2003), that information can be embedded in physical objects according to the HMS (Holonic Manufacturing System) paradigm, in order to ensure the traceability of customized products, goods for manufacturing issues and services for logistics issues. Such a holonic approach requires aggregating separate object views and constructs of the IEC/ISO 62264 standard in order to define the relevant holons field, lack of established standards, which sometimes happen after the fact, and the rapid and unstable growth of the basic technology with a lack of commonly supported global strategy.

4.3 Concerning Enterprise Integration of Applications and ICT

Enterprise Integration (EI) is also becoming a reality for many companies, especially networked companies or enterprises involved in large supply-chains (extended or virtual enterprises). Some major projects for enterprise integration have been conducted in Europe (AIT Initiative) or in the US (EIF, NIIIP, NGM). The problem is that EI is both an organizational problem as well as a technological problem. The organizational problem is still partially understood so far. The technological problem has been the focus of major advances over the last decade, mostly concerning computer communications technology, data exchange formats, distributed databases, object technology, Internet, object request brokers (ORB such as OMG/CORBA), distributed computing environments (such as OSF/DCE and MS DCOM), and now J2EE (Java to Enterprise Edition and Execution Environments), .NET, and Web services. Some important projects having developed integrating infrastructure (IIS) technology for manufacturing environments include (Weston, 1993; Goranson *et al.*, 2002; Morel *et al.*, 2003):

- CIMOSA IIS: Integrating Infrastructure of CIMOSA
- AIT IP: This is an integration platform developed as an AIT project and based on the CIMOSA IIS concepts
- OPAL: This is also an AIT project that has proved that EI can be achieved in design and manufacturing environments using existing ICT solutions
- NIIIP (National Industrial Information Infrastructure Protocols)

More recently, research programs such as the ATHENA Integrated Project and the INTEROP Network of Excellence funded by E.C. in Europe look for innovative solutions regarding interoperability between legacy systems. However, EI suffers from inherent complexity of the field, lack of established standards, which sometimes happen after the fact, and the rapid and unstable growth of the basic technology with a lack of commonly supported global strategy. Nevertheless, EI must be seen as a goal, not a solution, to continuously progress towards a more integrated enterprise.

5. CONCLUSIONS

It is the author's opinion that Enterprise Modeling and Integration is slowly but surely becoming a reality. However, these technologies would better penetrate and serve any kind of enterprises if:

- There was a standard vision on what enterprise modeling really is and there was an international consensus on the underlying concepts for the benefit of business users (Goranson *et al.*, 2002)
- There was a standard, user-oriented, interface in the form of a unified enterprise modeling language (UEML) based on the previous consensus to be available on all commercial modeling tools (Chen *et al.*, 2002b; Panetto *et al.*, 2004)
- There were real enterprise modeling and simulation tools commercially available taking into account function, information, resource, organization, and financial aspects of an enterprise including human aspects, exception handling, and process coordination. Simulation tools need to be configurable, distributed, agent-based simulation tools (Vernadat and Zeigler, 2000)
- There were design patterns and model-based components available as (commercial) building blocks to design, build, and reengineer large scale systems (Molina and Medina, 2003)

- There were commercially available integration platforms and integrating infrastructures (in the form of packages of computer services) for plug-and-play solutions (Chen and Doumeingts, 2003)

Future trends in enterprise integration and enterprise modeling would be toward loosely-coupled interoperable systems rather than high-cost monolithic solutions and low-success holistic integration projects.

REFERENCES

Berio, G., Vernadat, F. (1999) New Developments in Enterprise Modeling Using CIMOSA. *Computers in Industry*, **40**(2-3), 99-114.

Bernus, P., Nemes, L. (1996) Framework to define a generic enterprise reference architecture and methodology. *Comp. Integrated Manuf. Systems*. **9**(3) 179-191.

Browne J., and Zhang J., (1999). Extended and virtual enterprises - similarities and differences, *International Journal of Agile Management Systems*, 1/1, pp. 30-36.

Camarinha-Matos LM, Afsarmanesh H, Garita C. (1998). Towards an architecture for virtual enterprises, *J of Intelligent Manufacturing*, 9 (2), 189-199.

Chen, D. and Doumeingts, D. (2003) European Initiatives to develop interoperability of enterprise applications - basic concepts, framework and roadmap. *Journal of Annual reviews in Control*, **27**(3), 151-160.

Chen, D., and Vernadat, F. (2004). Standards on enterprise integration and engineering – A state of the art. *International Journal of Computer Integrated Manufacturing*, **17**(3), 235-253.

Doumeingts, G. Vallespir, B. (1995). A methodology supporting design and implementation of CIM systems including economic evaluation. *Optimization Models and Concepts in Production Management* (P. Brandimarte and A. Villa, eds.), Gordon & Breach Science Publishers, New York. pp. 307-331.

Fox, M.S., Gruninger, M. (1998) Enterprise Modeling. *AI Magazine*, Fall. 109-121.

Goranson, T., Jochem, R., Nell, J., Panetto, H., Partridge, C., Sempere Ripoll, F., Shorter, D., Webb, P., Zelm, M. (2002). New Support Technologies for Enterprise Integration. *IFIP International Conference on Enterprise Integration and Modeling Technology (ICEIMT'02)*, Kluwer Academics Publisher, Valencia, Spain, 24-26 April 2002, pp. 347-358.

IFAC-IFIP Task Force (1999). GERAM: Generalized Enterprise Reference Architecture and Methodology, Version 1.6.3, IFAC-IFIP Task Force on Architecture for Enterprise Integration.

Molina, A., Sanchez, J.M., Kusiak, A. (1999). *Handbook of Life Cycle Engineering: Concepts, Models and Technologies*, Kluwer Academic Press.

Molina, A., and Flores, M., A. (1999). Virtual Enterprise in Mexico: From Concepts to Practice. *Journal of Intelligent and Robotics Systems*. 26: 289-302.

Molina, A. and Medina, V. (2003). Application of Enterprise Models and Simulation Tools for the evaluation of the impact of best manufacturing practices implementation. *IFAC Annual Reviews in Control*. 27 (2), 221-228.

Morel, G., Panetto, H., Zaremba, M-.B., Mayer, F. (2003). Manufacturing Enterprise Control and Management System Engineering: paradigms and open issues. *IFAC Annual Reviews in Control*. 27 (2), 199-209.

National Research Council. (1998). Visionary Manufacturing Challenges for 2020, Visionary manufacturing challenges for 2020 / Committee on Visionary

Manufacturing Challenges, Board on Manufacturing and Engineering Design, Commission on Engineering and Technical Systems, National Academy Press. (http://search.nap.edu/readingroom/books/visionary/index.html)

Panetto, H. (2001). UML semantics representation of enterprise modeling constructs. *Proceedings of the Workshop on Knowledge Management in Inter- and Intra-organizational Environments*, EADS – Paris, December 5-7, 2001.

Panetto, H. (2002). UEML constructs representation. *Proceedings of the workshop on Common Representation of Enterprise Models*. IPK, Berlin, February, 20-22

Panetto, H., Berio, G., Benali, K., Boudjlida, N., Petit, M. (2004). A Unified Enterprise Modeling Language for enhanced interoperability of Enterprise Models. Preprints of the 11th IFAC INCOM2004 Symposium.

Panetto, H., Berio, G., Petit, M., Knothe, T. (2003a). Enterprise integration semantics issue. Proc. International NIST/NSF Workshop on Semantics Distance, NIST, November 10-12, 2003, Gaithersburg, Maryland, USA.

Panetto, H., Bernard, A., Kosanke, K., (2003b). Product and Process models for Enterprise Integration. Proceedings of the ISPE-CE2003 conference on Concurrent Engineering: Research and Applications, Madeira, Portugal, 26-30 July, Enhanced Interoperable Systems, 747- 835.

Vernadat, F.B. (1996). Enterprise Modeling and Integration: Principles and Applications, Chapman & Hall, London.

Vernadat, F. (2002). UEML: Towards a unified enterprise modelling language, *International Journal of Production Research*, **40** (17), 4309-4321.

Vernadat, F.B. (2002). Enterprise Modeling and Integration (EMI): Current Status and Research Perspectives. *Annual Reviews in Control*. **26**, 15-25.

Vernadat, F.B., Zeigler, B.P. (2000). B.P. New simulation requirements for model-based Enterprise Engineering. Proc. IFAC/IEEE/INRIA Int. Conf. on Manufacturing Control and Production Logistics (MCPL'2000), Grenoble,.

Weston, R.H. (1993). Steps towards enterprise-wide integration: a definition of needs and first-generation open solutions. *International Journal of Production Research,* **31**(9), 2235-2254.

Williams, T.J. (1994). The Purdue Enterprise Reference Architecture. *Computers in Industry*, **24** (2-3), 141-158.

Whitman, L. and Huff, B. (2001). On The Use Of Enterprise Models. Special Issue on: Business Process Design, Modeling, and Analysis, International Journal of Flexible Manufacturing Systems, **13**(2), 195-208.

Whitman, L.E., D. Liles, B. Huff, and K. J. Rogers. (2001). A Manufacturing Reference Model for the Enterprise Engineer. *Special Issue on Enterprise Engineering, The J. of Engineering Valuation And Cost Analysis*, 4(1), 15-36.

Whitman, L., B. Huff, and A. Presley. (1997). Structured models and dynamic systems analysis: The integration of the IDEF0/IDEF3 modeling methods and discrete event simulation. Proc. Winter Simulation Conference, Atlanta, GA.

Yoshikawa, H. (1995). Manufacturing and the 21st Century – Intelligent Manufacturing Systems and the renaissance of the manufacturing-industry. *Technological Forecasting and social change.* **49**(2), 195-213.

26. Enterprise Integration Approaches in Healthcare: A Decade of Trial and Error

V. Jagannathan

West Virginia University & MedQuist, Inc.
juggy@csee.wvu.edu

This paper chronicles the different approaches for enterprise integration used in the field of healthcare over the past decade, and which approach succeeded and which failed. It ends with the new approach just launched through the Health Level 7 standards organization with support from the Health and Human Services in US.

1. INTRODUCTION

Integration and interoperability issues are not new to healthcare. They are the very reason why large standards body such as Health Level 7 (HL7) and EU CEN TC 251 exists with the goal of simplifying integration and promoting interoperability. The problem of interoperability however takes on massive connotations in healthcare, unlike most other disciplines. One needs to look no further than the National Healthcare Information Infrastructure (NHII) goals in US and similar efforts around the world to understand what is at stake. Very simply stated, the goals of these initiatives is to provide access to any pertinent information which furthers the treatment of any individual at any location and point in time by any authorized care giver. Now if one overlays the scope of information of individuals – also dubbed their longitudinal electronic health record – which tracks what was done to individuals from cradle to grave, and any relevant medical treatment information the scope becomes a bit more clearer. Now, if we extrapolate to hundreds of millions to billions of the general populace, privacy and confidentiality considerations, and the tens of thousands of information systems that are deployed in the real world, it is clear that this is a problem that is not going to be solved in a hurry. This document chronicles the different approaches tried to date and the new approach being attempted by the NHII initiative to address interoperability and integration in healthcare.

2. APPROACHES

2.1 Classification of approaches

Standards clearly play a part when diverse systems need to be integrated. One of the traditional approaches to integration, which has been very successful in financial communities, is the Electronic Data Interchange (EDI) standards. In healthcare, this approach is facilitated by Health Level 7 (HL7) standard. The HL7 standard is primarily a message-oriented approach to integration (Hinchley, 2003).

The latter part of last decade, gave credence to another approach that can be best described as "Service Oriented Architectures". Proponents of this approach include the Object Management Group and their Healthcare Task force (HDTF, 2004).

Another approach, which at the time it was proposed was unique to healthcare, is the notion of "Visual Integration". Here, the integration happens at the customer/client user end of the applications (CCOW, 2004).

Lastly, but not the least, there are approaches which hinge around ontology and agent-based architectures and peer-to-peer communication protocols. The impact of these types of architectures is limited in healthcare to date though gaining in ground given the focus on evidence-based practice of medicine (AgentCities, 2002).

All of the above approaches and their strengths and weaknesses are highlighted in the next section.

2.2 Message-Oriented Integration

HL7 is a well established standard and is the primary mechanism by which most applications in healthcare communicate with each other. HL7 has a very rigorous approach to creating message structures that are based on a general purpose Reference Information Model of healthcare that supports variety of workflows in healthcare (Hinchley, 2003). For instance, there are specific messages one can use to communicate that a patient has been admitted, discharged or transferred. Hundreds of message definitions exist and they serve the purpose of communicating between various applications in healthcare.

The biggest advantage with this approach in healthcare is that it is well established. However, since every system behaves like a complete application and stores whatever information it needs, there tends to be significant duplication of information. Almost every clinical application deployed in a hospital for instance, needs to have some patient demographic information. This is typically acquired during the hospital registration process. What is typically done is this information is communicated individually in a separate message to every application that needs this information – typically fifteen to twenty applications in just one hospital. Every one of these system will store this common information – leading to significant duplication of some common information. This approach has given rise to an entire category of applications called "Interface Engines" whose sole purpose in life is to make the process of getting the sending and receiving applications to handle the vagaries of the messages that are being communicated.

A variation of this approach that was attempted in healthcare, include one where the messages, instead of being an ASCII string, was instead an object with rich structure. Sending and receiving messages as objects save the sending and receiving applications from having to parse through the syntactic structure of the message. It made the process a bit more efficient at run time at the expense of the complexity of the standard itself. Andover Working Group and ActiveX for Healthcare and HL7 Special Interest Group on Object Brokering Technologies (SIGOBT) promoted this approach in 1998-2000 time frames (Jagannathan *et al, 1998)*. This particular effort failed in its entirety and was abandoned. The reason this failed was because there was not much perceived value for "object version of messages", as most vendors had already invested in parsing HL7 message standards and they were unwilling to replace something that worked really well.

Another standard which supports this paradigm and well established in healthcare is the Digital Imaging and Communication in Medicine (DICOM) standard. This standard is used to communicate digital images such as XRays, MRIs, CT Scans etc and enjoys uniform support and conformance from most imaging solution vendors (DICOM, 2004).

In message-oriented approaches there is an implicit assumption that information is being sought only in the context of ONE application and the information that particular application provides is all that will be accessible to the user. It is simply untenable to assume that there is going to one application that warehouses all patient information from cradle to grave for all patients. Purely message oriented approaches to integration will not satisfy the NHII goals.

2.3 Service-Oriented Architectures

An alternative approach is the one advocated by the Object Management Group (OMG) and others. The notion here is simply that each application provides a programmatic interface to which other applications can send messages and get back responses. A request-response paradigm for integration is dubbed as "Service-Oriented Architectures" (SOA). OMG's Healthcare Task Force between 1996 and 2002, came up with a series of specifications in very specific healthcare application domains that support this model.

OMG's Common Object Request Broker Architecture (CORBA) provided an interoperable framework that allowed clients and servers running on different platforms and different language bindings to interoperate. This was and still is a mature technology and it had proved itself in numerous other domains such as telecom and finance. It also separates specifications into horizontal and vertical segments. Horizontal specifications are those that are applicable to all industries and include infrastructure pieces such as security and event notification and communication. Vertical specifications are industry specific.

Though some of the specifications developed by the Healthcare Domain Task Force (DTF) (HDTF, 2004) enjoyed some adoption, the OMG effort in healthcare failed for a number of reasons. These include: 1) A perception that this approach of using CORBA competed with the use of DCOM and Microsoft technologies, 2) The level of complexity and the bar for entry was significant for application developers in healthcare, and 3) This effort competed with HL7 charter for providing interoperability standards and specification in healthcare. HL7 had much broader representation than the technically oriented OMG.

The European Committee for Standardization CEN, Technical Committee 251 is focused on healthcare informatics[76]. At various times they have investigated SOA architectures and currently the WG1 is focused on specifying such solutions as part of their "Healthcare Common Information Services (HCiS) effort.

Though the OMG effort failed, there is a resurgence of SOA, due to the advancement in XML-related technologies in general and Web Services in particular. This aspect of integration is discussed in section 3.0 below. It is also implicitly assumed that to realize the goals of NHII a combination of SOA and message oriented solutions will need to be deployed. And it is quite conceivable that

---

[76] European Union - CEN TC 251 - Healthcare Informatics Technical Committee. http://www.centc251.org

the specifications developed by OMG might re-appear in a morphed form to address the goals of the EHR and NHII (See section 4.0).

2.4 Visual Integration

One unique approach to integration that has taken some hold is a focus on providing value to the users by visually integrating the client applications. The approach here, embodied in a standard named Clinical Context Object Workgroup (CCOW) is to take the client side of multiple applications and get each of them to switch context in tandem to user actions. The following time-sequence diagram and analysis explains this concept in greater detail. In the following, we assume that there are two vendor applications – labeled "App A" and "App B", one user, labeled "User P" and a context management application labeled as "CM". The numbered items are interaction between applications and user and various applications.

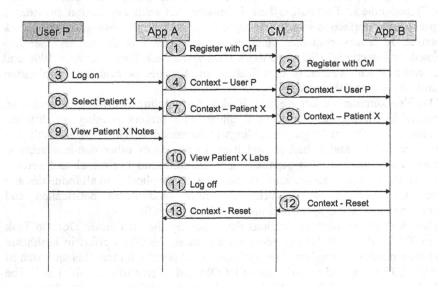

1. Application A registers with the context manager (CM). This allows CM to notify application A when it needs to.
2. Application B registers with the context manager (CM). This allows CM to notify application B when it needs to.
3. User P logs on to Application A.
4. Application A sends a notification to CM stating the user context is P – essentially indicating that P has logged on.
5. CM propagates (notifies) the context to Application B. Now application B knows user P is logged on and resets the context of the application to reflect that.
6. User P interacts with application A and selects patient X.
7. Application A sends this contextual information to CM.
8. CM propagates that Patient X is the context to Application B.

9. User P interacts with Application A and reviews say Patient X clinical notes.
10. User P interacts with Application B and reviews the lab results for patient X. Note that the user did not have to log on to Application B nor select Patient X. Application B already knows the context by receiving context information from CM.
11. User P logs off on Application B.
12. Application B sends notification to CM to reset the context – equivalent to not having context or nobody is logged on.
13. CM propagates this to Application A and hence Application A logs the user off and resets the context. By virtue of logging off on Application B, user is logged off from Application A as well.

Application A and B in above are referred to as Context Participants (CP) in the CCOW standard. CCOW standard enjoys modest support in healthcare. However, when it is used for purely providing a single-sign-on capability between multiple applications, there are more general and simpler approaches available. The simplest of these approaches include solutions provided by password management applications. Here a browser keeps track of (caches) username/password pairs for multiple web-based applications and simply use them to log on to multiple applications. Other efforts to provide single sign on solutions include efforts within the Oasis group such as Secure Access Markup Language (SAML, 2004).

2.5 Agent and Ontology-based Architectures

Agent-based and artificial intelligence based techniques have been around for a while. They gained major significance in late 80's with the advent of fifth generation computing in Japan. However, the technologies failed to live up to the expectations of the users and the techniques promoted took a back seat to the evolution of internet and web in the 90s. There is currently a resurgence of interest in knowledge and agent-based solutions. Agentcities.com is a group focused on promoting agent technologies. A subgroup of this group, work on healthcare applications (AgentCities, 2002) with hopes of developing and promoting solutions for integration using multi-agent architectures.

Part of the basis for this effort is knowledge ontology of health care concepts. The "Protégé" project at Stanford provides a tool that will allow users to create their own ontology (Li *et al*, 2000). It has been used extensively by a number of different groups to develop ontologies in healthcare. Another related effort in the tools arena is the "OpenGalen" project. This project seeks to build an open framework for supporting a terminological server that will allow documentation using standard terminologies[77].

Medical vocabularies are complex and hitherto have lead to a number of different standards with varying specificity and goals. The International Classification of Diseases (ICD), developed by the World Health Organization is a widely used classification scheme. ICD 9, the 9th version, is used in US for billing and reimbursement purposes. ICD 10 is a better classification of diseases and is yet to find widespread adoption in US. Current Procedural Terminology (CPT) developed by the American Medical Association, is a classification of procedures

---

[77] OpenGalen Foundation, Nijmegen, the Netherlands (http://www.opengalen.org/index.html)

performed by physicians and is used very heavily in US to pay for physician services. ICD and CPT and various variations of it have as their primary goal of supporting the reimbursement process (Buck, 2002) and have less relevance for actually providing care. In order to support the goal of reducing medical errors and providing evidence-based decision support at the point of service, other vocabularies and coding schemes are more relevant. The Unified Medical Language System (UMLS) developed by the National Library of Medicine (NLM) is a scheme to allow the classification and indexing of medical knowledge (UMLS, 2004). The Systematized Nomenclature of Medicine: Clinical Terms (SNOMED CT) is a systematic classification of medical terms and shows the promise of representing medical conditions and treatments with aim towards supporting clinical decision support. SNOMED is now made available freely in US as part of the NHII promotion effort[78]. Ontology and knowledge driven integration of healthcare applications is a necessity if the goals of safety and quality in healthcare delivery has to be realized.

3. IMPACT OF XML TECHNOLOGIES ON HEALTHCARE INTEGRATION

3.1 Special Interest Group in XML in HL7

In 1997, HL7 established a Special Interest Group (SIG) on SGML – the precursor to HTML and XML. This group recognized early on that a mark-up language will simplify encoding clinical content. When XML standard was established, this SIG switched their efforts to XML. The lobbying effort by this group resulted in the wholesale adoption of XML as the core technology to be used for specifying all new standards specification from HL7. The Reference Information Model (RIM) which uses Unified Modelling Language (UML) for its core representation, supports translating actual message definitions into XML. One of the core specifications that were adopted early on was encoding of clinical content using XML. This specification was called Clinical Document Architecture or CDA for short. Currently, the second generation of these specifications are in the works. The CDA was a departure from the usual types of specifications that HL7 supported. The CDA was not designed as a message – but rather as a free standing content document. It could be delivered as a payload in a message or it could be delivered as a response to a request. How the content is to be used and delivered was left to the implementers of systems (CDA, 2004).

3.2 ebXML and Healthcare

Electronic Business XML (ebXML) has traction particularly in Europe and the primary goal of this effort is to enable electronic transactions between business partners. This standard is supported by the Oasis group (ebXML, 2001) and (Kotok and Webber, 2002). This standard is being explored by Centers for Disease Control (CDC) and Federal Drug Agency (FDA) to allow organizations to report to them on communicable and syndromatic disease outbreaks and drug-drug interactions.

---

[78] Systematized Nomenclature of Medicine Clinical Terms (SNOMED CT) : http://www.snomed.org

3.3 Integration Demonstrations

HL7 orchestrated a sequence of demonstration using XML technologies starting in 1999. These demonstrations progressively became more and more complex over the years. In 2004, the demonstrations joined hands with a number of other organizations such as HIMSS and RSNA to put together what was dubbed the "Integrated Healthcare Enterprise (IHE)"[79]. IHE actually started out as a demonstration of integration technology focused in the radiology domain in 2001. The IHE demonstration this year (2004) combined all the elements discussed in this paper to this point – HL7 messaging, DICOM standards, CCOW standards, CDA standards and use of XML-web services to access CDA documents. Clearly the healthcare pendulum has swung in the direction of using XML and web services. Nearly 25 different vendors and numerous standards group participated in this demonstration showcasing that fairly significant interoperability is indeed feasible. However, these efforts are nowhere close to supporting the true goal of NHII – that of a longitudinal electronic health record. This leads logically to the next effort – discussed in the next section.

4. FUNCTIONAL MODEL OF ELECTRONIC HEALTH RECORD – DRAFT STANDARD FOR TRIAL USE (DSTU)

4.1 NHII Imperatives

The Institute of Medicine came up with a series of reports which raised the national conscience on the state of healthcare. Their report, "To err is human" (Institute of Medicine, 2001) showed that there were 98,000 preventable errors in healthcare annually in US. A follow on report, showed that serious quality concerns persist in healthcare (Institute of Medicine, 2001b). To address all of these, the National Healthcare Information Infrastructure (NHII) initiative was born. NHII is an initiative that is currently managed by the Department of Health and Human Services in US. NHII is "… the set of technologies, standards, applications, systems, values, and laws that support all facets of individual health, health care, and public health" (NCVHS, 2000)[80].

4.2 EHR – Draft Standard for Trial Use

Institute of Medicine under the direction of Health and Human Services (HHS) released a new report in summer 2003 titled: "Key Capabilities of an Electronic Health Record System" (Institute of Medicine, 2003). This report became the seed for a fast-track effort at HL7 to define a functional model for what constitutes an EHR – driven by HHS. The goal that HHS has is to provide financial incentives through their CMS (Medicare/Medicaid) wing to further adoption and implementation of EHR starting in 2004. In order for them to provide such an incentive, there needs to be a definition in place what constitutes an EHR and what functionality it needs to support. This is what HL7 is currently trying to do. An initial effort which resulted in a functional model with over 1000 elements was rejected by the HL7 membership in September 2003 as being overly complex. A

---

[79] Integrating Healthcare Enterprise (IHE): http://www.rsna.org/IHE/index.shtml.
[80] National Healthcare Information Infrastructure (NHII): http://aspe.hhs.gov/sp/nhii/

revamped functional model was balloted again and has now been adopted as EHR Functional Model – Draft Standard for Trial Use in July 2004 (HL7, 2004).

The revamped model summarizes the functionality of the EHR in three categories: 1) Direct Care, 2) Supportive functions and 3) Infrastructure. Direct care functions include applications that support capturing orders, medication and clinical documentation. Supportive functions include providing decision support, alerts, drug-drug interaction warnings etc. Infrastructure functions include addressing security and privacy concerns, Public-Key Private-Key infrastructure and the like.

The functional model is a blue print for what EHR functionality needs to be supported. The specification and mapping of specific standards on how this functionality needs to be supported so there can be a broader integration and interoperability is a future task.

5. CONCLUSIONS

This paper chronicles the efforts to develop interoperable solutions in healthcare over the past decade. The current status is also presented. The problem of enterprise integration and interoperability has been around for a while in healthcare and standardized solutions can dramatically improve the quality of healthcare delivered. There are some similarities to the approach taken in healthcare to efforts in other industries. The HL7 standards are developed with a healthcare model as its core foundation.

In the non-healthcare manufacturing world, the Open System Architecture for Computer Integrated Manufacturing (CIMOSA) has an excellent model driven approach to dealing with integration[81]. CIMOSA effort dates back to late eighties and continues to date with a single minded focus on modeling enterprise systems and deriving integration strategies from enterprise models.

The adoption of SOA has been slow in healthcare. It appears that the climate for that is changing and can lead to true integration and interoperability. The functional EHR model and the DSTU gives hope to that belief.

REFERENCES

Andrew Hinchley (2003). Understanding Version 3. A primer on the HL7 Version 3 Communication Standard, Alexander Moench Publishing, ISBN 3-933819-18-0, Munich, Germany.

HDTF (2004) Object Management Group, Health Domain Task Force (OMG HDTF, formerly CORBAmed). http://healthcare.omg.org/Healthcare_info.htm (accessed 2004)

V. Jagannathan, K. Wreder, R. Glicksman, Y. Alsafadi (1998). "Objects in Healthcare – focus on standards," ACM Standards View, Summer 98.

AgentCities (2002) – Working group on Healthcare Applications. http://wwwcms.brookes.ac.uk/hcwg/LivDoc.pdf (accessed 2004)

CCOW (2004) HL7 Context Management Standar, Clinical Context Object Workgroup, http://www.hl7.org/special/Committees/ccow_sigvi.htm (accessed 2004)

---

[81] Computer Integrated Manufacturing – Open System Architecture (CIMOSA). http://cimosa.de

DICOM (2004) Digital Imaging and Communication in Medicine (DICOM), National Electrical Manufacturers Association, Rosslyn, VA (http://medical.nema.org/)

Q. Li, P. Shialane, N. F. Noy, M. A. Musen. (2000) "Ontology Acquisition from On-Line Knowledge Sources. AMIA Annual Symposium, Los Angeles. (http://smi-web.stanford.edu/pubs/SMI_Abstracts/SMI-2000-0850.html)

C. J. Buck (2002) "Step-by-step Medical Coding," W. B. Saunders Company, An Imprint of Elsevier Science, Philadelphia, PA, ISBN: 0-7216-9333-4.

UMLS (2004) Unified Medical Language System, National Library of Medicine, National Institutes of Health, USA (http://www.nlm.nih.gov/research/umls/archive/2004AC/umlsdoc_2004ac.pdf, accessed December 2004)

SAML (2004) Secure Access Markup Language (SAML), Organization for the Advancement of Structured Information Standards (OASIS), Billerica, MA (http://www.oasis-open.org/committees/tc_home.php?wg_abbrev=security)

CDA (2004) Clinical Document Architecture (CDA), Health Level Seven, Inc. http://www.hl7.org/special/Committees/structure/struc.htm

ebXML (2001) Electronic Business XML (ebXML), UN/CEFACT and Organization for the Advancement of Structured Information Standards (OASIS), Billerica, MA http://www.oasis-open.org/committees/tc_home.php?wg_abbrev=ebxml-bp

Kotok, D.R.R.Webber (2002) ebXML: The New global standard for doing business over the internet. New Riders Publishing. ISBN: 0-7357-1117-8.

Institute of Medicine (2001) To Err is Human. Building a safer health system. ISBN# 0-309-06837-1. The National Academy Press, 2001.

Institute of Medicine (2001b) Crossing the Quality Chasm: A new health system for the 21st Century. ISBN# 0-309-07280-8. The National Academy Press, 2001.

Institute of Medicine (2003) Committee on Data Standards for Patient Safety: Key Capabilities of an Electronic Health Record System: Letter Report. Washington, D.C., The National Academy Press, 2003.

HL7 (2004) EHR System Functional Model: A Major Development Towards Consensus on Electronic Health Record System Functionality. Available from www.hl7.org.

27.   An Enterprise Modelling Approach to Team Systems Engineering

Nikita Byer [1], Richard H Weston [2]

1 Loughborough University, MSI Research Group, Email [n.a.byer@lboro.ac.uk]
2 Loughborough University, MSI Research Group, Email [r.h.weston@lboro.ac.uk]

Teams are engineered by dependent processes involving a spectrum of activities commencing with the initial identification of need, extending through to the realisation of that need and in some cases dissolution of the team. A new model of the team systems engineering life cycle is described in this paper which includes four main groupings of activities corresponding to: 'design', 'build', 'operate' and 'maintain' (DBOM) life phases through which a typical team system progresses. The paper illustrates how Enterprise Modelling concepts and the DBOM model can be innovatively deployed in order to systematically capture published knowledge about teams; thereby providing an analytic basis on which teams can be designed, built, operated and maintained. Here EM modelling constructs were used to document and visually represent relatively enduring aspects of team systems. This paper illustrates the approach by creating a semi-generic model of project teams.

1. INTRODUCTION

This paper reports on new understandings gained when applying Enterprise Modelling (EM) techniques in a novel way in order to facilitate the systematic reuse of published knowledge about human teams. The prime purpose of so doing is to enable best practice to be achieved when engineering specific cases of this class of complex system in unique manufacturing enterprise (ME) settings. The paper reports on research contributions made during the first author's PhD study. The full set of research arguments, modelling methodologies proposed and results obtained from case study applications of the proposed modelling methodologies are described in Byer 2004.

Teams and teamworking are topics that have received very significant research attention, primarily by human and behavioural scientists and practitioners. It follows that there exists a massive body of literature on these topics. Teams and teamworking are known to be topics of concern to manufacturing industries worldwide. When a ME gets its teamworking 'right' significant benefits can accrue. Team experts, consultants and academics advance a plethora of theories and techniques, which are derived from the literature on teams and teamworking, to inform and facilitate the successful design, development and implementation of teams. Despite these efforts the literature is populated with examples of teams that fail to produce desired results. Evidently there is a gap between the team approaches that are conceived and tested in academia, and deployed by human factors, specialists and consultants, and those that are widely understood and reused within specific industrial settings.

The authors observed a lack of homogeneity within the existing body of literature of teams and teamworking. Although similar theories and concepts are shared, in general these theories and concepts are ill defined with similar terms commonly used to mean significantly different things. This is as might be expected because people are individuals and when attributed to organisational groupings, are known to constitute complex entities that can be viewed from many different perspectives; even then they can only be partially understood. Many related issues (such as emergent behaviours, team culture and motivation) are by nature soft and difficult to quantity. However evidently there is commonality amongst teams (e.g. commonality of purpose, processes, composition, inputs, outputs, performance measures) of similar ilk. The authors also observed (a) significant variation in the perceived importance of teams and teamworking in different MEs and (b) that existing knowledge about teams is currently used in a fragmented, and ad hoc manner, typically driven by an in-house human factors expert or hired teamworking consultant. This again is not surprising because MEs are in general even more complex than is a specific team system and require multi-perspective understandings by people systems which are also concerned about soft and hard issues, which may be of a relatively enduring, or relatively transient nature.

Hence the authors decided to deploy EM techniques to achieve the following objectives, in new and improved ways:

1. Document general understandings about teams and teamworking so as to (a) improve the homogeneity of those understandings, (b) facilitate understanding and interpretation of existing knowledge on teams and (c) encourage the computer-based formulation and reuse of that knowledge.
2. Document semi-generic understandings about different types of team used in MEs, thereby facilitating the capture of reference models of teams and successful scenarios of team systems engineering.
3. Provide generic and semi-generic means of informing, and lending structure, to team working aspects of EE projects; to facilitate decision making about team type selection, team system implementation, enabling team systems operation, and so forth.

2. REVIEW OF TEAM SYSTEMS LITERATURE

With aims (i) through (iii) in mind the authors reviewed existing literature on teams and team working. The literature on team systems reported variously on four distinctive system life phases (Byer and Weston, 2004), namely: i) Team System Design; ii) Team System Building; iii) Team System Operation; and iv) Team System Maintenance. Those phases are described in outline in the following.

2.1 Team System Design

Teams are commonly designed in a top-down manner so as to implement specified business strategies (Chiesa, 1996 a, b; Dunphy and Bryant, 1999; Schilling and Hill, 1998). Prasad and Akhilesh (2002) state that virtual teams (one of many possible team types) are required to achieve stated objectives and their structures and context have to support these objectives. Thus it is very important to maximise the fit between team design and their stated intent. Team system design, they continue, uses the stated objectives of the team to determine the size, team composition and team structures. Prasad and Akhilesh (2002) suggest that proper team design in itself would be of no avail if it does not finally lead to the delivery of top performance. Thus teams need to be structured to achieve maximum effectiveness.

Hence it is observed that team system design requires adequate problem definition to define the problem and to use that definition to determine appropriate team characteristics. Also the characteristics defined should be related to some specific team type (such as a project team or a self-managed team) with stereotypical abilities to meet the problem defined. Hence team design can be viewed as choosing the right team type, i.e. a team type with known characteristics, capacities and capabilities to perform the task(s) at hand within defined constraints.

2.2 Team System Building

Team system building has been viewed as the process of selecting a collection of individuals with different needs, backgrounds, and expertise and transforming them into an integrated and effective work unit (Thaimhain and Wilemon, 1987). In such a transformation the goals and energies of individual contributors have to merge so as to support the objectives of the team (Thaimhain and Wilemon, 1987). It follow that effective team system building can be advantageous in adding to the diverse viewpoints and the needed widespread contacts for anticipatory learning (i.e. learning that anticipates change) (McDonald and Keys, 1996).

Partington (1999) states that the quality of the human resources, which make up teams, is a critical determinant of team performance. Castka *et al* (2003) observations concur with those of many team experts by identifying team composition as being key to top team performance. Oakland (1993) states that no one person has a monopoly of good characteristics, because characteristic behaviours that need to be developed by teams (and within teams) are often contradictory (i.e. good listener vs. fluent communicator).

2.3 Team System Operation

Woodcock (1979) stated that any team has two prime areas of concern, namely: i) teamworking development, 'the way the team plays'; and ii) task realisation, 'the direction of individual skills towards a united effort needed to complete tasks and attain goals'. Since Woodcock's publication, many authors have supported his view that teams will concurrently carry out sub-processes concerned with (i) and (ii).

Teamworking Development involves 'taking care of the team members' and is centred on team behaviour, roles, work assignment, communication and so forth (Woodcock, 1979; Stickley, 1993). While task realisation is about 'getting the job done'. Task realisation should lead to attainment of goals and completion of the purpose for which the team was developed (Woodcock, 1979; Stickley, 1993). These two prime areas of concern characterize the actual processes (i.e. activities that need to occur over time) during the team system operation stage

2.4 Team System Maintenance

Groesbeck and Van Aken (2001) suggest that achieving effective team design, building and operation represents only a start towards achieving increased competitiveness through teams. Teamworking is the process by which people work together more effectively. Teams are not objects to be installed and left to run without further attention or support, rather, team systems must be monitored and supported (Polley & Ribbens, 1998). This can differentiate team systems markedly from existing forms of technical systems (e.g. machine and software systems).

Given potential difficulties of developing and sustaining effective team system processes, Polley and Ribbens (1998) advocate a 'wellness' approach to deal with chronic and potential team system problems. Wellness has been described as being two interrelated sub-processes, such as: i) Monitoring includes diagnosing past performance and assessing group processes at key checkpoints to assess a team's process health; and ii) Maintenance includes assuring that teams receive the support and coaching needed to develop productive processes for the way they work together and perform core task routines.

3. FORMULATION OF THE DBOM TEAM SYSTEMS ENGINEERING MODEL

This research advances the notion that a four-phase DBOM (design, build, operate and maintain) model can represent team systems engineering activities from team 'conception' to team 'grave'.

During team system design, the D phase, information about the objectives of the team is deployed. Here information about the task and task characteristics is interpreted to determine needed team characteristics, such as its composition, size, structure and so on. Team system building, the B phase, involves constructing the team system by selecting and bringing together individuals with appropriate skills, knowledge and expertise. Research of the authors of this paper has shown that skill-requirements of teams can be usefully determined from the task definition and task characteristics (Byer 2004). Both functional roles and team roles should be allocated during the team system building phase. Team system operation, the O phase, concerns two distinct but related sub-processes, namely: i) teamworking development and ii) task realisation. Whereas team system maintenance, the M phase, is characterised by two main concerns, i) monitoring and ii) maintaining team system processes.

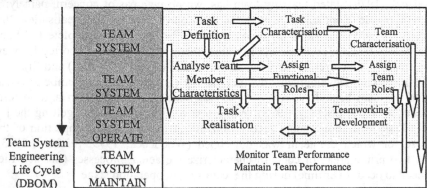

Figure 1.   The DBOM Life Phases of Team Systems

Figure 1 was constructed to illustrate the DBOM phases described above. It groups the team system engineering activities that must be performed from identifying task requirements, through designing the team and allocating resources, to achieving teamworking development and realising the task, and terminating with completion of the task or dissolution of the team. It also illustrates relationships and dependencies between these activity groupings. For example it was determined that team system design comprises three main activities, namely: task definition, task characterisation and team characterisation, which are interrelated. The task characterisation activity receives inputs from the task definition activity, and task characteristics are used to inform the determination of team characteristics.

Information contained in the task definition and about task characteristics is used to inform decisions made about needed member characteristics during the team system build phase. It follows that DBOM engineering activities are inter-related and impact collectively on team system effectiveness.

4. DECOMPOSITION OF THE DBOM TEAM SYSTEM ENGINEERING PHASES

During the process of designing a team system, task descriptions and task requirements are referenced as they provide details pertaining to: the team's purpose; time required for task completion; leadership requirements; team size and team composition (in terms of functional skills). Figure 2 illustrates team system design in terms of its primary input events and output results. It also lists its key sub-processes.

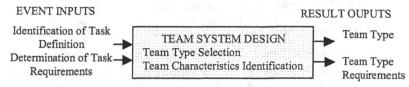

Figure 2. Graphical Model of Team System Design

Team system building is illustrated by figure 3. This constitutes a process of building the team by: selecting the most appropriate team members based on their technical competences; allocating team roles to selected members centred on their behavioural competences; providing the team with an organising 'structure'; and finally releasing the team to achieve the task. Thus it was observed that the process of team system building can be further decomposed into four sub-processes, namely: member selection; role allocation; structure development and team release. During the team system building process, team member selection is determined with reference to the functional skills required and matching this to the technical skills,

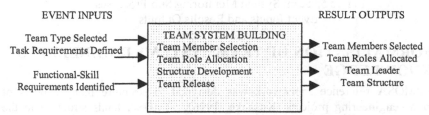

Figure 3. Graphical Model of Team System Building

experience and education of candidate team members. Once selected, team members are assigned team roles. The team leader is chosen and the team structure can subsequently be developed. Figure 4 depicts the elements of the team system operation process. This also identifies sub-processes, event inputs and output results. Team system operation, the O phase, is centred on teamworking development and task realisation. Team System Operation is concerned with how the team develops from a group of individuals to an efficient functioning unit while successfully realising the task assigned and delivering the task goals.

Team system maintenance incorporates two interrelated processes: monitoring and maintenance. In this paper the authors emphasis the importance of the team system monitoring sub-process, as a vital source of information pertaining to the performance and effectiveness of the D, B and O life phases. Team system monitoring naturally provides feedback with regard to team system effectiveness. In the first instance it can function to provide feedback with respect to teamworking development and task realisation performance. Secondly, it can provide feedback regarding the suitability of the team members selected, team members roles allocated and the structure developed and deployed. Finally it can feed back quality (i.e. fitness for purpose) about the type of team selected. Figure 5 shows a graphical decomposition of the team system monitoring sub-process.

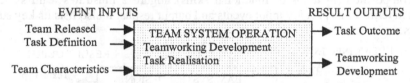

Figure 4. Graphical Model of Team System Operation

In general it is assumed that the team system maintenance sub-process can utilise feedback information generated by the team system monitoring process in a wide variety of ways. For example, maintenance sub-processes might analyse, reflect on, predict and improve the performance of (1) operational teams and team members or (2) future teams (when used to achieve similar tasks or when similar team types are to be used to realise new tasks).

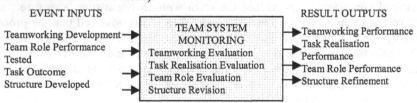

Figure 5. Team System Monitoring Sub-Processes,
Event Inputs and Results Outputs

5. FORMALISATION OF TEAM SYSTEMS ENGINEERING KNOWLEDGE

The CIMOSA reference framework was designed to structure many aspects of enterprise engineering projects (Kosanke, 1996). CIMOSA lends structure to the activities of enterprise modellers as they seek to generate models with different levels of generality, i.e. generic, partial or particular enterprise models. Monfared *et al* (2002) states that users can generate enterprise models at any of these three levels of generality and that other models at different levels of generality can be derived by referencing captured models, thereby reducing the modelling effort required.

In this study it was considered appropriate to develop partial models of team systems, and the processes and sub-processes needed to engineer teams during their lifetime. A particular focus of attention was to be placed on semi-generic application domains in which enterprise teams are commonly deployed. One such domain analysed concerned cases where project teams are assigned a new task. With respect to this domain it was decided that partial models could usefully be captured formally by using the following representational forms (Byer, 2004):

- Overall context diagram for the 'engineer new project team system' domain
- Overall interaction diagram for the 'engineer new project team system' domain
- Context diagram for 'team system building' (B phase of DBOM)
- Interaction diagram for 'team system building'
- Structure diagrams for 'team role allocation', 'team structure development' and 'team release' (sub-processes of team system building)
- Context diagram for 'team system operation' (O phase of DBOM)
- Interaction diagram for 'team system operation'
- Structure diagrams for 'task realisation', 'teamworking development' and 'team operation progression' (sub-processes of team system operation)
- Context diagram for 'team system monitoring' (M phase of DBOM)
- Interaction diagram for 'team system monitoring'
- Structure diagram for 'team role performance', 'teamworking development performance' and 'structure refinement' (sub-process of team system monitoring)

Only the first four of these representational forms are illustrated in this paper.

5.1 Overall Context Diagram for "Engineer New Project Team System"

Figure 6 depicts the overall context diagram for the target 'engineer new project team system' domain process. This shows that engineering a new project team system involves interactions between activities that 'belong to' the 'design team system', 'build team system', 'operate team system', and 'monitor team system' domains. In this case the 'design team system' domain was designated a non-CIMOSA domain and was not considered in detail within the authors' research.

5.2 Overall Interaction Diagram for "Engineer New Project Team System"

Next it was considered necessary to formulate a detailed analysis of elements that constitute the complete 'engineer new project team system' process. To accomplish this existing information about team systems that was coded into the *ad hoc* graphical models described in previous sections of this paper, was reformatted in the form of a high-level CIMOSA conformant interaction diagram so as to characterise primary interactions in the 'engineer new project team system' domain, see figure 7. This figure documents the primary flows of information, human resources and physical resources between the sub-domains of the 'engineer new project team system' domain.

Detailed descriptions of task requirements, task goals and teamworking information are passed from the 'team system building' domain process (designated DP 1) to 'team system operation' domain process (designated DP 2). Through the team systems life cycle this information can include task definitions; task requirements; task goals; teamworking goals; initial team operating procedures; initial organisational structures; and constraints and boundaries. Physical resources such as task descriptions; drawings and other types of technical specifications; machines; equipment; and computer software to support the transfer and deployment of this information. A team developer is also designated responsibility for regularly monitoring team system operation.

The 'team system operation' domain process (DP 2) generates extensive information for deployment by the 'team system building' domain process (DP 1). Teambuilding proposals and recommendations are typically derived from a number of project team meetings. Each change request from the 'team system operation' domain process (DP 2) requires the 'team system building' domain process (DP 1) to respond with change assessment information. Information about 'team system operation' is derived with respect to task realisation and teamworking development outcomes.

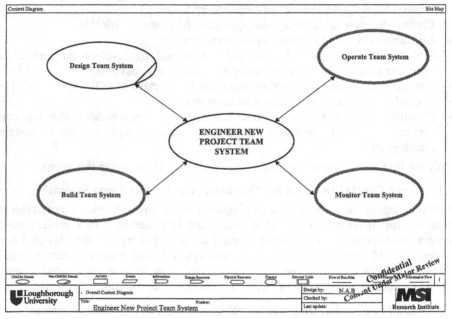

Figure 6. Overall Context Diagram for 'Engineer New Project Team System'

5.3 Context Diagram for "Team System Building"

Figure 8 illustrates the context diagram developed to formally represent the 'team system building' domain. This CIMOSA domain can be sub-divided into four constituent CIMOSA sub-processes, namely: 'team member selection'; 'team role allocation'; 'structure development'; and 'team implementation'. The reader should note that the analysis presented in this paper focuses predominantly on team aspects rather than the task aspects of the team system

5.4 Interaction Diagram, Team System Building

Initially this domain was considered from a broad perspective; hence an interaction diagram was created for the 'team system building' domain with the aim of visually identifying how its four sub-processes play a role in the overall 'team system building' domain process (DP1). As illustrated by figure 9, it can be observed that this domain can be decomposed into four sub-domain processes, namely: i) Team Member Selection (designated DP11); ii) Team Role Allocation (DP12); iii) Structure Development (DP13); and iv) Team Release (DP14).

Error! Objects cannot be created from editing field codes. Figure 7. High Level Interaction Diagram for "Engineer New Project Team System"

This figure shows that information concerning task definition and task requirements from the non-CIMOSA domain 'team system design' constitutes a key input of the 'team member selection' (DP11) domain process. Physical outputs from this domain were observed to include 'member's technical competence lists', which is passed both to the 'team role allocation' domain process (DP12) and 'team structure development' domain process (DP13).

Also observed was that the team release process (DP14) should receive information on task operating instructions, teamworking instructions, members' team role preference and team role allocation. Further this domain (DP14) receives information, physical resources and human resources from all three of the other domains and is required to use this information to facilitate the progression of the team system, from a group of individuals to a unit working towards a common goal.

6. CONCLUSION AND FURTHER WORK

This paper explains that teams can be viewed as being complex systems with associated life phases and processes that characterise them in life cycle engineering terms. Literature on team systems was reviewed and organised with reference to four phases of a newly proposed generic model of team systems engineering, comprising team system design; team system building; team system operation; and team system maintenance. It was presumed that a formal characterisation and representation of existing literature on teams could be achieved using EM concepts and that resultant enterprise models could promote an effective reuse of that knowledge. The life phases observed were distinctive in terms of the events, activities and processes that occur. Further, these DBOM phases encompass groupings of activities performed from initial identification of task definitions and requirements through designing the team, allocating resources, achieving teamworking development, realising the task and terminating at the completion of the task or dissolution of the team.

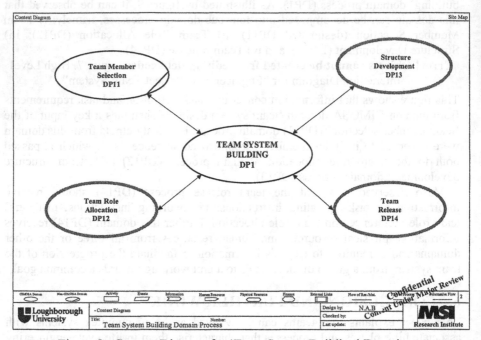

Figure 8. Context Diagram for 'Team System Building' Domain

CIMOSA diagramming templates were used to capture, represent and formalise static (relatively enduring) views of team system processes. It was observed that best-in-class EM techniques could usefully facilitate the capture and reuse of both semi-generic and particular models of team system engineering. This paper shows in outline how CIMOSA diagramming notations were deployed to formally document various semi-generic perspectives on team systems engineering related to the 'project engineering' team class.

It was envisaged that semi-generic and particular models of teams created using EM concepts can be reused for a variety of purposes; such as to inform the life cycle engineering of specific project teams that have been assigned a new task, or to provide an action plan for design, operation and maintenance (as a template) for modelling many, possibly most, types of teams used in industry.

It was observed that many domain models could usefully be created to inform the lifecycle engineering of common team types found in MEs. Here it was envisaged that semi-generic DBOM models could be populated and used to inform the creation and reuse of particular DBOM models which could be related to models of specific ME processes. Thereby it was envisaged that EE practice could be informed and advanced by using EM to organising and facilitating the reuse of the massive body of existing knowledge on teams. However, it is understood that much work remains to determine all needed semi-generic models of team types and to populate and validate an industry-wide use of those models. Also it was observed that not all aspects of team systems engineering can be formalised and enabled via EM. Some softer aspects of team systems engineering need to remain essentially ad hoc and human centred.

The authors consider their DBOM proposal to be of potentially high significance because it provide a semi-formal definition of terms needed to facilitate an interchange of conceptual ideas and proposals between human scientists and ME practitioners. EM and simulation modelling can facilitate an organised and explicit reuse of human systems knowledge and can explicitly articulate ME needs in terms understood by human scientists.

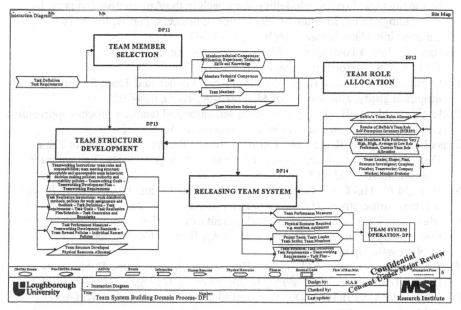

Figure 9. Interaction Diagram for 'Team System Building' Domain Process

REFERENCES

Ajaefobi, J.O. (2004) Human systems modelling in support of enhanced process realisation. PhD Thesis, Loughborough University, Leics., UK.

Byer, N. A. (2004) Team Systems Engineering and the Role of Enterprise Modelling Technologies. PhD Thesis, Loughborough University

Byer, N.A., Weston, R.H. (2004) A Life Cycle Model of Team System Engineering. In Proceedings of the Institute of Mechanical Engineers, Part B, *Journal of Engineering Manufacture*, Submitted in 2004.

Castka, P., Bamber, C. J. and Sharp, J. M. (2003) Measuring Teamwork Culture: the use of a modified EFQM model. *J. of Management Development*, 22(2), 149-170

Chiesa, V. (1996) Managing the Internationalisation of R&D Activities. *In IEEE Transactions on Engineering Management*, 43(1)

Chiesa, V. (1996) Strategies for Global R&D. *Research Technology Management*, 39(5)

Dunphy, D., Bryant, B. (1999) Teams: Panaceas or Prescriptions for Improved Performance? *Human Relations*, 49(5), 677 – 699

Groesbeck, R., Van Aken, E.M. (2001) Enabling Team Wellness: Monitoring and Maintaining Teams After Start-up. *Team Performance Management: An International Journal*, 7(1/2), 11 – 20

Kosanke, K. (1996) Process Oriented Presentation of Modelling Methodologies. In Modelling Methodologies and Enterprise Integration, P. Bernus, L. Nemes (Eds) London : Chapman and Hall

McDonald, J.M., Keys, J.B. (1996) The Seven Deadly Sins of Teambuilding. *An International Journal of Team Performance Management*, 2(2), 19-26

Monfared, R.P., West, A.A., Harrison, R., Weston, R.H. (2002) An Implementation of the Business Process Modelling Approach in the Automotive Industry. In Proceedings of the Institute of Mechanical Engineers, Part B, Journal of Engineering Manufacture, 216(B11), 1413-1427

Oakland, J. (1993) Total Quality Management: the Route to Improving Performance. Oxford: Butterworth-Heinemann

Partington, D. and Harris, H. (1999) Team Role Balance and Team Performance: an empirical study. *The Journal of Management*, 18(8), 694-705

Polley, D., Ribbens, B. (1998) Sustaining self-managed teams: a process approach to team wellness. *Team Performance Management*, 4(1), 3 – 21

Prasad, K., Akhilesh, K.B., (2002). Global Virtual Teams: What Impacts Their Design and Performance. *Team performance Management: An International Journal*, 8(5/6), 102-112

Schilling,M.A., Hill,C.W.(1998) Managing the New Product Development Process: Strategic Imperatives. *IEEE Engineering Management Review*, Winter, 55-68

Stickley, A. (1993) Selection and Development of Effective Teams within Design Function Deployment. In Quality and Its Application Conference at University of Newcastle Upon Tyne

Thaimhain,H.J., Wilemon,D.L. (1987) Building High Performing Engineering Project Teams. *IEE Transactions on Engineering Management*, 34(3), 130-137

Woodcock, M. (1979) Team Development Manual. Farnborough, UK: Gower Publishing

28. Improving Supply Chain Performance through Business Process Reengineering

Andréa Wattky, Gilles Neubert
University of Lyon2, Bron, FRANCE
[andrea.wattky@wanadoo.fr; gilles.neubert@univ-lyon2.fr

Customer satisfaction and service reliability are not any more the assets but the unavoidable condition for a company to be accepted as a supplier of a product or service. The creation of value added in a company concerns all functions and specifications that are involved in delivering a product or service to the customer. Part of that value enhancing chain is the Supply Chain Management (SCM) conception which is defined as all management principles by which the supply chain is considered as a whole.

1. INTRODUCTION

Our environment has changed during the past few decades and is more complex today then ever. It seems quite often to be very disorderly and not foreseeable and requires better economic performance. Market globalization had severe consequences for business; competition is fierce, faster and more reactive every day.

Cultural and socio-economic changes have been favored by new information and communication technologies and new relationships within and between organizations. However, in spite of technology investments, productivity did not have the expected significant effects (Grover and Malhorta, 1997). The concept of business process reengineering (BPR) thus appeared; literature witnesses this phenomenon and several books confirm this tendency (Hammer andChampy, 1993) and (Davenport, 1993). This change has been largely recognized (Davenport, 1993), (Stevens, 1989), etc. and today companies try to decompartmentalize their departments, services, and functions in order to end the existing silo organization and to establish integrated and transverse organizations (Kramer and Tyler, 1995).

The Supply Chain Management (SCM) concept is considered as a major stake to gain a competitive advantage over their competitors (Porter, 1985) and (Lynch, 2000). It describes the manufacturing and movement of a good or service starting from the origin of an order until its distribution to the final customer and thus represents an evolution from an intra-company functional integration over an internal corporate logistics integration to at last, an external integration in a logistic network, that is extended upstream to suppliers and downstream to customers.

SCM puts forward the process engineering and reengineering for the company reorganization, therefore companies seek for new ways and solutions in order to respond to this organizational challenge (new forms of management...) concerning the internal level (federate applications...) as well as the external level (synchronized processing...). A new challenge for companies concerns also the creation of alliances and partnerships (work in network and through projects, assume interdependency, and develop process rhythms...) (Mentzer, 1999). Exchanging and sharing of information to attest performance towards the different actors, detecting the expectation progress, etc. becomes thus essential for a company today.

Business environment is not a stable situation anymore (pipeline, supply chain) but it belongs to a supply chain network (extended and transverse enterprise). In the following section, we will try to analyze and propose different concepts for process modelling.

2. PROCESS MODELLING TECHNIQUES

The supply chain is the combination of multiple processes, which contribute to the creation of value added for an external customer. It includes and is closely related to the concept of process management and process reengineering.

There are different techniques and methods for modelling and remodelling company processes. When choosing one a or number of such techniques, one needs to keep in mind that a process consists in multi-actor activities, which are carried out through time and space; the links between these activities and the activities them-selves may belong to different functions and even to different organizations (extended enterprise). The process is consequently considered as the means by which the organization reaches its objectives.

Companies model and remodel their processes mostly through reference models (standard of the ASLOG, the Supply Chain Council's SCOR-model, Framework of Zachman, the model of M. Cooper...), also called standards or architectural frameworks. They are used to represent a problem and to extract useless details in order to provide a better comprehension of that problem in the whole. Those structures and architectural frameworks include standards, languages, and techniques (Bal and Jay, 2004). Their primary objective is to indicate, which information should be captured and by which means this is possible.

Then, there are enterprise modelling tools (Aris from IDS Scheer, MEGA from MEGA International...) that include concepts, meta-models and semantics. The objective of these modelling tools is to support architectural frameworks. There are also modelling techniques (Use Case de UML...), that are often associated with the before mentioned modelling tools. Graphical modelling techniques are particularly interesting because they are very appropriate for a good visualization and communication. Traditionally, techniques like flow charts and data flows are used to model applications.

Finally, one should not forget to mention the description techniques, which support reference models and which are used by organization modelling tools. These are modelling and analysis techniques that provide sometimes implementation examples. The ISO standards are among the most known and put forward the main principles and vocabulary of the management system for quality etc., main guidelines for performance improvement, inter-enterprise relationship requirements, and regulation requirements, which are applicable in order to continually improve enterprise performances.

In order to understand the hierarchy of the concepts, mentioned above, one needs to distinguish between the terms methodology and methods. The definitions of the (*Merriam-Webster dictionary, 2004*) are the following

- Methodology:
  Etymology: New Latin *methodologia,* from Latin *methodus* + *-logia* –logy
  1 : a body of methods, rules, and postulates employed by a discipline : a particular procedure or set of procedures
  2 : the analysis of the principles or procedures of inquiry in a particular field
- Method
  Etymology: Middle French or Latin; Middle French *methode,* from Latin *methodus,* from Greek *methodos,* from *meta-* + *hodos* way
  1 : a procedure or process for attaining an object: as **a** (1) : a systematic procedure, technique, or mode of inquiry employed by or proper to a particular discipline or art (2) : a systematic plan followed in presenting material for instruction **b** (1) : a way, technique, or process of or for doing something (2) : a body of skills or techniques
  2 : a discipline that deals with the principles and techniques of scientific inquiry
  3 a : orderly arrangement, development, or classification; **b** : the habitual practice of orderliness and regularity
  4 : *capitalized* : a dramatic technique by which an actor seeks to gain complete identification with the inner personality of the character being portrayed.

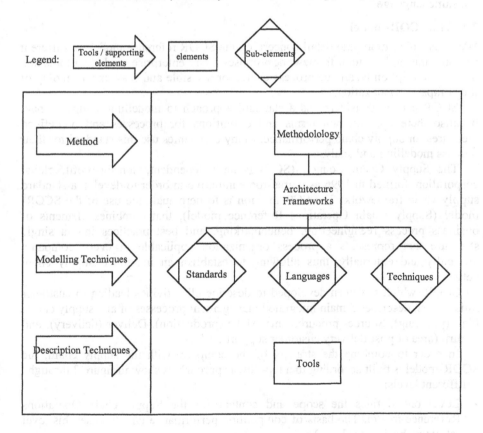

Fig. 1 State of the art overview

One can assume that the term methodology contains one or more methods. The state of the art concerning the architectural frameworks, methods, methodologies etc. can be represented like in Fig. 1.

In the next section, we will tempt to present two different modelling techniques, which put the processes in the heart of the company's transformation actions thanks to the process reengineering approach. We will try to explain how the use of a standard framework and business process reengineering (BPR) can help defining, organizing, and implementing optimized processes in a supply chain network.

3. TWO MODELLING TECHNIQUES IN DETAIL

A process is a continuation of supple steps and is based on knowledge management. It needs to be built up in a manner to best serve customer needs and requirements at the lowest cost for the company. Thus one can say that the difficulty of the process implementation does not only lie in the complexity of its realization and execution, but also in the conducting of transverse change.

In order to create a mutually accepted and common understanding, one needs to talk the same language.

3.1 The SCOR-model

We chose to go more into detail concerning the SCOR reference model for different reasons: this architectural framework proposes many enterprise modelling tools and uses as description technique process charts for a visible and easy understanding of the company organization.

SCOR can be considered as a standard approach to modelling a supply chain because there are common terms and definitions for processes and predefined measures for supply chain performance. Many companies use this standard for their process modelling and analysis.

The Supply Chain Council (SCC) is an independent, non-for-profit, global corporation, formed in 1996 as a grassroots initiative in order to develop a standard supply chain framework. The SCC mission is to perpetuate the use of the SCOR-model (Supply Chain Operations Reference model), that combines elements of business process reengineering, benchmarking, and best practices into a single structure that represents a process organization applicable in every company, internally and externally, thus allowing to establish an integrated supply chain network.

SCOR, which has been developed to describe all activities leading to customer satisfaction, describes 5 main integrated management processes of any supply chain: Plan (planning), Source (procurement), Make (production), Deliver (delivery), and Return (area of post-delivery customer support).

In order to combine the strategy of the company with the supply chain, the SCOR-model is built according to a top-down approach as show in figure 2 through 3 different levels:

- Level one defines the scope and content for the Supply chain Operations Reference-model. The basis of competition performance targets is set this level (strategy, business rules...).

- At level two, a company's supply chain can be "configured-to-order" from core "process categories." Companies implement their operations strategy through the configuration they choose for their supply chain (product environment...).
- Companies "fine tune" their Operations Strategy at level three. This level defines a company's ability to compete successfully in its chosen markets, and consists of: process element definitions, information inputs and outputs, process performance metrics, best practices, system capabilities required to support best practices, and systems/tools.

Companies implement specific SCM practices at level four, which is not in the scope of SCOR because it concerns the unique circumstances of each company and cannot be part of a structure which is applicable to any company.

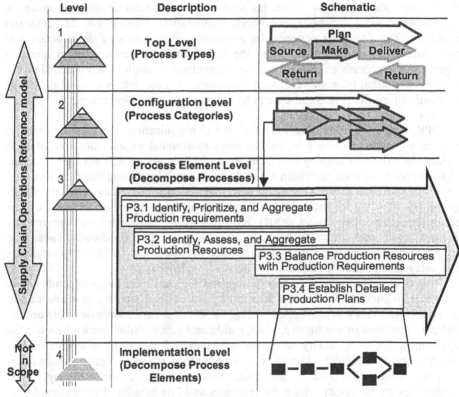

Figure 2: Three Main Levels of SCOR

Even though SCOR considers technology as a tool that can support a better company organization, the management processes are the ones which are classified as most important sp that a true interdependency chain of great value between its actors is described and the internal and external company relations are dealt with in an optimal way.

However, the SCOR-model does not have all the answers. There are processes which cannot be described by SCOR. For example, training, marketing or quality are processes out of scope of the SCOR-model.

3.2 BPR methodologies

There are different methodologies for modelling and remodelling processes, called business process reengineering (BPR). BPR is about radical change in the way in which an organization performs its business activities and it involves the rethinking of the business processes followed by their redesign to enhance all or most of its critical measures (cost, service quality, staff dynamics...) (Grover and Malhorta, 1997).

BPR can be done following different viewpoints: process, information technology, organization according to (Gilmour, 1999); process, market or channel according to (Bowersox and Daugherty, 1985) and (Clinton and Closs, 1997).

One can comprehend BPR through different ways: one can analyze what work is done, who does it and how, when this work is done and who the decision maker is. A process includes functions, behaviors, organization, information, decisions and resources. The functions concern the activities and elements of the process. The behaviors focus on the "when" and the "how" of the process. The organization represents the process execution and the mechanisms through which interaction and transfer of content takes place. Information represents the details or entities that are manipulated by the process; they can be data and relationships associated with the process.

BPR, hence the transformation of a vertical organization (traditional hierarchy) into an organization, based on its processes (horizontal organization), requires to rethink the existing company processes. Nevertheless, one should not forget that the different actors of a supply chain are at different work levels and that therefore they do not consider the process the same way. The vision of these actors is thus neither global nor the same for each of them. Their performance objectives are not the same either (Nathalie Fabbe Costes, 2004). For instance, for those who are concerned by physical flows, cost control is the most important performance indicator. Those who are concerned by quality have totally different priorities like the delivery reliability and delay optimization to improve and secure the physical flows.

This means that the process functioning and the used technology depends totally on the motivation of people and their ability and will of learning and adapting to change. Difficulties can also appear during the information transmission between the different functions of a company. Thus, a clear and simple collaboration between the different actors is necessary to create an integrated and transverse organization (Kramer and Tyler, 1995). Therefore, one needs to describe processes, which allow to share information and to accelerate the reactivity of the company. Good interactions are necessary within the processes and this in spite of the different time and culture horizons and different objectives of all actors. That's also why simplifying and standardizing have become key words today because more a product or an information circulate with different functions to follow a process flow, more there are risks of delay, bad interpretations, and value destruction.

BPR concerns the reorganization of tactical and strategic processes, made in light of people, process, and technology and it is always guided by an "as is – to be" methodology of understanding the current process situation before making any changes.

In the following two sections, we will try to clarify both process modelling techniques presented above through two case studies: one using the SCOR-model, and one using BPR for analyzing and optimizing business processes.

4. FIRST CASE: PROCESS IN SCOR SCOPE

Being considered as a reference model of international standards, SCOR can be considered as a process guide that includes performance indicators, best practices, and benchmark information. The following process analysis will rest on three modelling stages: the SCOR process charts are used for the process description and visualization; common notations and definitions help to measure supply chain performance and get to the desired process situation; and thus the final process description and validation.

4.1 AS IS process description & SCOR comparison

First of all, the material flow and the information/work flow of a process are described in an AS IS situation and compared with the flow proposed by SCOR (figure 3). When this is done, the actual process situation and SCOR allow determining all barriers in the process structure and its fluidity bottlenecks.

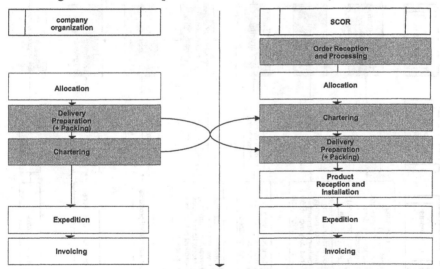

Figure 3: AS IS company and SCOR comparison

4.2 SCOR implementation & KPI/BP analysis

SCOR being a guide of process and sub-process identification, and optimization opportunity identification, this step concerns the SCOR implementation in the AS IS situation in order to compare all data (inputs, outputs, sub-processes…).

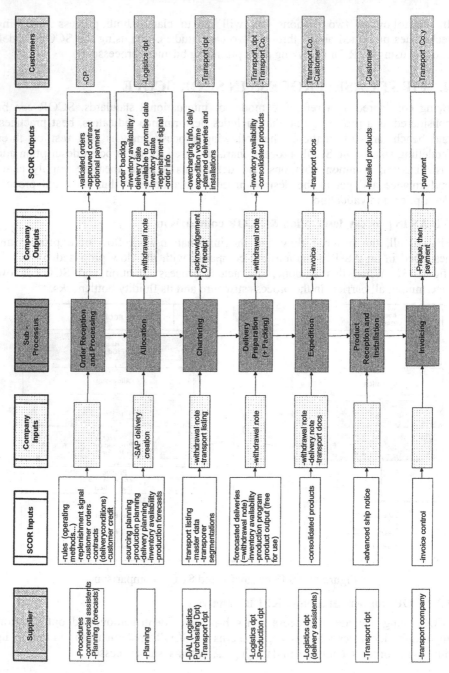

Figure 4: SCOR Implementation in an AS IS Process

As shown in Fig. 4, the model defines many different sub-process and their inputs and outputs so that the company can check and compare those with the existing ones and analyze their performance. In order to optimize this process, it is of primary importance to include measurement of the internal and external process performance because it allows making different adjustments (design, marketing...).

So as to facilitate the choice of pertinent performance indicators (KPI), SCOR recommends four basic performance attributes (service, cost, reactivity, and resource usage) on which performance measures are built upon.

In order to realize the TO BE status of the new process, action plans have to be determined. At this stage of the process analysis, SCOR also proposes a useful tool: the best practices (BP), which aim at evaluating the maturity of a process to help defining the actions, which are to be undertaken, in order to optimize the process in the best possible way.

4.3 TO BE process validation & process GO LIVE

The last step is then to describe and communicate the TO BE process as shown in figure 5. The process flow is optimized and fluid, and all possible adjustments for the TO BE process have been made.

SCOR is helpful concerning the interactions between the individuals in charge of the sub-processes and other multiple tasks within the process and facilitates the final implementation of the optimized process (GO LIVE).

5. SECOND CASE: PROCESS OUT OF SCOR SCOPE

Business process reengineering methodologies can help to improve and optimize processes that are not in the scope of the SCOR-model because it's a matter of an approach that puts the processes in the heart of the company's transformation actions. This helps to remove any work surplus in the processes and to automate the majority of tasks. It is a matter of exploiting efficiency and productivity opportunities across multiple inputs into the supply chain; establishing hence an integrated supply chain process.

5.1 AS IS description & pointing out of dysfunctions

One first describes the AS IS situation of the process. No comparison with SCOR is possible at this point. Through audits, all interactions between the actors and the different sub-processes are put forward. All process dysfunctions within these are highlighted as one can see in figure 6.

The critical path concerning these dysfunctions can thus be analyzed, stating the main problems in the process.

5.2 TO BE description, analysis, & action plan implementation

The next step concerns the description of the TO BE process by putting in place different action plans, which emphasize for each actor where his improvement areas are, concerning which activities. This is done through regular meetings with all actors for a simple process, and as for complex processes through meetings with all group pilots.

5.3 TO BE process, action plan validation & process GO LIVE

At the next stage of process optimization, priorities for all action plans are set as shown in figure 7.

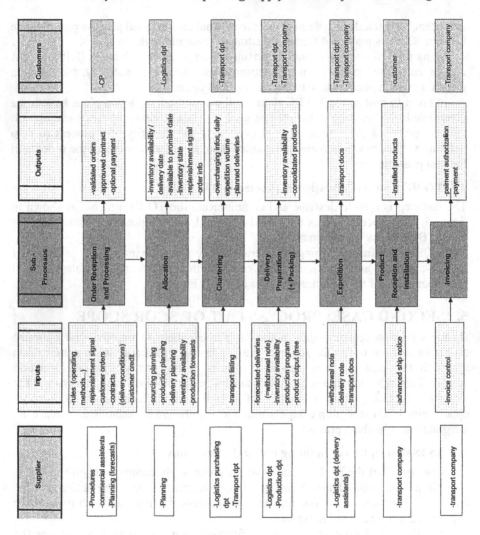

Figure 5: TO BE Process

In order to follow-up these plans of action, the regular meetings help to put forward new questions and/or dysfunctions concerning the process and to indicate the plans of action, which succeeded (GO Live). During the last piloting meeting, every plan of action is reviewed and validated thus determining the success of their implementation (GO LIVE).

6. LESSONS LEARNED

To guarantee the lasting quality and the efficiency of rewritten and optimized processes, it is imperative that the company's reorganization is in the scope of strategic company processes. It is also essential to appoint the individuals in charge of each process and sub-process so that their continuous improvement is assured. In order to reach the performance optimum of the "new" processes, it is useful to gather the operational units that depend hierarchically on the person in charge of the process. That way, neither the different functions (production, marketing, etc.) nor the different divisions within a company (products, markets) constitute an obstacle for these individuals to use their different resources and competencies in an optimal way. At this point of company reorganization SCOR and BPR help gathering all efforts and contributions of each individual of the company so as to combine them in a common process of value added.

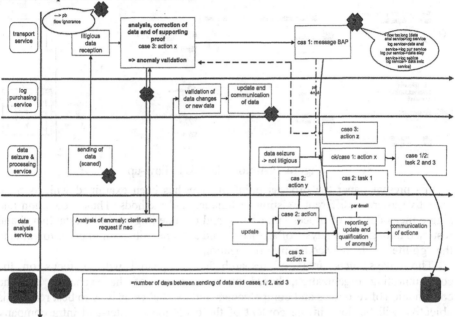

Figure 6: AS IS Process Flow

However, a company should not embark on such a process reengineering analysis without having profoundly thought about the reasons and the possible obstacles concerning this company and process reorganization. Using SCOR and BPR for the process optimization, one needs to simplify and base the process configuration on two levels: the material flow and the information/work flow. Once, the processes described and analyzed, relevant and adapted indicators and BP should be used to measure the success of the efforts made.

Organizations, which are based on processes, seem to be most capable to learn and generate continuous progress because they are open minded aiming at external objectives (customer satisfaction, service level, etc.). Companies want to become best-in-class in their industry sectors and markets and seek to become most agile thanks to multifunctional entities.

Process reorganization needs to be made in light of people, process and technology. Not only the tactical and strategic processes need to be optimized, but also the individual users motivation and the technology they use to operate. Without all three molded into a final methodology, continuous improvement cannot take place.

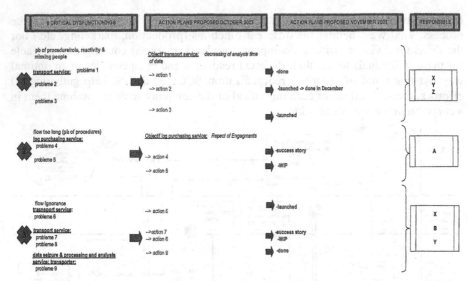

Figure 7: Action Plans & Follow-up

In this present paper the process point of view has been examined with as main objective to cross different existing techniques and methods. Thus a common and single methodology of process modelling and remodelling could be established and resumed through the analysis of an "as is" and a "to be" situation in order to validate the "go life" of the new company organization.

These days, new areas of research have appeared with a focus on the communicating organization, based on mechanisms of the exchange and the coordination of resources and competencies. Future studies on how to best reach this objective will be done in the context of the evolution of inter-and intra company relationships. Information exchange and information sharing will be the main focus concerning the organizational change, from simple contractual relationships until collaborative supply chains creating supply chain networks (alliances and partnerships).

REFERENCES

Grover,V., Malhorta,M.K (1997). Business process reengineering: A tutorial on the concept, evolution, method, technology and application. *Journal of Operations Management*. no. 15.

Hammer,M., Champy,J. (1993). Reengineering the corporation: A manifesto for business revolution. Harper Collins. New York.

Davenport,T.H. (1993). Process Innovation: Reengineering Work Through Information Technology. In Harvard Business School Press. Boston.

Stevens, G. (1989). Integrating the supply chain. *Physical Distribution & Materials Management*. no **19** 8. pp.3–8.

Kramer, R.M., Tyler, T.R. (1995). Trust in Organizations: Frontiers of Theory and Research. Berkeley, CA, Sage Publications.

Porter, M.E. (1985). Competitive advantage: creating and sustaining superior performance. New York, Free Press. Chapter 2.

Lynch, D.F. (2000). The effects of logistics capabilities and strategy on firm performance. *Journal of Business Logistics.* **21** 2. pp.47-67.

Mentzer, M.S. (1999). Two heads are better than one if your company spans the globe. In the Academy of Management Executive: no 13. pp89–90.

Bal, J. (2004). Process Analysis Tools for Process Improvement. Proc International Manufacturing Centre, University of Warwick. pp1-3

Merriam-Webster *dictionary.* (2004). http://www.m-w.com/netdict.htm, March 29th

Gilmour, Peter. (1999). Benchmarking supply chain operations. *International Journal of Physical Distribution & Logistics Management.* no **5** 1. pp.283–290.

Bowersox, DJ.J, Daugherty, P.J., (1985). Emerging Patterns of Logistical Organization. *Journal of Business Logistics:* no **8** 1. pp.46-60.

Clinton, S.R., D.J. Closs. (1997). Logistics strategy: does it exist? *Journal of Business Logistics*: no **18** 1. pp.19-44.

Fabbe Costes,N. (2004). Maîtriser le temps des processus logistiques pour créer de la valeur? Proc Carrefours Logistiques, Paris, France. pp3-7.

29. Toward the Knowledge-based Enterprise

Raffaello Lepratti[1], Jing Cai[2], Ulrich Berger[1] and Michael Weyrich[2]

[1]*Brandenburg University of Technology at Cottbu, Chair of Automation Technology*
Email:[lepratti;berger.u@tu-cottbus.de]
[2]*DaimlerChrysler AG, Information Technology Passenger Car*
Email:[jing.cai;michael.weyrich@daimlerchrysler.com]

In order to support European industry in its transition process towards the knowledge-based enterprise a set of novel information-based tools for enabling knowledge, skill and data transfer is needed. Their design depends on the organic and functional enterprise infrastructure features and relations between the heterogeneous agents involved across the whole value added chain. This paper presents two approaches aiming at overcoming interoperability barriers arising in communication process among humans and machines. First one is an ontological approach, which focuses on computer-supported human collaboration and human-machine interaction by means of natural languages, enabling semantic independence of shared knowledge and data contents. The second one proposes an approach for machine data exchange and sharing, applying standards as highly extruded common knowledge.

1. INTRODUCTION

European industry is in transition process from a mass production industry towards a knowledge-based customer- and service-oriented one, which aims at a production model on demand, mass customization, rapid reaction to market changes and quick time-to-market of new innovative products.

In this transition, it faces the challenge to produce according to a *lot-size one paradigm* at low cost and high quality. A customizing in final products leads to a strong individualization of product features, which influences the normal course of the product life cycle making risky investments and resource plans in production.

Following this vision, networked, knowledge-driven and agile manufacturing systems stand out as necessary key elements towards this future production scenario, which shall allow European industry long-term competitiveness improvements above all by added values in product-services.

The context of collaborative engineering and manufacturing has witnessed a striking expansion in all fields of the value added chain. In spite of a successful employment of a set of information-base tools, knowledge, skill as well as data transfer it shows many inefficiencies and hurdles (Goossenaerts *et al.*, 2002) and still represents the major problem toward the achievement of a suitable and efficient infrastructure ensuring the establishment of the knowledge-based enterprise.

Therefore, research efforts and technology development on information infrastructures are ongoing, addressing a.o. information architecture, methodologies, ontologies, advanced scenarios, standard machining tools and services. These should contribute in providing a holistic solution for a knowledge-based engineering and manufacturing architecture, which must feature a systemic dynamic learning behavior, where innovation emerges from new complex interaction forms between integrated technologies, human resources, management and organizations in all phases of the valued-added chain, i.e. (i) production preparation, (ii) planning and programming as well as (iii) process execution. Hence, new solution approaches shall allow above all operational knowledge acquisition and knowledge feedback in computer-based collaborative engineering and manufacturing.

Both, already existing and arising industrial know-how should be gathered together either manually from human experiences (usually by experts) and by use of intelligent cognitive sensing systems, or automatically derived from human interventions (e. g. short process corrections at shop-floor level) or other machine equipments. Through knowledge retrieval mechanisms, also machines acquire intelligence reaching the necessary challenging level of efficiency and robustness. Such visionary workflow architecture is shown in Figure 1.

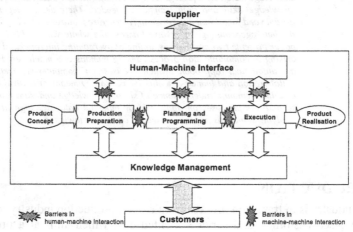

Figure 1: Basic buildings block structure of the knowledge-based enterprise

2. INTEROPERABILITY ISSUES

If, on the one hand, the enterprise structure proposed in Figure 1 represents an adequate solution to meet the growing market challenges, on the other hand, it shows to be also ambitious in connection with its functional requirements.

A smooth global information flow between all actors involved in this process is the most important aspect for ensuring the correct process behaviour. However, still too many complications evolve when trying to find standard criteria for interoperability across the entire heterogeneous human qualifications and machine programming languages setting.

As already stressed in (Lepratti and Berger, 2003), on the one hand, possible understanding problems arise, while two persons try to communicate with each other as consequence of discrepancies in their cultural and/or professional backgrounds. They are incline to cognitive perceive and mentally process same situations of the real world in subjective ways, referring these to different models (so called mental models). Thus, also two interaction partners, even though speaking the same language and using identical terminologies, can misunderstand each other, since vocabulary terms represent merely etiquettes of cognitive categories.

On the other hand, in machine-to-machine communication, incompatibilities in data structure or code languages (different syntax and semantic rules) are major reasons of impediments in transferring information from a software system to another one, which uses distinct technology solutions.

3. STATE-OF-THE-ART

Some standards enabling interoperability in engineering and manufacturing have been already successfully employed. Some relevant examples are here described:

The KIF (Knowledge Interchange Format) (Genesereth and Fikes, 1992) as well as the KQML[82] (Knowledge Query and Manipulation Language) allow interchange of information and knowledge among disparate software programs - either for the interaction of an application program with an intelligent system or for two or more intelligent systems - to share knowledge in support of co-operative problem solving with the possibility to structurally represent knowledge at a meta-level.

The STEP ISO 10303 (STandard for the Exchange of Product data) (Fowler, 1995) addresses to the exchange and sharing of information required for a product during its life cycle (such as parametric data like design rationale, functional specification and design intent). STEP is nowadays a well-known standard for real world product information modeling, communication and interpretation. Some examples are STEP AP-203 (Application Protocol) (Configuration Control for 3D Design of Mechanical Parts and Assemblies), STEP AP-214 (Core Data for automotive Mechanical Design Processes), STEP AP-224 (Mechanical Parts Definition for Process Planning Using Machining Features), STEP-240 (Machining Process Planning) and STEP-NC (ISO 14649 Industrial automation systems and integration Physical device control) (Richard et. al., 2004).

Finally, CORBA (Common Object Request Broker Architecture) (Object Management Group, 1995) and COM/DCOM83 (Distributed Component Object Model) provide neutral - both platforms and languages independent - communication between remote applications based on object oriented distributed technology, allowing different clients and/or servers connected within a network to live as individual entities able to access to the information they need in a seamless and transparent way. Both solutions aren't standards but are widely used e. g. for the development of agent-based systems.

While shown standard and further solutions have been successfully proved and employed, they are often too strong task-oriented in their applications or remain just one-off solutions. At present, a generic knowledge management concept for architecture as shown in Figure 1 has not been developed. The extent of knowledge and skill transfer in engineering and manufacturing is often strong limited.

New requirements for innovative and generic holistic knowledge-based enterprise architectures such capability in upgrading different heterogeneous systems, transparent data exchange among them, distributed open environments and improved information sharing go therefore beyond the actual state-of-the-art. To

---

[82] http://www.cs.umbc.edu/kqml/
[83] http://www.microsoft.com/com/default.asp

foster the transition process of European enterprises, big efforts in the research of further suitable concepts are still needed.

In the next Section two approaches, which aim at improving interoperability, will be presented. However, while the first one bases on the use of ontologies and addresses mostly the semantic standardization of computer-supported human-human communication as well as human-machine interaction by the use of natural languages, the second one focuses on overcoming complications in data exchange among heterogeneous software applications of machines and equipments.

4. THE ONTOLOGICAL APPROACH

4.1 The role of ontology

In today's production systems the development of communication and production technologies becomes not only more efficient but also more complex. The employment of such technologies represents challenges facing professional and cultural requirements of personnel, which has to work with. This stresses the importance of a novel knowledge management solution able to archive semantic standardization of knowledge contents and provide task-oriented as well as user-based redistribution of stored information. A corresponding building block knowledge management architecture concept is illustrated in Figure 2.

However, it is difficult to identify a unified knowledge form, when considering the different nature of tasks needed across the whole value added chain. According to Figure 1, three different knowledge forms are identified: (i) The so called 1-D interaction form, i. e. textual, is for instance still the most common way used for information exchange in scheduling tasks during both product preparation and planning phase. (ii) 3-D technologies of the Digital Factory have gained importance in the last years above all with regard to process & planning activities and represents the most profitable way to design production environments (e. g. planning of human and machine activities and machine programming). Finally, (iii) graphical (2-D) technologies such as interactive platform systems support user-friendly on-line process corrections at shop-floor level.

Under these circumstances the need of standard procedures for an efficient processing of knowledge contents, which are able to acquire, filter and retrieve data of different multi-dimensional sources, is assuming more and more an essential role.

In Figure 2 the core of the architecture is represented by the Ontology Filtering System (OFS). It plays this important role enabling semantic autonomy of different information contents independently of their nature of being. All multi-dimensional data sources mentioned above could be processed in an equivalent manner, i. e. knowledge contents of different forms are stored in the OFS knowledge data base in a standard data structure according to a pre-defined set of semantic definitions and relation rules. This offers a number of advantages: On the one hand, it supports knowledge retrieval and representation in a task-oriented manner according to the specific user requirements and, on the other hand, it facilitates the computer-supported knowledge exchange among humans or between human and machine avoiding possible semantic ambiguities of knowledge contents.

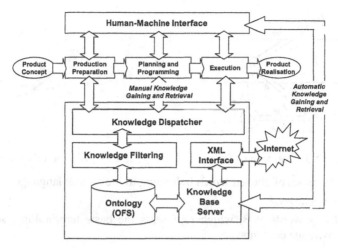

Figure 2: Ontology-based knowledge management architecture

In the next Section, a mathematical description of the applied ontology is presented. It has been developed and already experimentally demonstrated within a research initiative focused on the use of natural languages in the automation technology domain (see (Lepratti and Berger, 2004)).

4.2 The Ontological Filtering System (OFS)

Although the use of natural languages still represents an hazard solution approach due to possible misinterpretations, which could arise during the interaction process as consequence of syntactical, lexical and extensional ambiguities connected to their domain of use, they represent the most familiar and understandable communication form for human beings. Following Winograd's theory (Winograd, 1980), assuming that there is no difference between a formal and a natural language, one finds proper reasons for all the efforts to formalize knowledge expressed by natural languages.

The so called Ontological Filtering System (OFS) removes possible ambiguities in natural languages by means of a semantic network, in which words are chained together hierarchically per semantic relations. This network consists, on the one hand, of a set of words selected for a specific domain of application and used as key words, in order to standardize information contents for the machine data processing. On the other hand, it encloses a set of additional words, which could be used from different persons in their natural communication, since there are more ways to express the same knowledge meaning. These words could have different abstraction degrees in their meaning (so called *granularity*). Thus, some words are more general in their expression than others, while others can go very deep with their meaning. A simplified example of this semantic network is given in Figure 3.

According to their specification level all words – key words and additional words - are linked together by means of semantic relations such as hypernymy, hyponymy, synonymy or antonymy. A parser within the OFS processes knowledge contents and leads back words meanings to these ones belonging to the set of pre-defined key words. In this way, one can say, the OFS provides a semantic filtering function. A mathematical description could better explain how it works.

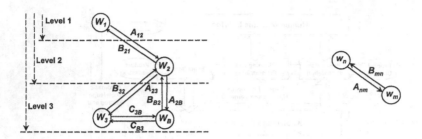

Figure 3: Example of OFS   Figure 4: Simple relation in OFS

Considering W as set of chosen words belonging to the natural language:

$$W_{NL}=\{w_1, w_2, ..., w_n\} \tag{1}$$

and a set of key words WB, which represents the basis terminology, selected to formalize knowledge contents:

$$W_B=\{w_{1B}, w_{2B}, ..., w_{nB}\} \tag{2}$$

Using following set R of semantic relations of natural language: hypernymy (A), hyponymy (B), synonymy (C) and antonymy (D):

$$R=\{A, B, C, D\} \tag{3}$$

one can define the OFS network as following ordered triple:

$$OFS=<W, R, S> \tag{4}$$

where W represents the addition $WNL \cup WB$ and S takes into consideration the specification level of the elements of W. According to Figure 3, relations between the elements of W can be included in a relation matrix $\Re$:

$$\Re = \begin{bmatrix} 0 & A_{12} & 0 & 0 \\ B_{21} & 0 & A_{23} & A_{2B} \\ 0 & B_{32} & 0 & C_{3B} \\ 0 & B_{B2} & C_{B3} & 0 \end{bmatrix} \tag{5}$$

Multiplying R by the transposed vector WT.

$$\Im = W^T \cdot \Re = \begin{bmatrix} W_1 \\ W_2 \\ W_3 \\ W_4 \end{bmatrix} \cdot \begin{bmatrix} 0 & A_{12} & 0 & 0 \\ B_{21} & 0 & A_{23} & A_{2B} \\ 0 & B_{32} & 0 & C_{3B} \\ 0 & B_{B2} & C_{B3} & 0 \end{bmatrix} \tag{6}$$

one attains the system of equations $\Im$, which reflexes the structure of OFS:

$$\Im = \begin{cases} W_1 = A_{12} \cdot W_2 \\ W_2 = B_{21} \cdot W_1 + A_{23} \cdot W_3 + A_{2B} \cdot W_B \\ W_3 = B_{32} \cdot W_2 + C_{3B} \cdot W_B \\ W_B = B_{B2} \cdot W_2 + C_{B3} \cdot W_3 \end{cases} \tag{7}$$

Considering Figure 4 one deduces the simple semantic relation (8)

$$
\begin{cases}
W_n = A_{nm} \cdot W_m \\
W_m = B_{mn} \cdot W_n
\end{cases} \Rightarrow A_{nm} \cdot B_{mn} = \gamma
\tag{8}
$$

where γ represents an empty element, since paths from W_n to W_m and vice versa over A_{nm} and B_{mn} are equivalent. Similarly, it counts also for:

$$
C_{nm} \cdot C_{mn} = \gamma
\tag{9}
$$

Resolving (7) as functions of W_B using (8) and (9) one obtains following results:

$$
\mathfrak{I} = \begin{cases}
W_1 = A_{12} \cdot A_{23} \cdot C_{3B} \cdot W_B + A_{12} \cdot A_{2B} \cdot W_B \\
W_2 = A_{23} \cdot C_{3B} \cdot W_B + A_{2B} \cdot W_B \\
W_3 = C_{3B} \cdot W_B + B_{32} \cdot A_{2B} \cdot W_B \\
W_B = W_B
\end{cases}
\tag{10}
$$

Every equation of (10) gives the number of different semantic paths, which lead a specific element W_n to the corresponding key word W_B.

The structure of the Ontology Filtering System presented in this paper could be easily extended to any other languages or data structures. As in (Guarino, 1998) accurately treated, when considering a further logical language L with a specific vocabulary of symbols V, one can rearrange the definition used above assigning elements of a specific application domain D to symbols of V and elements of R to predicate symbols of V.

5. ENGINEERING DATA EXCHANGE APPROACH

In Section 4, the ontological approach for knowledge management shows how knowledge contents expressed in specific language can be computer-processed, i. e. standardized in their meaning. Also data exchange among heterogeneous software applications of machines and equipment in engineering represents an important issue towards the development of the holistic architecture of Figure 1.

5.1 Standard for Engineering Data Interoperability

Nowadays in engineering domains, the data communication process represents a crucial aspect within the digital product creation process. On the one hand, heterogeneous set of software tools is applied. Different data formats and structures describing same engineering object lead to incompatibilities. Furthermore, with development of new information technology, the more digital simulation tools have been used for complicated scenarios, the more complex data become. It ranges from plain text to 2-D, 3-D geometries with semantic information. On the other hand, the data communication within the extended enterprise makes the exchange of data with customers, partner or supplier for a specific engineering object more complex. Therefore, data compatibility of various engineering tools in the extended enterprise represents the essential requirement in exchanging data in different applications.

The following second approach bases on the use of knowledge-derived engineering standards, with which the encompassing architecture of Figure 1 should be composed, its components functions specified and validated in real scenarios.

5.2 Knowledge-derived Standard-based Data Exchange Architecture

The High Level Architecture (HLA, IEEE 1516) for enterprise-wide and in external supply cooperation respectively, describes the test platform for performing distributed simulations. As to the data exchange requirements, i. e. engineering data compatibility, engineering knowledge retrieval and application, a corresponding architecture should be built up. The architecture of engineering data exchange using knowledge-based standard is depicted in Figure 5. Its components are:

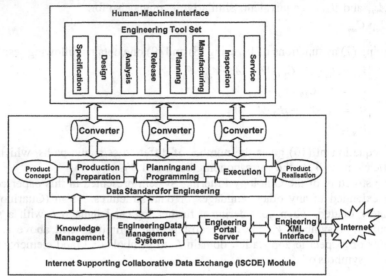

Figure 5: Standard based knowledge management architecture in product creation

1. Engineering tool set: A typical HMI, which is an aggregation of engineering IT tools and computer integrated manufacturing (CIM) machine for engineering i.e. specification, design, analysis, planning, manufacturing, inspection, services, etc. This interface for computer application must support like: (i) access to data, (ii) exchange of information and (iii) Multiple views of product data.

2. Converter: The interface between the software (machine) and knowledge-based engineering standard back-bone. The standard should fit for the entire digital product creation process, i.e. product concept, planning and programming, execution, and finally product realization. For each milestone in this process chain, a universal converter should be available.

3. Engineering data management: It synchronizes i.e. saves/provides engineering data for corresponding tools and realizes standard-specified data management, using data bank functions, i.e. data configuration and interface accessibility.

4. Connection mechanism for external integration: It is based on the net interface to the engineering data management system. This connection includes the engineering portal and engineering XML interface for the Internet application. For building an extended enterprise, this connection is an important element.

5. Knowledge management: (described in the Section 4) Its connection to the data standard for engineering is described in Section 5.3.

5.3 Standard as Common Knowledge Representation

A standard is a consistent definition or behaviour established by custom, authority, or consensus. In another word, standard is a higher knowledge level as greatest common denominator syndrome. Standards must be taken into account regarding too many variants or options to maximize performance. The data standardisation is a process of knowledge accumulation, sharing, description, application guiding. Each entity in standard can represent the knowledge in all relevant engineering fields and be directly used in engineering tool set for the intelligent application.

Connection between engineering data standard with knowledge management i.e. ontology-based knowledge management architecture, can be described as follows: i) Knowledge management retrieves the knowledge from the engineering domains, converts the knowledge into ontology through the knowledge dispatching system, then knowledge filtering system, into the ontology data management system. ii) The standard structures and data formats are defined only after standardization of ontologies, especially the ones in the entire supplier chain. Finally, the conversion of standardized ontology into specific standard data formats is to be done.

5.4 Scenario Process Planning

In engineering domain, rapid system functional development of Computer-Aided Design (CAD) and Computer-Aided Manufacturing (CAM) are progressed. Computer Aided Process Planning (CAPP) system working with 3D-Geometry is far from the engineering interest, because: (i) A CAPP system has the upstream (CAD) and downstream (CAM) application which are at dynamic unbalanced technology levels, this brings the complexity to be integrated. (ii) CAPP itself is very complex. Process planning uses not only planer know-how but also a variety of heuristic rules and logical decision. Conventional computer algorithmic programs can do little with regard to the logic inference. (iii) The most important reason is that there is still missing standard description of manufacturing process. The standard for CAx domain is usually the STEP format. STEP 3D product data exchange has been achieved in an industry-practical way (ProSTEP, 2004). However, STEP is still a development activity and has its limited consideration fields (Michael, 2001). For CAPP system such as AP 240 process planning and STEP standard definition for machine tool is in development. (iv) Due to the application of heterogeneous CAx systems in supplier chain, the process data exchange is blocked (Zhang and Alting, 1994). Thus, knowledge and standard for process planning should be acknowledged through knowledge-derived standard-based data exchange architecture.

The scenario for the process planning can be described as in Figure 6. The OEM planning engineering cooperates with the machine supplier using own planning systems for process definition, and supplier integrates the entire process and orders the equipment and cutting tool from its own sub-suppliers. The Internet supporting collaborative data exchange module (ISCDE) is applied. The connection to the engineering data management system with extension to the supplier using engineering portal server and engineering XML interface is necessary to exchange data with the supplier, finally to build up the extended enterprise.

As to the knowledge management for data standardization of process planning is based on this defined approach is still in progress. This approach promises the interoperability in the collaborative process planning in the extended enterprise. Also this knowledge-based standard makes the entire CAx-engineering in up /downstream application continuously. The next step is to connect the knowledge system to retrieve the data standard for the specification of the engineering data management systems in the extended enterprise for the process planning.

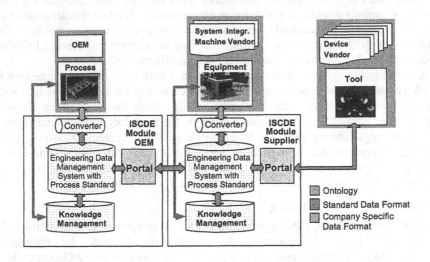

Figure 6: Scenario of collaborative process planning using ISCDE module

6. CONCLUSIONS

In product creation process, knowledge acquisition, retrieval and application can be found in every manufacturing and engineering phase. The knowledge management should embrace the entire enterprise structure, giving the necessary structure flexibility to need the growing market challenges such as rapid product creation with higher quality and short time-to-market strategy. As shown in Section 4, ontologies help in standardizing knowledge and data semantic contents in the communication among humans and in the human-machine interaction, while in Section 5 knowledge-derived standard based data exchange architecture is defined. The components in this architecture are described. An application scenario regarding the process planning is developed, further work to realise this architecture is concluded.

REFERENCES

Fowler, J. (1995) STEP for Data Management, Exchange and Sharing. Technology Appraisals.

Genesereth, M. R. and Fikes, R. E. (1992) Knowledge interchange format, version 3.0. Reference manual. Technical report, Logic-92-1, Computer Science Dept., Stanford University.

Goossenaerts, J. B. M.; Arai, E.; Shirase, K.; Mills, J. J.; Kimura, F. (2002) Enhancing Knowledge and Skill Chains in Manufacturing and Engineering. In: Proc. of DIISM Working Conference 2002.

Guarino, N. (1998) Formal Ontology and Information System. In: Proceedings of FOIS'98, Trento, Italy, 6-8 June 1998, Amsterdam, IOS Press, pp3-15.

Lepratti, R.; Berger, U. (2004) Enhancing Interoperability through the Ontological Filtering System. In Processes and Foundations for Virtual Organisations.

Proceeding of the 5th IFIP Working Conference in Virtual Enterprises (PRO-VE 2004), Kluwert, Boston-London, 2004, (in print).

Michael J. P. (2001) Introduction to ISO 10303 - the STEP Standard for Product Data Exchange, *Journal of Computer Information Science Engineering.* 1(1), pp102-103.

N. N. (2004): OpenDESC-CAD Data Conversion made easy, http://www.prostep.com/en/solutions/ opendesc/funktion, ProSTEP Webpage,

Object Management Group (1995) The Common Object Request Broker Architecture and Specification (CORBA). Revision 2.0, July, 1995.

Richard, J.; Nguyen, V. K., Stroud, I. (2004) Standardisation of the Manufacturing Process: IMS/EU STEP-NC Project on the Wire EDM Process. In: Proceedings of the IMS International Forum 2004, May 17-19, Cernobbio-Italy, pp1197-1204

Winograd, T. (1980) What does it mean to understand language? *Cognitive Science* 4, 1980, 209-241.

Zhang H.C., Alting L. (1994) Computerized manufacturing process planning systems. Chapman & Hall, ISBN: 0412413000, pp188- 198.

30.   Strategic Process Integration

Juan Carlos Méndez, [1],

1 Affiliation Email:jcmendezb@adn.com.mx,
www.adn.com.mx

The advent of a global economy is forcing companies to improve
competitiveness more than ever and to increase collaboration by providing for
ICT based interoperability. These needs generate the necessity to focus on
company core processes and increase operational flexibility to satisfy customer
requirements. The paper is aimed on strengthening the enterprise adaptation to
changing markets focusing on the integration between strategic planning and
business processes, using enterprise modeling as documented in CEN/ISO
19439 and 19440.

1. INTRODUCTION

Global markets require companies to improve competitiveness and increase inter-organizational collaboration. In most enterprises their mission, mission-objectives and quality policy have been defined independently of enterprise functionality or the corresponding enterprise model. Reason for this lack of integration between strategic planning and enterprise modeling is the missing formal support for objectives in enterprise process models. This situation arises because the process definition is not well understood, let alone sufficiently standardized in most supporting modeling tools.

Strategic management has evolved from an early emphasis on planning to become a comprehensive management approach that helps organizations align organizational direction with organizational goals to accomplish strategic change (Vinzant *et al*, 1999). Strategic management process is the full set of commitments, decisions, and actions required for a firm to achieve strategic competitiveness and earn above average return. (Rumelt *et al.*, 1994)

In summary, the lack of integration between the entity strategies to the enterprise model leaves great question marks on what functionality is needed to improve competitiveness of the organization and its interoperation with external organizations. The CEN/ISO work on standardization of enterprise modeling, (CEN/ISO 19439 and 19440) has established an enterprise model and modeling language definitions that allow integration of the strategies into the business model, achieving an enhanced vision of enterprise integration. The goal we must seek is to develop the skills of the work force to overcome opposition and to create a unified system of global governance. (Alexander the Great, 330 B.C.)

Practitioners of the strategic planning as well of the enterprise modeling have to their disposal tools like Balanced Score Cards (BSC) (Smith, *et al*, 2002) and Hoshin Kanri (Tennant, *et al*, 2001) to unfold the strategies throughout the company. These tools look for things to do and to establish actions that impel the strategies throughout the organization. The aim of this paper is to have strategic planning become integral part of the enterprise model and, as a consequence, any changes to the strategy will result in changes to the enterprise model accordingly.

Nowadays a valid, robust framework exists that is free of ambiguity and that can manage the integration of enterprise functionality and enterprise strategies. With that, it is possible to clearly identify how the processes are going to contribute to the mission fulfillment in an enterprise model.

The following benefits can be obtained:

- Mission-based company structure.
- Process-based company operation, with each process fulfilling one strategic objective. With clearly assigned process ownership eliminating multiple objective owners.
- Increased operational flexibility due to fast operation reorganization through enterprise model adjustments according to strategic changes made as a result of changing markets.
- Identification of the needed capabilities to improve the competitiveness of the business operation.

This improvement process will apply to the areas of information technologies, human resources and other resources participating in the enterprise processes. We must always remember that behind every successful company, there is a superior strategy. (Markides, 1999)

1.1 Missing a Strategy - Process Integration

Strategy is an integrated and coordinated set of commitments and actions designed to exploit core competencies and gain a competitive advantage. (Hitt *et al.*, 2003) Consequence of the missing link between strategic planning and business modeling is the lack of support for the mission objectives from enterprise processes. .

This is clearly observed when the company personnel has to make a lot of unplanned efforts in order to meet their given objectives and their customer requirements. Another example is the lack of upper level management support for new information technologies implementations when the board does not see any contribution to the company base line.

This missing link leads to very similar ICT implementation across an industry. But with all competitors working with similar technologies, there is no competitive advantage to be gained with ICT. The question is, how do we determine which capability must be enhanced or added to improve our position against our competitors? This question comes up constantly through out company life, since one must continuously determine what technology is going to contribute the greatest value to the organization. But poor integration with the mission will generate a weak vision of the required capabilities.

The same holds for all other operational resources: What capabilities in machinery have to be increased? What skills of the workforce have to be enhanced through training and education to enable them to perform their functions?

In summary, the lack of integration between the mission and its objectives to the processes generates great uncertainties of what functionality is needed to improve the competitiveness of an organization.

2. ENTERPRISE MODELING – FROM MISSION TO OPERATION

Modeling enterprise operation based on its mission and its objectives allows identification of organizational weaknesses and directs leaders of process improvement to follow the strategic path as a solid source to improve competitiveness. It avoids the waste of resources that do not contribute to the value add of the organization.

2.1 The Mission and its Objectives

With a solid mission statement, the company defines what the company is, in order to enable the organization to direct its efforts to fulfill it. With a well defined mission employees will not be confused on why the company is in a particular product segment and consequently, they will focus their energies in fulfilling the mission. One important aspect of the mission is that identifies the primary goals or objectives to be fulfilled by all levels of the organization. "Goals make the mission specific and direct the company toward the future" (Christopher, 1993). These goals or objectives are the pillars on which the company mission is going to rest; allowing the business to obtain its strength and its flexibility.

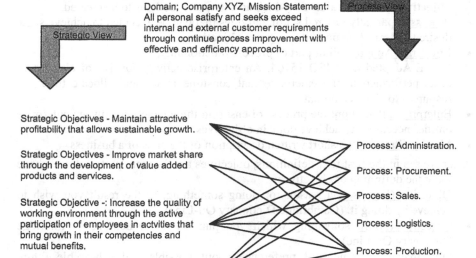

Figure 1: Two views with uncertain relations between objectives and processes.

Great disadvantages for competitiveness materialize when the mission statement does not have a clear link to the company processes, from which they're supposed to receive its strength. (see Figure 1) What we have found in most of the organizations visited in the course of our practice it's that strategic objectives are not related to the operational processes, generating great confusion about how those processes are going fulfill the company's mission and its objectives. The mission only defines a generic path, but does not guide the processes in "how to follow the path".

The main role of strategic objectives should be to identify the functionalities needed for mission fulfillment. Therefore the practitioners of strategic planning defining strategic objectives should think in operational processes, because these objectives will become the objectives of the specific processes or even domains of the enterprise. If they do this the integration of processes and mission becomes a natural step.

Generalizing, with the mission being the reason for the being of the company, we could conclude that the objectives are the reason for the being of the domain and its processes.

2.2 CEN/ISO 19439 and the Strategic Process Integration

Until this moment we have seen the disadvantages of the lack of integration between the mission and its objectives and the processes of an organization. Now we are going to establish the relation that can exists between the mission and the enterprise model by starting from the definitions of GERAM (Generalized Enterprise Reference Architecture and Methodology) (IFAC/IFIP Task Force) that are shaped into a language in CEN/ISO 19439 and 19440.

Let's consider these definitions of CEN/ISO 19439:

- Domain: that part of the enterprise considered relevant to a given set of business **objectives** and constraints for which an enterprise model is to be created.
- Process: a partially ordered set of activities that can be executed to achieve some desired end-result in pursuit of a given **objective**.
- Enterprise Activity: all or part of process functionality,
  NOTE Adapted from ISO 15704. An enterprise activity consists of elementary tasks performed in the enterprise that consume inputs and allocate time and resources to produce outputs.
- Enterprise Integration: the process of ensuring the interaction between enterprise entities necessary to achieve domain **objectives**.
- Mission Statement: a short written description of the aims of a business

After we see in the first four definitions, objectives are the main part of it, let's have a look at the definition of objectives:

- Objective (Purpose): a reason for doing something, or the result you wish to achieve by doing it. (Cambridge Dictionary On-Line)
- Objective (Aim): something which you plan to do or achieve. (Cambridge Dictionary On-Line)
- Objective: a statement of preference about possible and achievable future situations that influences the choices within some behavior. (CEN/ISO 19439)

If both the domain and the process pursuit objectives, and the strategic planning define objectives, these two tasks – strategic planning and enterprise modelling – have to be linked in order to integrate the enterprise strategies into the operational processes.

Therefore, we can establish that the functional description of an organization – the enterprise model - must be completely related to the mission and its objectives. With CEN/ISO 19439 and 19440 the practitioners of strategic planning and enterprise modelling now can and must start their models from the mission of the organization. With this relation we have a strategic direction into our models and thereby will define the functionality according to the basic needs identified in the mission statement.

3. STRATEGIC PROCESS INTEGRATION (SPI)

For incorporation of strategic planning to processes modelling there exist many techniques that allow unfolding strategic targets throughout the organization. The intentions here are not to propose yet another technique, but to adapt the existing international agreements and standards into the strategic planning methodologies. The paper proposes to change from unfolding the strategic planning, to make the strategic planning part of the design of the processes.

Strategic competitiveness is achieved when a firma successfully formulates and implements a value-creating strategy. When a firm implements such a strategy and other companies are unable to duplicate it or find it too costly to imitate, this firm has a sustained (or sustainable) competitive advantage (called competitive advantage). (Maritan, 2001; Helfat, 2000; Barney, 1999)

In order to obtain the Strategic Process Integration (SPI) the following sequence of steps will support practitioners to accomplish the integration between strategic planning and enterprise modelling: Mission, Strategic Objectives, Decomposition of Objectives and Indices, Domain, Process Decompositions and Enterprise Activities, and Complementing Enterprise Model Views.

When these steps were lectured in several companies in Mexico participants were in a dilemma. Either follows what they had learned in the strategic planning or what enterprise modelling was dictating them. Although they all agreed that the effort must be directed by the mission, they found that it was not possible to tie the objectives to the processes.

What they decided to do is to redefine their strategic objectives, respecting the definitions of the enterprise modelling. These objectives redefinition gives the opportunity to define them now oriented to processes. The result was the board of directors' defined new strategic objectives (SO), and managers define new key processes base on those SO with more clarity on how those processes are going to support the strategic.

Two examples of the benefits resulting from this approach: 1) identification of the key processes that fully support the mission; when defining the strategic objectives base on processes, they found three key processes instead of five. 2) the organizational structure is now oriented fully to the processes; with three key processes they reduced the number of directors of first level from six to three, giving each of the new organizations complete control to fulfill their strategic objective.

3.1 The Steps of SPI

Mission

Starting point is the company mission, which must be defined first. It is the departure point of all the processes in enterprise modelling aimed on improving the enterprise operation; especially its competitiveness and interoperability. The mission can be defined with its essentials: Business Type, Strategy, Goals, Performance Measures and Values. Strategic mission is a statement of a firm's unique purpose and the scope of its operations in product and market terms (Ireland and Hitt, 1992). The practitioners may use whatever tool or methodology is best suited for him.

Strategic Objectives

The strategic objectives make the mission specific and direct, orienting the future of the organization. It is important that when defining the strategic objectives one thinks of the generic process that is going to support the specific objective. It should be avoided that the strategic objectives must be shared by different processes, because the dividing of an objective hampers its fulfillment. (see Figure 2)

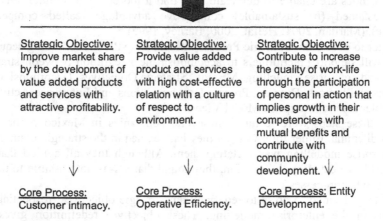

Domain; Mission Statement: All personal satisfy and seeks exceed internal and external customer requirements through continues improvement in process effective and efficiency.

Strategic Objective: Improve market share by the development of value added products and services with attractive profitability.	Strategic Objective: Provide value added product and services with high cost-effective relation with a culture of respect to environment.	Strategic Objective: Contribute to increase the quality of work-life through the participation of personal in action that implies growth in their competencies with mutual benefits and contribute with community development.
Core Process: Customer intimacy.	Core Process: Operative Efficiency.	Core Process: Entity Development.

Figure 2: Decompose mission statement using the SPI approach

Decomposition of Objectives and Metrics

Each strategic objective is decomposed into functional objectives and transformational objectives. The functional objectives are satisfied by the processes or sub-processes and the transformational ones by the enterprise activities. This decomposition simplifies the unfolding of the mission and its fulfillment. The elementary level of the decomposition of objectives is the transformation objectives. (see Figure 3) At the end each objective must have his metrics to control his performance. For this the control system must be define.

Domain

Mission is related to an enterprise domain. The domain therefore is defined by its mission, and then the name of the domain could be based on mission statement definition. (see Figure 2)

Process and Enterprise Activity Decomposition

Using the decomposition of objectives (strategic or functional), lets do just the same at the process level. Assign strategic objectives to processes, functional objectives to sub-processes and transformational objectives to one enterprise activity each. The main change to the actual definition of process decomposition is that we no longer decompose processes; we decompose objectives then name the processes from their objectives. (see Figure 3)

Complementary of Enterprise Model Views

The by now constructed process and activity decomposition allow working with the enterprise model using its views: Functional, Information, Resources and Organization. It is important before starting this aspect remember to verify that objective has his goals or indices. Only with these indices it is possible to find out if the enterprise model has the performance needed for his competitiveness and his interoperability.

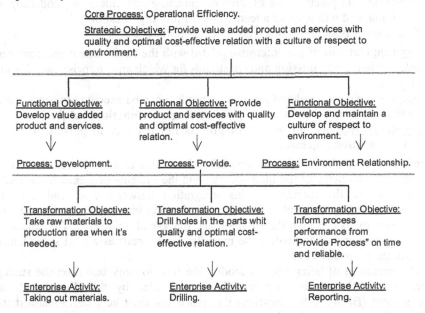

Figure 3: Decomposition and integration of Processes and Enterprise Activities

Developed through these steps of the Strategic Process Integration, the model may be used make all kinds of analysis to determine the improvements needed in the enterprise operation and organization to improve its competitiveness and interoperability. It is needed to improve enterprise activity performance to make real better competitiveness. Two major approaches can be found to make attainable this competitiveness; improving the quality of inputs and improving capabilities. To perform the analysis practitioners must focus on what is the objective that must be accomplish, and make a list of weaknesses the activity has.

After this practitioners can find weaknesses in IT or Machinery or Knowledge or Infrastructure. Let identify the set of weaknesses we can get more cost-effective result, and the put the resources needed in order to straight them. Always remember modifications in the strategic objectives will modify enterprise model; respecting the enterprise model phases as is defined in CEN/ISO 19439.

4. CONCLUSION AND FURTHER WORK

The enterprise modelling serves to capture the reality of an organization. With the explicit integration of the strategic planning in the modelling we strengthen the processes of the company and develop the strategic flexibility. Strategic capability is a set of capabilities used to respond to various demands and opportunities existing in a dynamic and uncertain competitive environment (Harrigan, 2001). Also, the personnel working in the company is going to know clearly what s(he) must improve and how they can directly or indirectly contribute to the mission.

Definitions of domain, process and enterprise activity in CEN/ISO 19439 must be adapted in order to contemplate in an explicit way the strategic planning. With these adjustments practitioners of strategic planning and enterprise modelling will find their link and will work as a team in their entity organization.

Examples of possible adjustments to the definitions are:

- Domain: that part of the enterprise related with the environment and considered relevant to a given **mission** and constraints for which an enterprise model is to be created.
- Process: a partially ordered set of activities that can be executed to achieve some desired end-result in pursuit of a given **strategic or functional objective**
- Enterprise Activity: all or part of process functionality in pursuit of a given **transformation objective**

In reference to methodologies for unfolding the strategies as Balanced Score Card, we can take advantage of the methodology and the systems that exist for unfolding and monitoring the indicators. This integration between BSC and enterprise modelling until this moment is not complete, but the investigation has shown that the strategic maps can give the strategic and functional objectives. The adjustment of the functionality of the enterprise modelling will occur as a next step from the adjustments in the objectives.

The paradigm of being able to modify the functionality based on the strategic necessities of the market and the clients is guided by the Business Process Management (BPM). BPM mentions that processes must be able to readjust their functionality to adapt to the changes in the requirements of the customers and to the situations of the market.

It has to be theme of further investigation harmonizing the definitions of enterprise modelling based on strategic planning, the development of the adjustments in the methodologies of strategic planning, unfolding indicators and administration of business processes, integrating all of these into enterprise models.

With this paper and future investigations entity organizations are going to be able to better clarify their future needs for resource capabilities and to support their decisions to get benefits from improving competitiveness and interoperability. Also SPI will fortify the relationship among stakeholders that is used to determine and control the strategic directions and performance of organizations (Corporate Governance) (Hillman *et al.*, 2001).

Acknowledgments

Kurt Kosanke and James Nell, who received me into this community of scientists, convinced me that Enterprise Integration is the present and the future of world-class entity organizations. Arturo Molina for allowing me to participate in enterprise integration events. This events give me the opportunity to strengthen my knowledge and abilities in Enterprise Integration.

REFERENCES

Tennant,Ch., Roberts,P. (2001) Hoshin Kanri: A Tool for Startegic Policy Deployment. *Knowledge and Process Management*. 8(4) 262-269

Rumelt, R., Schendel, D., Teece, J. (eds.) (1994), Fundamental Issues in Strategy, Boston: Hardvard Business School Press, 527-530.

Ireland, R., Hitt, M. (1992) Mission statements: Importance, challenge, and recommendations for development. *Business Horizons*, 35(3):34 – 42.

Maritan, C. (2001) Capital investment as investing in organizational capabilities: An empirical grounded process model, *Academy of Management J.*, 44: 513-531.

Helfat, C. (2000) Evolution of firm capabilities, *Strategic Management Journal*. 21(special issue): 955-959.

Barney, J. (1999) How firms' capabilities affect boundary decisions. *Sloan Management Review*. 40(3): 137-145.

Hitt, M., Ireland, D., Hoskisson, (2003) Strategic Management: Competitiveness and Globalization, Fith Edition, Thomson South-Western.

CEN ISO 19439, (2002) Enterprise Integration – Framework for Enterprise Modelling, CEN TC'310 WG1 together with TC 184 SC5 WG1

CEN ISO 19440, (2002) Language Constructs for Enterprise Modelling. CEN TC'310 WG1 together with TC 184 SC5 WG1

Smith,H., Fingar,P. (2002) Business Process Management: the third wave, The breakthrough that redefines competitive advantage for the next fifty years. Meghan-Kiffer Press, Tampa

Cambridge On-Line Dictionaries, http://dictionary.cambridge.org/

Christopher, W. (1993) The Starting Point: Company Mission. In Handbook for Productivity Measurement and Improvement, William F. Christopher *et al* (Eds) Portland : Productivity Press 2-2.1 – 2-2.6

IFAC/IFIP Task Force (1999) Generalized Enterprise Reference Architecture and Methodologies, Annex to ISO 15704

ISO 15704: (1999) Requirements for Enterprise Reference Architecture and Methodologies ISO TC 184/SC5/WG1

Vinzant, J., Vinzant, D. (1999) *Journal of Management History*. 5(8) 516-531.

Harrigan, K., (2001) Strategic flexibility in old and new economies, in M. A. Hitt, R. E. Freeman & J. R. Harrison (eds), Handbook of Strategic Management, Oxford, U.K.: Blackwell Publishers, 97-123.

Hillman, A., Keim, G., Luce, R. (2001) Board composition and stakeholders performance: Do stakeholder directors make a difference? *Business and Society*, 40:295-314

Part II –

DIISM 04

**Manufacturing and Engineering in the Information Society:
Responding to Global Challenges**

31. Manufacturing and Engineering in the Information Society: Responding to Global Challenges

Jan B.M. Goossenaerts[1], Eiji Arai[2], John J. Mills[3], and Fumihiko Kimura[4]

1 Dept. of Technology Management, Eindhoven University of Technology, the Netherlands
j.b.m.goossenaerts@tm.tue.nl
2 Dept. of Manufacturing Science, Graduate School of Eng., Osaka Univ., Japan
3 Dept. of Mechanical and Aerospace Eng., The Univ. of Texas at Arlington, TX, USA
4 Dept. of Precision Machinery Eng., The Univ. of Tokyo, Japan

This introductory paper to the DIISM'04 volume explains the DIISM problem statement and applies principles of architecture descriptions for evolutionary systems (IEEE 1471-2000) to the information infrastructure for engineering and manufacturing. In our vision, knowledge and skill chains depend on infrastructure systems fulfilling missions in three kinds of environments: the socio-industrial domain of society and its production systems as a whole, the knowledge domain for a scientific discipline, and the sectorial domain, which includes the operational entities (companies, organizational units, engineers, workers) in engineering and manufacturing.

The relationships between these different domains are captured in a domain paradigm. An information infrastructure that enables responses to global challenges must draw on a wide range of both industrial and academic excellence, vision, knowledge, skill, and ability to execute. Responses have a scope, from the company, the factory floor and the engineering office to external collaboration and to man-system collaboration. In all scopes a system can offer services to different operational levels: operations, development or engineering, and research. The dimensions of scope and service level are briefly explained in relation to the architecting of an infrastructure. Papers are grouped according to their contribution to an infrastructure scenario or to an infrastructure component.

*Keywords: architecture,   engineering,   information   infrastructure,
manufacturing*

1. INTRODUCTION

The context of engineering and manufacturing has witnessed a striking expansion: from the product at the workshop during the workday of the craftsman, towards the portfolio of products and services, the resource base, and the business processes of the globally operating virtual enterprise. Simultaneously, the *set of information-based tools,* supporting the knowledge and skill chain has expanded: from the paper, pen and ruler to computer-and-communications aided applications for a growing range of functions ('CCAx'), with their impacts ranging from the core manufacturing process, over intra- and inter-enterprise integration, to the supply chain and the total life time of the extended product.

Computer-and-communications applications do well support many of the engineering, manufacturing and business functions that are key to manufacturing excellence and product success. But still, the engineering and manufacturing knowledge and skill chain shows many inefficiencies and hurdles. Therefore research and technology development on information infrastructure is ongoing, addressing a.o. information architectures, methodologies, ontology, advanced scenarios, tools and services. This research is driven by the insight that throughout an integrated life cycle of products and enterprises, the manufacturing knowledge and skill chain sources information from globally distributed offices and partners, and combines it with situational awareness, local knowledge, skills and experience to initiate decisions, and to deliver solutions. Hence the top-level objective of the information infrastructure: responding to global challenges by enhanced knowledge and skill chains.

However, how to design the information infrastructure that manages knowledge, information, data, and related services and tools that are shared by the different autonomous entities collaborating and seeking solutions in the socio-economic fabric in a finite global environment? Because the collaborators are part of different enterprises and economies, the information infrastructure is not regarded as a long-term differentiator in the business strategy of any enterprise. The infrastructure rather is a common enabler for the globalizing enterprise networks and professionals. For these entities, the common services matter at different levels of aggregation: for the external collaboration, for the teams and machine devices working in the factory or office, and for each person working in one or more enterprises. Hence the scope of this volume: information infrastructure systems and services for any level of aggregation in the engineering and manufacturing knowledge and skill chain.

2. AN INFRASTRUCTURE PROBLEM?

A series of IFIP TC5 WG 5.3/5.7 working conferences has been dedicated to the design of the information infrastructure systems for manufacturing (Yoshikawa and Goossenaerts, 1993; Goossenaerts *et al.*, 1997; Mills and Kimura, 1999; Mo and Nemes, 2001, Arai *et al.*, 2005). At this 6th working conference, building on recent research results and the results reported at and discussed at the previous conferences, contributions demonstrated a combination of breadth and depth, academic focus and industrial relevance. While multiple and more capable components are being developed, global challenges are being articulated, as well as roadmaps to overcome them. The Millenium Development Goals and the Kyoto Protocol are two examples. The connectedness of the global fabric is widely recognized but is in contrast with our inability to enact concerted practices that deliver the required results. Unless a sound information infrastructure gets deployed, the chaining of the problem solving scenarios will meet problems of quality, of interoperability of data, and of the scaling and combination of knowledge. How to offer continuity of service, the ubiquitous reuse of data and knowledge, and continuous interoperability while responding to new challenges, as companies compete, stakeholders evolve and new technologies emerge?

Contributions to this volume address components and scenarios of future knowledge and skill chains, as seen from the viewpoints of expert researchers in engineering, manufacturing and information technology. Traditionally, in industry, the integration of such components and scenarios is performed at companies. Today, and for the future, the globality and connectedness of the economic fabric and its problems oblige the research community to also address these chains supportive of improving the state of 'manufacturing industries as a whole'.

3. ARCHITECTING THE INFRASTRUCTURE

Architecture is defined in IEEE 1471-2000 (IEEE, 2000) as 'the fundamental organization of a system embodied in its components, their relationships to each other, and to the environment, and the principles guiding its design and evolution'. Every system has an architecture that can be recorded by an architectural description (AD) consisting of one or more models. The viewpoints for use selected by an AD are typically based on consideration of the concerns of the stakeholders to whom the AD is addressed.

Modeling techniques support communication with the systems stakeholders, prior to system implementation and deployment. Methodologies and tools come available for the model driven building and deploying of information systems and information infrastructures.

The relevance of architecting for the infrastructure addressed in DIISM derives from its life cycle focus: architecting is concerned with developing satisfactory and feasible systems concepts, maintaining integrity of those system concepts through development, certifying built systems for use and assuring those system concepts through operational and evolutionary phases. This is important as the domain of engineering and manufacturing is immensely complex, diverse and evolving. Where infrastructure sub-systems fulfill missions in different scopes, these systems should co-evolve and their architectures be aligned. Their AD's should be based on stable viewpoints.

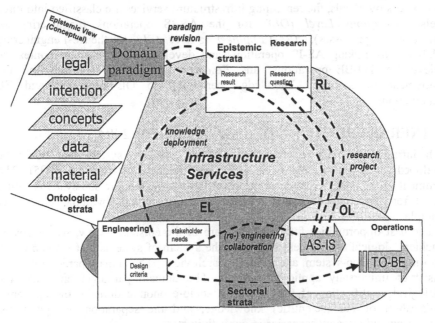

Figure 1. Three operational levels to serve

The four different scopes for which scenarios must be supported are the *natural &
socio-economic domain* (DP – domain paradigm), the *external collaboration* (EC)
between enterprises, the *factory floor* (FF), and the *man-system collaboration* (MS).
In each scope systems evolve under a result focus: outcomes are defined, problems
and stakeholder needs are observed and analyzed in the AS-IS, requirements update
and design deliver an extended or new specification, development and
implementation deliver the TO-BE operational system which is monitored for the
occurrence of new problems. The assets involved in system evolution include
natural capital, knowledge, data and models, human capital, social capital and
financial capital.

Each of the four views in Figure 1 offers services to the above scenario of
systems evolution. The *epistemic view* offers an *ontological stratification* that
structures the design space within which intentions, models and operational systems
evolve. The *research view* offers *epistemic stratification* (one strata per scientific
discipline such as logistics, mechanics, chemistry, and ergonomy) that structures the
discipline knowledge and derived design criteria (constraints) that must be met in
modifying or creating the operational system. The *engineering view* merges
constraints and contributions from ontological and epistemic strata to obtain new
operational capabilities. In the *operations view* repeating tasks are performed, in
accordance with the models developed. Operations must comply with the hard laws
of nature (as studied in the natural sciences), and the soft laws of the socio-economic
fabric (social sciences), while deploying the technology at hand. Both the
engineering and operations view show *sectorial stratification* that is evident in the
industrial differentiation of the modern society.

Assuming that a stable (meta-) model of the epistemic view exists, and that it
rarely needs overhauls, the remaining infrastructure services are classified into three
levels: *Operations Level (OL):* for the AS-IS operations (engineering or
manufacturing processes); (Re-) *Engineering Level (EL):* for the (re-) engineering
collaborations linking AS-IS operations and development for certain context to
achieve the TO-BE operations; and *Research Level (RL):* research and the
deployment of scientific knowledge pertaining to OL processes and EL
collaborations.

4.   INFRASTRUCTURE DESIGN AT DIISM 2004

Each infrastructure sub-system is a software intensive system that could be
developed using the widely used 4+1 view model of (Kruchten, 1995). The
alignment of the architecture descriptions of these infrastructure sub-systems would
benefit from a maximal reuse across those views, in accordance with the subsidiarity
principle.

The best opportunities for such reuse are in the epistemic view, which covers
Kruchten's logical and process views for the system of systems that we can call a
socio-industrial eco-system, and in the research view. The domain paradigm would
consist of universally applicable models. The domain paradigm embodies the
ontological stratification of the natural & socio-economic domain, the epistemic
stratification of our (scientific) knowledge, and the separation of operations,
engineering and research scenarios in our activities.

Two papers address this conceptual architecture and generic infrastructure
components. These contributions address viewpoints or services that in principle can

be shared by all scopes (society, external collaboration, factory floor and man-system collaboration).

Shu Qilin and Wang Chengen address a framework of product lifecycle model that comprises three parts: product information model, process model based on product life cycle, and extended enterprise resource model. They then describe the relationship and formation of product models at different stages and propose an integrated information architecture to support interoperability of distributed product data sources. Gonsalves and Itoh propose a technology-neutral integrated environment for system performance estimation during the requirement analysis and design phases, i.e. much before the implementation phase. The authors use a generic core life cycle of system development, consisting of three phases: system modelling, performance evaluation and performance improvement.

With the availability of reusable domain-level infrastructure components, the focus in the scopes of EC, FF and MS is on their differentiating aspects and scenarios. This volume contains contributions on External Collaborations, the Factory Floor Infrastructure and the Man-System Collaboration.

Wiesinger addresses engineering level services for external collaboration. He presents the software solution "Workbench" for the planning of large logistics networks as well as for the network structures of the facilities in an enterprise. The "Workbench" ensures a better information flow and provides a basis for Factory planning. It enables planners who lack expert planning knowledge.

Three papers address engineering and operation level services for the factory floor. Muljadi *et al.* describe an ontology for the development of a feature library. Requirements are derived by considering both the designer's intention and the extraction of manufacturing information for process plans generation.

Kato *et al.* propose a planning method for linear object manipulation, especially knotting. Topological states of a linear object are described and transitions between states are defined. Possible sequences of state transitions are generated, from which, one can choose an adequate path from the initial state to the objective state. Furthermore, a method to determine the grasping points and a planning method are proposed. A system based on the proposed methods is demonstrated.

Using the concept of Activity-Based Costing, Narita *et al.* propose an accounting method of production cost for machine tool operation. The cost factors considered in the research are the electric consumption of machine tool components, coolant quantity, lubricant oil quantity, cutting tool status and metal chip. The cost prediction system is embedded into a virtual machining simulator.

Technical architecture and the infrastructure life cycle are addressed in two papers. Takata *et al.* describe an implementation of the Integrated Process Management System, which includes manufacturing process management for building parts, and also construction process management at construction site. To observe the flow of the building parts, RFIDs are stuck to all parts to be managed, and several checkpoints are introduced within the coherent process through part-manufacturing and building construction. The requirements of the RFID directory services are also discussed.

Sugitani *et al.* propose the effective tools of operation standardization for mass production of a new product. The cycle of operation standard consists of three stages of design, improvement and evaluation. It is divided into seven steps, that is, decision, communication and understanding, observance, supervision, notice, decision again, and evaluation. The proposed seven tools of operation standardization (OS7) correspond to these steps. These tools help to realize mass production of a new product and to stabilize a product quality much earlier.

5. POST CONFERENCE GAPS

To better respond to global challenges, business, engineering and manufacturing decision making must introduce new criteria and develop new tools for operations design, improvement and evaluation. DIISM 2004 has further explored the multiple issues and approaches to address them. Over the past decade, while globalization has been studied as a driver for competitiveness, the international community has articulated desirable outcomes, including social and environmental, and it has achieved consensus about global development goals, such as the Millenium Development Goals, and environmental targets, such as the Kyoto Protocol. Suddenly the pre-competitive and post-competitive phases of the knowledge production process (Yoshikawa, 1993) can be addressed in a much more mature socio-technical global environment. A new performance paradigm is being shaped. It recognizes the broad context within which production capabilities develop, and the enabling role of "manufacturing industries as a whole" in achieving development goals. Knowledge that is produced in the pre- and post-competitive phases is best considered a global public good. The result focussed management of this knowledge (see Kimura, 2005 for critical issues) by the multiple product life cycle stakeholders, is a major challenge requiring a dedicated collaborative effort of public-private partnerships.

REFERENCES

Arai, E., J. Goossenaerts, F. Kimura and K. Shirase (eds.) (2005) Knowledge and Skill Chains in Engineering and Manufacturing: Information Infrastructure in the Era of Global Communications, Springer.

Goossenaerts, J., F. Kimura and J.C. Wortmann (eds.) (1997) Information Infrastructure Systems for Manufacturing, Chapman & Hall, London, UK

IEEE (2000) Recommended Practice for Architectural Description of Software-Intensive Systems IEEE Std 1471-2000. July 2000.

Kimura, F. (2005) Engineering Information Infrastructure for Product Life Cycle Management. In Arai, E. *et al.* (2005), pp 13-22

Kruchten, P. (1995) , Architectural Blueprints - The '4+1' View Model of Software Architecture, IEEE Software, 12 (6)

Mills, J. and F. Kimura (eds.) (1999) Information Infrastructure Systems for Manufacturing II Kluwer Academic Publishers, Boston

Mo, J.P.T. and L. Nemes (eds.) (2001) Global Engineering, Manufacturing and Enterprise Networks, Kluwer Academic Publishers, Boston

Yoshikawa, H. and J. Goossenaerts (eds.) (1993) Information Infrastructure Systems for Manufacturing. IFIP Transaction B-14. Elsevier Science B.V. (North Holland)

Yoshikawa, H. (1993) Intelligent Manufacturing Systems: Technical Cooperation that Transcends Cultural Differences. In: Yoshikawa, H. and J. Goossenaerts (1993) pp. 19-40

32. Considering Designer's Intention for the Development of Feature Library of a Process Planning System

H. Muljadi[1], K. Ando[2], H. Takeda[1] and M. Kanamaru[1]

1 National Institute of Technology Email:hendry@nii.ac.jp
2 Shibaura Institute of Technology

In this paper, the creation of ontology of manufacturing features for the development of a feature library is described. The designer's intention described in functional data of the feature constructing face elements is considered for the creation of the ontology. The creation of the manufacturing feature ontology is intended to make the feature library be useful for the extraction of manufacturing information for process plans generation.

1. INTRODUCTION

In the production stage, dynamic changes such as increased production, part type changes, machine breakdowns etc are ordinary occurrences. To deal with these dynamic changes, we presupposed the need to integrate design, manufacturing and scheduling activities. Our research puts its goal in the generation of a CAPP system that can integrate process planning, scheduling and manufacturing activities (Sakurai , 2000). Figure 1 shows the overview of the proposed CAPP system. The system consists of 3 steps.

- Step 1: feature sets creation from the product design data (CAD data).
- Step 2: generation of process plan of a part based on the created feature sets.
- Step 3: determination of optimal set of process plans for product mix.

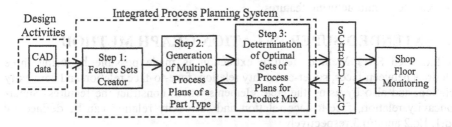

Figure 1: Overview of Integrated Computer-Aided Process Planning System

The optimal set of process plans obtained in Step 3 is used for the shop floor scheduling. During the shop floor monitoring, re-scheduling may occur to handle the dynamic changes in the manufacturing stage. In the re-scheduling stage, we can return to Step 3 to determine the optimal set of process plans for the present shop floor or production planning condition.

In order to bring this integrated process planning system to realization, we have proposed a Feature Sets Creator that can lead to the generation of multiple process plans (Muljadi *et al*, 2005). For the development of the Feature Sets Creator, we implemented Super Relation Graph (SRG) Method (Kao *et al*, 1995). We did some modification to the SRG Method and proposed the Modified SRG Method. The Feature Sets Creator uses the Modified SRG Method to extract manufacturing features from the product design information. We have further modified the Modified SRG Method, and proposed the Extended SRG Method that is able to extract not only single depression features, but also protrusion and compound features (Muljadi *et al*, 2003). Protrusion and compound features can also be extracted by the Extended SRG Method since these features can be represented by the Extended SRG. We store manufacturing features and their Extended SRG representations in a feature library.

For the development of a feature library for the CAPP system, we collect and store manufacturing features, their corresponding Extended SRG representations and the manufacturing information needed to create the shape of the manufacturing features. However, we found that instances of same type of manufacturing features may require different manufacturing methods. For the automated extraction of manufacturing information, the task will become easier if we have instances of a feature class in the feature library refer only to same possible manufacturing methods.

In this paper, for the development of the feature library, we propose the creation of ontology of manufacturing features by considering the designer's intention described in the functional data of the face elements that construct the features. The goal of this ontology creation is to make the feature library be useful for the automated extraction of proper manufacturing information to create the manufacturing features that are extracted by the Extended SRG Method.

The structure of this paper is as follows. In order to make this paper self-content, the Extended SRG Method is described briefly in Section 2. In Section 3, we discuss the creation of ontology of manufacturing features for the feature library. In Section 4, a case study is used to show the validity of the proposed manufacturing feature ontology to enable feature library to extract proper manufacturing information from the extracted manufacturing features.

2. EXTENDED SUPER RELATION GRAPH METHOD

In Extended SRG Method, feature extraction is made possible by using three relations between faces, super-concavity relation, face-to-face relation and convexity relation, and also by using the edge elements which construct the features. Super-concavity relation, face-to-face relation and convexity relation can be defined by Eq.1, Eq.2 and Eq.3 respectively.

$$n_{f_i}^+ . n_{f_j}^+ \neq -1; \ f_i \cap S(f_j)^{|+|} \neq \emptyset \text{ and } f_j \cap S(f_i)^{|+|} \neq \emptyset \tag{1}$$

$$n_{f_i}^+ . n_{f_j}^+ = -1; \ f_i \subset S(f_j)^{|+|} \text{ and } f_j \subset S(f_i)^{|+|} \tag{2}$$

$$n^+_{f_i} \cdot n^+_{f_j} \neq 1; n^+_{f_i} \cdot n^+_{f_j} \neq -1; f_i \cap S(f_j)^{|+|} = \emptyset;$$

$$f_j \cap S(f_i)^{|+|} = \emptyset; E_{f_i} \cap E_{f_j} \neq \emptyset \tag{3}$$

where $n^+_{f_i}$ is the positive face normal of face f_i (Figure 2(a)), and the strict positive half space of face f_i, $S(f_i)^{|+|} = \{ x \mid n^{+^T}_{f_i} x > k \}$ is the positive half space which exclude the embedding plane of face f_i, $P(f_i) = \{ x \mid n^{+^T}_{f_i} x = k \}$ (Figure 2(b),(c)), E_{f_i} is the set of edges of face f_i. $n^+_{f_j}$, $S(f_j)^{|+|}$ and E_{f_j} are defined similarly as above.

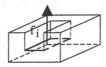

(a) Positive Face Normal of Face f_i

(b) The Embedding Plane of f_i

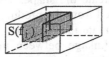

(c) Strict Positive Half Space of f_i

Figure 2: Explanation of terms used in Extended SRG Method

Figure 3 shows the Extended SRG representation of a stepped-hole feature. A node with one circle in the Extended SRG corresponds to a plain face of the feature. A double circle node corresponds to a curve face. Dotted links are used to represent face-to-face relations. Solid links are used to represent super-concavity relations and face-to-face relations. To distinguish these two relations, 0 is used as the attribute of the solid links to represent super-concavity relations and 1 to represent convexity relations. Solid links with no attribute are used to represent the face-edge relations.

Plain edges are represented by e_n and curve edges are represented by e^+_n. The Extended SRG Method has the ability to extract not only single depression features, but also protrusion and compound features, since protrusion and compound features can have their Extended SRG representations too.

In the development of feature library, a collection of manufacturing feature types, their corresponding Extended SRG representations and the possible manufacturing information to create the manufacturing features is stored. In the next section, we discuss the creation of manufacturing feature ontology to make the feature library be useful for the extraction of proper manufacturing information of manufacturing features extracted by Extended SRG Method.

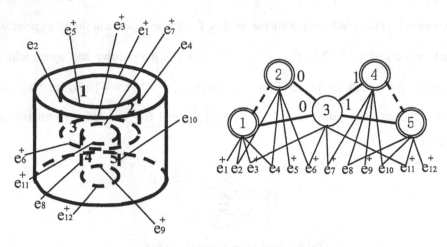

Figure 3: Stepped-hole feature and its Extended SRG representation

3. FEATURE LIBRARY

3.1 Feature Library for the Generation of Process Plans

A manufacturing feature can be defined simply as a geometric shape that has its manufacturing information to create the shape. This definition means that when a manufacturing feature is extracted from the product design information, the possible manufacturing information to create the shape is also extractable at the same time (Kanamaru *et al*, 2004). The linkage between manufacturing features and their corresponding possible manufacturing information is stored in the database, which we call as a feature library. Thus, a feature library plays a big role for the generation of process plans based on the recognized manufacturing features. As manufacturing features are extracted, the manufacturing information can also be extracted from the feature library.

3.2 Representation of Designer's Intention

Manufacturing features are extracted from the product design information. However, in order to extract proper manufacturing information to create the manufacturing features, we need to recognize the functions of the manufacturing features. To do so, we have to know why the designer designs the geometrical shapes, or in other words, we have to understand the designer's intention. So, we can say that understanding the designer's intention is very important to extract the proper manufacturing information to create the manufacturing features. However, since normally designer does not design a part using manufacturing features, we suppose that it is better to understand the designer's intention by considering the functions of the face elements that construct the manufacturing features.

The functional data of face elements can be described as basic function, mechanism utilized for realization of the basic function, and condition and direction of the motion. The detail explanation of the functional data of face elements is given in another report (Ando *et al*, 1989). Table 1 shows the contents of functional properties of face elements that are used for the creation of manufacturing feature ontology. The scope of the functional properties of face elements shown in Table 1 is limited to machined products.

Table 1 - Contents of Functional Properties

Basic Function	Mechanism utilized for realization of the basic function	Condition and direction of the motion
Transmission of motion	1: friction-mech. 2: gear-mech. 3: link-mech. 4: cam-mech.	1: liner 2: smooth-liner 3: very-smooth-liner 4: round 5: smooth round 6: very smooth round
Constraint of motion	1: rigidity-mech. 2: ball-bearing-mech. 3: sliding-mech.	1: liner 2: weak-radial 3: strong-radial 4: weak-thrust 5: strong-thrust
Fixation of motion	1: bolt-and-nut 2: bolt-only 3: friction-mech. 4: bearing-fit 5: key-fit 6: river-fit 7: shrinkage-fit	1: stationary-object 2: revolutionary-object

3.3 Creation of Manufacturing Feature Ontology

Figure 4 shows the manufacturing feature ontology and the functional data ontology. New classes for manufacturing feature ontology are created to have their relation with the functional data ontology. The relation between the class of the manufacturing feature ontology and the functional data ontology represents how the manufacturing features should be manufactured to fulfill the required functions. Each new class of the manufacturing feature ontology will refer to a collection of possible manufacturing information that can be used to create the shape of the instances of the feature class.

In Figure 4, a "precise drilled thru hole" class is created to relate the thru hole feature type in the manufacturing feature ontology with the "transmission by friction in liner motion" class of the functional data ontology. This is done since the "precise drilled thru hole" can fulfill the required mentioned functions. And for the "precise drilled thru hole" feature class, a collection of possible manufacturing information for the instances of the "precise drilled thru hole" feature class should be prepared so that when a manufacturing feature extracted by the Extended SRG Method falls to this class to fulfill the required functional data as intended by the designer, a proper manufacturing information can be extracted automatically

Thus, by creating the manufacturing feature ontology, we can collect and manage the knowledge of process planners to create manufacturing features, and also that the manufacturing feature ontology will make the feature library be useful for the extraction of proper manufacturing information that can lead to the generation of process plans of a part.

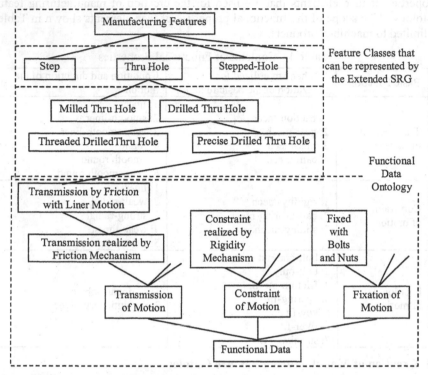

Figure 4   Ontology of manufacturing features and functional data

4. CASE STUDY

Using a sample part shown in Figure 5, we confirm the effectiveness of the proposed manufacturing feature ontology to allow the extraction of manufacturing information from the feature library. Figure 5 shows a sample part with the required functions of the face elements of the part.

First, Extended SRG Method is applied to extract manufacturing features from the sample part. As illustrated in Figure 6, there are 5 thru hole features extracted. Then using the functional data shown in Figure 5 as the input, the extracted features find the matched feature class from the feature library. As illustrated in Figure 7, one thru hole feature falls to the "grinded thru hole" feature class, and four thru hole features fall to the "threaded drill thru hole" feature class. Then, manufacturing information for each thru hole features are extracted by referring the instances of the feature classes. The thru hole feature that falls to the "grinded thru hole" feature class extracts a manufacturing method where cylindrical grinder is required to manufacture the shape. The four thru hole features that fall to the "threaded drill thru hole" feature class extract a manufacturing method where threading is required to manufacture the shape. Thus it shows that the manufacturing feature ontology is effective to make the feature library be useful for the automated extraction of proper manufacturing information to create the manufacturing features that are extracted by the Extended SRG Method.

Basic Function: Fixation
Mechanism: bolt
Motion: Stationary-Object

Basic Function: Constraint
Mechanism: rigidity
Motion: weak radial

Basic Function: Fixation
Mechanism: bolt
Motion: Stationary-Object

Basic Function: Fixation
Mechanism: bolt
Motion: Stationary-Object

Basic Function: Fixation
Mechanism: bolt
Motion: Stationary-Object

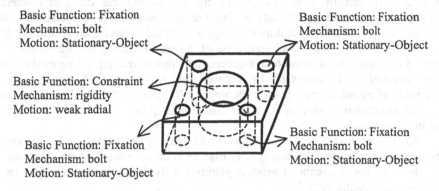

Figure 5: A sample part and the functions of the face elements

Thru Hole Feature

Thru Hole Feature

Thru Hole Feature

Thru Hole Feature

Thru Hole Feature

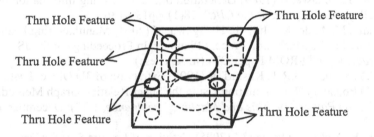

Figure 6: Extracted Manufacturing Features

Thru Hole Feature
Feature Class: "Threaded Drill
Thru Hole"

Thru Hole Feature
Feature Class: "Grinded
Thru Hole"

Thru Hole Feature
Feature Class: "Threaded Drill
Thru Hole"

Thru Hole Feature
Feature Class: "Threaded Drill
Thru Hole"

Thru Hole Feature
Feature Class: "Threaded Drill
Thru Hole"

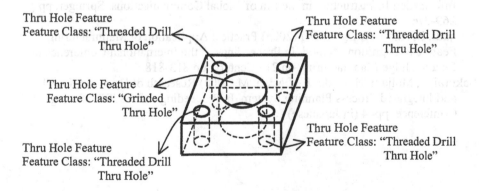

Figure 7: Extracted Manufacturing Features and Their Proper Feature Classes

5. CONCLUSION

In this paper, we presented the creation of ontology of manufacturing features for the development of a feature library by considering the designer's intention described in the functional data of the feature constructing face elements. New classes for manufacturing feature ontology are created to have their relation with the functional data ontology. Each new class of the manufacturing feature ontology will refer to a collection of possible manufacturing information that can be used to create the shape of the instances of the feature class. As shown in the case study, the creation of manufacturing feature ontology will make the feature library be useful for the automated extraction of proper manufacturing information for the generation of process plans.

For the automated generation of process plans, further works need to be done on how the extracted manufacturing information to create manufacturing features can be used for lower stream of process planning activities, such as setup generation, process sequencing etc.

REFERENCES

Ando, K, Yoshikawa, H (1989) Generation of Manufacturing Information in Intelligent CAD. *Ann. of the CIRP*, 38(1). pp133-136

Kanamaru, M, Ando, K, Muljadi, H, Ogawa, M (2004) Manufacturing Feature Library for the Machining Process Planning. In Proceedings of the JSPE Autumn Conference. CD-ROM Paper C74 (in Japanese)

Kao, C.Y, Kumara, S.R.T, Kasturi, R (1995) Extraction of 3D Object Features from CAD Boundary Representation Using the Super Relation Graph Method. *IEEE Trans. on Pattern Analysis and Machine Intelligence*, 17(2), December. pp1228-1233

Muljadi, H, Ando, K, Ogawa, M (2005) Creation of Feature Sets for Developing Integrated Process Planning System. In Arai, E, Goossenaerts, J, Kimura F, Shirase K (eds.) Knowledge and Skill Chains in Engineering and Manufacturing: Information Infrastructure in the Era of Global Communications, Springer. pp 269-276

Muljadi, H, Ando, K, Ogawa, M (2003) Practical Application of Manufacturing Feature Recognition Method. In Proceedings of the International Conference on Leading Edge Manufacturing on 21st Century. pp 813-818

Sakurai, T, Muljadi, H, Ando, K, Ogawa, M (2000) Research of Dynamic, Flexible and Integrated Process Planning System. In Proceedings of the JSPE Spring Conference. pp44 (in Japanese)

33. Manipulation Planning for Knotting Linear Objects with One Hand

Tsunenori Kato, Hidefumi Wakamatsu, Akira Tsumaya, and Eiji Arai

Affiliation: Dept. of Manufacturing Science, Osaka Univ.
Em:{kato, wakamatu, tsumaya, arai}@mapse.eng.osaka-u.ac.jp

A planning method for linear object manipulation, especially knotting is shown. At first, topological states of a linear object are described. Next, transitions between states are defined. Then, we can generate possible sequences of state transition, from which, we can choose an adequate path from the initial state to the objective state. Furthermore, a method to determine the grasping points is proposed. In the fourth, a planning method is proposed. Finally, our system based on proposed methods is demonstrated

1. INTRODUCTION

In production sites, a lot of deformable linear objects like wires, codes, and cables are used widely; for data transmission, object transportation, fixing or packing of objects, and so on. However, systematic approach for realizing those manipulative tasks aimed at such deformable objects has not established yet. Because the physical property of them is diversity, it is very difficult to adopt the method for manipulating rigid objects.

Focusing on linear objects, especially, those applications are accompanied by knotting manipulation usually.

Wolter *et al.* have proposed the method to describe the deformation process of linear objects qualitatively (J.Wolter, 2001). Leaf has described deformed shape of fabric geometrically (G.A.V.Leaf, 1960). Morita at el. have proposed a system for knot planning from observation of human demonstrations (T.Morita, 2002). Matsuno at el. have realized a task of tying a cylinder with a rope by a dual manipulator system identifying the rigidity of the rope from visual information (T.Matsuno, 2001).

When we make a knot, we manipulate a linear object by several fingers of both hands for bending, twisting, and holding the linear object. The way to make several knots depends on human makeup or experience, so it is not unique. We can generate manipulation plans suitable for equipment and facilities with unlike physical makeup of human if processes for knotting a linear object can be modeled. Then, in this paper, we propose a method for automatic planning and execution of linear object manipulation which includes knotting.

At first, we propose qualitative crossing state of linear object in three dimensional space. Secondly, we propose a manipulation process of a linear object can be represented as a sequence of crossing state transition. Thirdly, it is shown that any manipulation process can be realized by one hand and a planning method for one-handed manipulation is proposed. Finally, we demonstrate a knotting experiment of an overhand knot performed by a vision-guided manipulator system to examine the usefulness of our approach.

2. QUALITATIVE REPRESANTATION OF CROSSING STATES

In this section, we define how to represent the state of a linear object qualitatively in order to generate manipulation planning.

At first, we define the state of a linear object as its projection on a plane. Then, on this projection plane, a curve may cross with itself. Note that how to cross of the 2D curve depends on the projection plane. Next, we number crossing points of the linear object along it. Then, the state of the linear object is represented as a set of C_i ($i=1..., n$) standing for crossing points, E_l, and E_r standing for the left endpoint and the right endpoint respectively. Fig.1 shows an example of a linear object. It has 5 crossing points and their sequence is E_l-C_1-C_2-C_3-C_4-C_5-C_1-C_2-C_5-C_4-C_3-E_r. And so, we can define the state of the object as a sequence of its crossing points. And then, at each crossing point, we define the upper part C^u_i and the lower part C^l_i. Furthermore, we can distinguish two types of crossing; one of the two is the crossing that the upper part overlaps from the left side of the lower part to its right side and the other is opposite crossing. We define the former as the right hand helix crossing and the latter as the left hand helix crossing, and C^+_i which stands for the right hand helix crossing, C^-_i which stands for the left hand helix crossing. So, Fig.1 can be described as E_l-C^u_1-C^l_2-C^l_3-C^u_4-C^u_5-C^l_1-C^u_2-C^l_5-C^+_4-C^u_3-E_r.

Consequently, we can represent the state of linear objects including knotted ones as finite crossing states qualitatively, regardless of its length, thickness, or other physical properties.

Fig.1 Example of knotted linear object   Fig.2 The definition of two types of crossing

3. DEFENITON OF OPERATIONS FOR STATE CHANGING

In the previous section, we showed the states of linear objects can be represented by a sequence of crossing points. In this section, we consider how to transit the states of linear objects. In order to change the crossing state of a linear object, some operation must be performed on the object. Then, a state transition corresponds to an operation that changes the number of crossing points or rearranges their sequence. In this, to execute state transition of linear objects, four basic operations are prepared as shown in Fig.3. Operation type-I, II, and III are equivalent to Reidemeister move type-I, II, and III in the knot theory (C.C.Adams, 1994). By these operations, topology of the object state is not changed. But, type-IV operation in Fig.3 is regarded as an operation of changing topology. This operation is not included in Reidemeister, because in the knot theory, endpoints are not focused on. By operation type-I, II, IV, the number of crossing points is increased or decreased. Operation type-III does not change the number of crossing points but change their sequence. Let us define operations to increase crossing points as crossing operations CO_I, CO_{II} and C_{IV}, operations to decrease them as uncrossing operations UO_I, UO_{II}, and UO_{IV}, and an operation keeping the number of them as an arranging operation AO_{III}.

The number of possible crossing states after a crossing operation can be much larger than those after an uncrossing operation. So, in this paper, a manipulation process can be represented as a sequence of uncrossing operations.

Fig.4 shows an example of a required manipulation. The initial state in Fig.4 (a) represented as $E_l\text{-}C^{u\text{-}}{}_1\text{-}C^l{}_2\text{-}C^{l+}{}_3\text{-}C^{u+}{}_4\text{-}C^{u\text{-}}{}_5\text{-}C^{l\text{-}}{}_1\text{-}$ $C^{u\text{-}}{}_2\text{-}C^l{}_5\text{-}C^{l+}{}_4\text{-}C^{u+}{}_3\text{-}E_r$, and the objective state in Fig.4 (b) represented as $E_l\text{-}E_r$. Fig.5 is a derived graph from the initial state to the objective state used uncrossing operations. In this graph, 14 crossing states and 32 state transitions are included. The example in Fig.5 shows unknotting processes. After deriving the sequence of uncrossing operations, by following it backward, knotting manipulation processes can be generated.

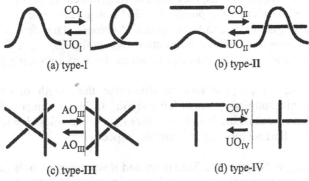

(a) type-I   (b) type-II

(c) type-III   (d) type-IV

Fig.3 Basic operations

4. PLANNING METHOD FOR ONE HANDED MANIPULATION

In this section, a planning method for one-handed manipulation for linear objects is proposed. A crossing state graph in Fig.5 includes sequences which consist of type-IV operation alone. All manipulation tasks can be achieved by iteration of type-IV operations. Therefore, in this paper, we show the ability of one-handed manipulation using type-IV operation alone. We define a grasping point and the approach direction of a manipulator for type-IV operation as shown in Fig.6. Fig.6 (a) shows them for CO_{IV} and Fig.6 (b) shows them for UO_{IV}. Fig.6 (a-2) and Fig.6 (b-2) shows the opposite crossing of the case illustrated in Fig.6 (a-1) and Fig.6 (b-1) respectively. We define the crossing shown in Fig.6 (a-1) and Fig (b-1) as the up-end crossing and that of shown in Fig.6 (a-2) and Fig.6 (b-2) as the down-end crossing. It is found that the upper part is selected as the grasping point, in both crossing. Furthermore, manipulator can access the objects from the front side of the projection plane in both cases. So, type-IV operation can be achieved by one-handed manipulator approaching from the front side. It implies that, we can realize manipulation for linear objects by one hand without turning over the whole or partial of it when it is laid on a table.

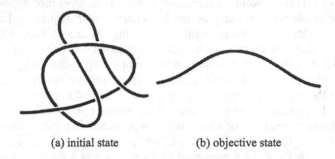

(a) initial state   (b) objective state

Fig.4 Example of state transition

In uncrossing operations, the positions of points to be uncrossed are given. However, in crossing operations, the positions of points to be crossed are unknown. So, to execute crossing operations, they must be determined. In the objective state, a knotted object with n crossing points has $2n+1$ segments and $2n$ upper/lower crossing points, that is, upper and lower points of crosses. Therefore, we divide the object into $2n+1$ segments in the initial state. We define D_i^* as a dividing point where subscript * and subscript i are equivalent to those of a crossing point created by crossing it.

In this section, we propose how to determine the length of each segment qualitatively. In this study, we assume left endpoint of linear object is fixed. Let us define some rules to determine length as follows. In these rules, L and S stand for long segment and short segment in comparison respectively.

Rule1.   The segment between the fixed point and it adjoining point is defined as L.

Rule2.   In the case of operating an up-end crossing, it is useful that the distance between a grasping point and the right endpoint is short. So, segments which exist in right side of the grasping point are defined as S.

Rule3.   In the case of operating a down-end crossing, segments existing between a grasping point and C*u are defined as L.

Rule4.   After repeating those rules above, segments which are not determined are defined as L.

Rule5.   After repeating those rules above, segments which are determined more than two times are defined as their product.

By using these rules, we try determining length of each segment. Fig.7 shows an example of knotting manipulation. In this example, the initial state and the objective state are expressed respectively as follows:

$$E_l - D_1^{l+} - D_2^{u+} - D_3^{l-} - D_5^{l+} - D_1^{u+} - D_2^{l+} - D_5^{u+} - D_4^{u-} - D_3^{l-} - E_r \qquad (a)$$

$$E_l - C_1^{l+} - C_2^{u+} - C_3^{l-} - C_5^{l+} - C_1^{u+} - C_2^{l+} - C_5^{u+} - C_4^{u-} - C_3^{l-} - E_r \qquad (b)$$

At first, the segment 1 in Fig.8 becomes L according to Rule1. And then, in the first operation in Fig.8, point D_1^{u+} is grasped, moved and crossed on point D_1^{l+} to realize this operation. In this time, the crossing operation to the up-end crossing is carried out. So, the segments **7, 8, 9, 10,** and **11** becomes S by Rule2. Next, in the second operation in Fig.8, point D_2^{u+} is crossed on point D_2^{l+}, but in this case, C_1^{u+} exists in the right side of D_2^{u+}. So, the crossing operation to the down-end crossing is carried out. Thus, segments **3, 4, 5,** and **6** become L by Rule3, and in this time, the distance between C_1^{u+} and C_1^{l+} had been determined already, so the rest segment, i.e. segment **2** becomes S. Next, in the third operation and the fourth operation, the crossing operation to the up-end crossing is used. So, segment **9, 10,** and **11** are S by Rule2. Finally, in the fifth operation, the crossing operation to the down-end crossing is used. So, segment **3, 4, 5** and **6** are L. Then, the length of each segment is determined as shown in Fig.8.

number of crossing points

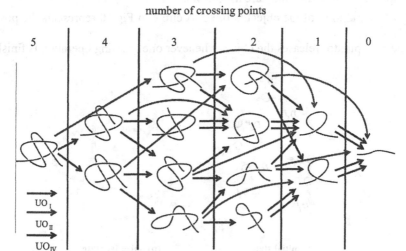

Fig.5 Result of manipulation process planning

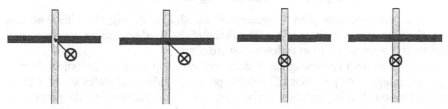

(a-1) up-end crossing (a-2) down-end crossing (b-1) up-end crossing (b-2) down-end crossing

(a) crossing operation (b) uncrossing operation

Fig.6 Grasping point for type-IV operation in crossing and uncrossing operation

Thus, if we substitute propriety numbers into L and S, we can determine the real length of each segment. In addition to this example, to realize that manipulation, two points which make a crossing point should be moved so that they create right hand or left hand helix crossings properly. If the tangents at these points are given, we can generate possible trajectories of the manipulator to create crossing points.

5. CASE STUDY

In this section, we demonstrate the validity of the method we have proposed in this paper. Fig.9 shows concise view of our pilot system. It consists of PC for controlling a manipulator and image processing, a 6 DOF manipulator, and a CCD camera. We attempt to plan and carry out one-handed knotting manipulation with this system. A linear object, twist yarn, is laid on a table and its shape is captured by the camera fixed above the table. The table corresponds to a projection plane.

Fig.10 shows a required manipulation. It corresponds to tying an overhand knot. The initial state and the objective state are shown in Fig.10 (a) and (b), respectively. They are represented as E_l-E_r and E_l-C^{l+}_1-C^{u+}_2-C^{l+}_3-C^{u+}_1-C^{l+}_2-C^{u+}_3-E_r, respectively. Assumptions of this case study are as follows:

- The left endpoint of the object is fixed. A circle in Fig.10 represents the position of fixture.
- The manipulator releases the object whenever one crossing operation is finished.

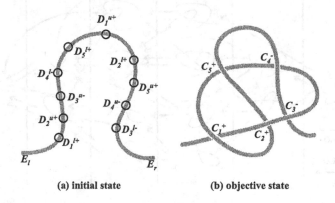

(a) initial state   (b) objective state

Fig.7 Example of knotting manipulation

Then, one manipulation plan is generated as shown in Fig.11. This knotting manipulation consists of three CO_{IV}. The first and the third operations act on up-end crossing; the second operation is down-end crossing.

Next, this system can recognize the current crossing state of the object from a gray-scale image. The positions of crossing points in individual states are able to be identified by substituting suitable numbers into the L and S derived by the proposed.

About grasping points, in this paper, we define the upper point of each crossing point as a grasping point. And, direction of the axes can be calculated from the tangent at the grasping point. As appropriate moving distance of a manipulator for a state transition is unknown, the system checks whether the crossing state of the linear object is changed or not after moving the object. Thus, the manipulator can approach, grasp, move and release the object according to the generated qualitative plan. Fig.12 shows the result of this manipulation.

Therefore, we think our proposed method we have proposed is effective for automatic planning and execution of linear object manipulation.

6. TOWARD DETAILED PLANNING

We can plan manipulation for linear objects qualitatively by applying our proposed method. But it may be not enough to make a more complex knot like a bowknot, because a crossing operation to a down-end crossing in Fig.6 is not certainty. So, in the case of carrying out a crossing operation to a down-end crossing, there is the necessity of preparations to make a down-end crossing certainly, for example pointing the right endpoint of the object to the grasping point in down-end crossing. If this situation comes true, quantitative analysis should be performed in order to check whether generated operations can be realized practically or not considering physical properties of a linear object. In quantitative analysis, the influence of the friction arisen by self-contact of the linear object is measurable, especially in knotting process. So, we had developed an analytical method to model the stable shape of a deformable linear object. Fig.13 shows the computed shape of an overhand knot, with/without the effect of friction.

Therefore, the manipulation strategy can be derived automatically by combining a qualitative planning proposed in this paper with the quantitative analysis.

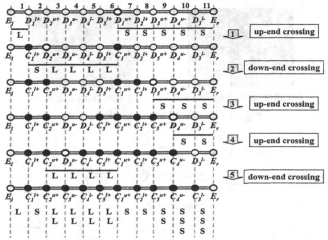

Fig.8 Result of length of each segment

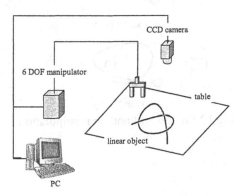

Fig.9 Overview of the experimental setup

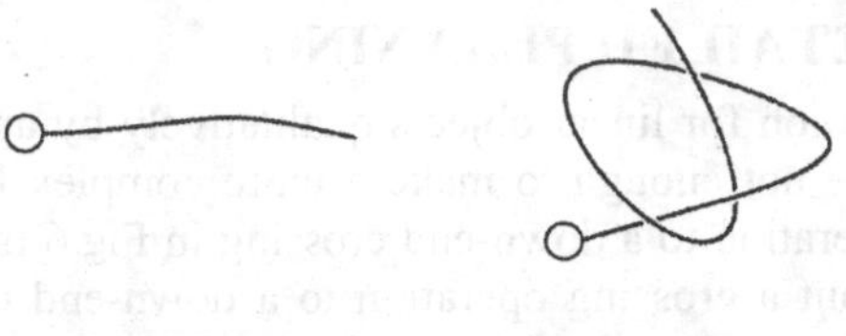

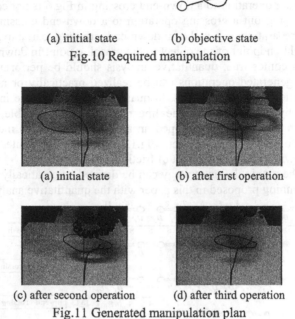

(a) initial state   (b) objective state

Fig.10 Required manipulation

(a) initial state   (b) after first operation

(c) after second operation   (d) after third operation

Fig.11 Generated manipulation plan

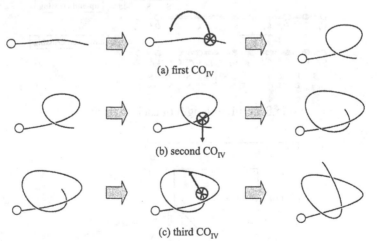

(a) first CO_{IV}

(b) second CO_{IV}

(c) third CO_{IV}

Fig.12 Process of knotting manipulation

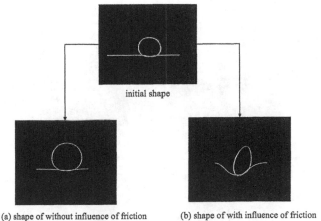

initial shape

(a) shape of without influence of friction   (b) shape of with influence of friction

Fig.13 Quantitative analysis of influence of friction

7. CONCLUSIONS

In this paper, a planning method for linear object manipulation including knotting was proposed. Especially, it was shown that any knotting manipulation can be realized by one hand.

Firstly, a representation of topological states of a linear object was proposed. Its topological state can be represented as finite crossing states, and to execute transition between them, four basic operations were introduced. A state transition corresponds to a basic operation which changes the number of crossing points or permutated their sequence. So, giving the initial state and the objective state of a linear object, possible manipulation processes can be generated. Secondly, a planning method for one- handed manipulation was proposed because it was found that any manipulation is realized by one hand. Furthermore, a method for determination of grasping point was proposed by suggesting how to determine the length of each segment in order to realize derived manipulation processes. Finally, in order to demonstrate the effectiveness of our method, planning and execution of linear object manipulation by one hand was carried out.

REFERENCES

Adams, C.C (1994) The Knot Book: An Elementary Introduction to the Mathematical Theory of Knots, W.H.FREEMAN AND COMPANY

Leaf, G.A.V (1960) Models of the Plain-Knitted Loop, Journal of the Textile Institute, Vol.51, No.2, 135-171

Wolter, J. and Kroll, E (2001) Toward Assembly Sequence Planning with Flexible Parts, Proc, IEEE Int. Conf. on Advanced Intelligent Macaronis, 677-682.

Matsuno, T., Fukuda, T. and Arai, F (2001) Flexible Rope Manipulation by Dual Manipulator System Using Vision Sensor, Proc. of International Conference on Advanced Intelligent Mechatronics

Morita, T., Takamatsu, J., Ogawa, K., Kimura, H., Ikeuchi, K (2002) Knot Planning from Observation, Proc. of IEEE Int. Conf. Robotics and Automation

34.   Cost Prediction System Using Activity-Based Model for Machine Tool Operation

Hirohisa Narita[1], Lian-yi Chen[2] and Hideo Fujimoto[2]

*1 Tsukuri Colloge, Graduate School of Engineering, Nagoya Institute of Technology, Japan
Email: narita@vier.mech.nitech.ac.jp
2 Omohi Colloge, Graduate School of Engineering, Nagoya Institute of Technology, Japan
Email: {chen, fujimoto}@vier.mech.nitech.ac.jp*

Production cost is one of the most important factors for manufacturing. The production cost associated with each machine tool is calculated from total cost of factory in general. The operation status of machine tools, however, is different, so accurate production cost for each product can't be calculated. Hence, accounting method of production cost for machine tool operation is proposed using the concept of Activity-Based Costing and is embedded to virtual machining simulator for the cost prediction.

1. INTRODUCTION

Production cost is one of the most important factors to decide the manufacturing process and manufacturing strategies. However, lots of machine tools are installed for production in recent factory and it is difficult to estimate and recognize accurate production cost due to machining operation. That is to say each operation status of machine tools is different, so conventional cost accounting method can not allocate accurate overhead costs and plant expenses to each product as production cost. Hence, the accounting method of production cost is proposed and is embedded to virtual machining simulator, which was developed to predict machining operation, for the cost prediction. So, the cost prediction system developed in this research can realize the automatic calculation of production cost from NC program generated by CAM.

Many research related to production cost prediction has been already carried out (e.g. Ohashi et. al., 2000). But, the difference of cutting conditions like depth of cuts, tool path pattern, feed rate, spindle speed can not be evaluated, so the calculated cost is not correct and this kind of system can not be used as general-purpose evaluation system. In this research, the cost accounting method is proposed using activity-based costing (ABC) (Brimson, J. A., 1997) concept. So far, some researches are carried out to account production cost using ABC concept (e.g. Fujishima, et. al. 2002, Sashio, et. al. 2004) and good results are achieved.

Hence, the cost prediction system which can solve conventional problems is developed based on the accounting method proposed and the feasibility of cost prediction system is shown through case studies in this paper.

2. OVERVIEW OF A PROPOSED SYSTEM

Figure 1 shows an overview of proposed system in this research. This consists of Estimator, Database and Analysis blocks. This system can evaluate NC programs generated by CAM. Here, the electric consumption of machine tool components, coolant quantity, lubricant oil quantity, cutting tool status and metal chip are cost factors in this research. Other factors in the figure mean the evaluation factors which are input by users according to needs like electric consumption of light, air conditioning, AGV's transportation, etc. and are ignored in this paper.

The analysis block can evaluate motions and activities related to machine tool and machining operation. The database block also consists of cost database and resource database. The cost database stores the production and disposal cost of each evaluation factors and the resource database stores machine tool specification data, cutting tool parameter, etc. for the estimation of machining process.

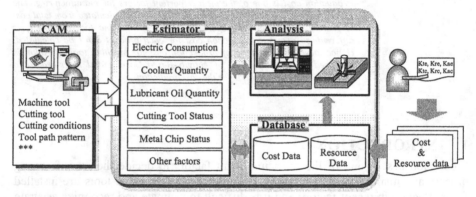

Figure 1. Prediction system of production cost for machining operation

3. CALCULATION METHOD OF PRODUCTION COST

Activity-based costing (ABC) concept is used for the calculation of production cost. Accounting method based on ABC can calculate and allocate production cost to each activity. ABC model is shown in Figure Figure 2.   All product costs can be classified to the activities used to manufacture them. Using this method, the system can identify the product cost factors which has direct implications on product cost.

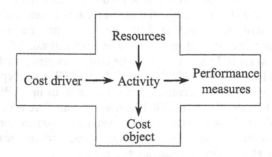

Figure 2. Activity-based model

In this research, cost driver corresponds to machining time, mean time of coolant update, etc., resource corresponds to work piece and cutting tool, cost object corresponds to product and performance measures correspond to analyzed results.

Total cost is calculated by the following equation. In this research, JPY (Japanese Yen) is used as currency.

$$Pc = Ec + Cc + LOc + \sum_{i \to N} (Tc_i) + CHc + OTc \tag{1}$$

where
    Pc: Cost of machining operation [JPY]
    Ec: Cost of machine tool electric consumption [JPY]
    Cc: Cost of coolant [JPY]
    LOc: Cost of lubricant oil [JPY]
    Tc: Cost of cutting tool [JPY]
    CHc: Cost of metal chip [JPY]
    OTc: Cost of other factors [JPY]
    N: Number of tool used in an NC program

In this paper, *OTc* isn't described. Calculation algorithms of *Ec*, *Cc*, *LOc*, *Tc* and *CHc* are described in detail as following.

Machine tool electric consumption (Ec)

The cost of electric consumption of machine tool is expressed by equation (2).

$$Ec = Ebc \times MT + ER \times CE \tag{2}$$

where
    Ebc: Basic rate of electricity [JPY/s]
    MT: Machining time [s]
    ER: Electricity bill [JPY/kWh]
    CE: Electric consumption [kWh]

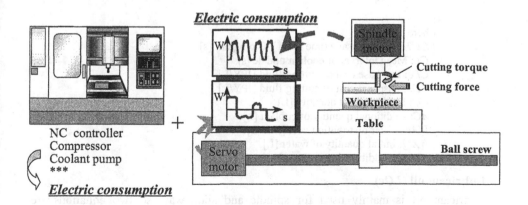

Figure 3. Electric consumption models of machine tools

CE in equation (2) is expressed by equation (3) and correspond to electric consumption of peripheral devices, servo and spindle motors shown in Figure 3. The electric consumption of peripheral devices can be predicted from machining time and each electric power. However, in order to estimate the electric consumption of the servo and spindle motors, cutting force in each axis and cutting torque are required. These values can be estimated by introducing cutting force model (Narita, et.al., 2000). This cutting force model can be applied to other cutting methods like drilling, turning and etc., so electric consumption model of machine tool proposed in this research can evaluate various machining operations.

$$CE = SME + SPE + SCE + CME + CPE + TCE1 + TCE2 + ATCE + MGE + VAE \quad (3)$$

where
    SME: Electric consumption of servo motors [kWh]
    SPE: Electric consumption of spindle motor [kWh]
    SCE: Electric consumption of cooling system of spindle [kWh]
    CME: Electric consumption of compressor [kWh]
    CPE: Electric consumption of coolant pump [kWh]
    TCE1: Electric consumption of lift up chip conveyor [kWh]
    TCE2: Electric consumption of chip conveyor in machine tool [kWh]
    ATCE: Electric consumption of ATC [kWh]
    MGE: Electric consumption of tool magazine motor [kWh]
    VAE: Vampire (Standby) energy of machine tool [kWh]

Coolant (*Cc*)

Coolant (water-miscible cutting fluid type) is generally used to enhance machining performance, and circulated in a machine tool by coolant pump until coolant is updated. During the period, some cutting oil is eliminated because of adhesion to metal chip and water escape as vapor, so additional quantity of coolant and water has to be considered. Hence, following equation is adapted to calculate the cost due to coolant.

$$Ce = \frac{CUT}{CL} \times \{(CPc + CDc) \times (CC + AC) + WAc \times (WAQ + AWAQ)\}$$

(4)

where
    CUT: Coolant usage time in an NC program [s]
    CL: Mean interval of coolant update [s]
    CPc: Purchase cost of cutting fluid [JPY/L]
    CDc: Disposal cost of cutting fluid [JPY/L]
    CC: Initial coolant quantity [L]
    AC: Additional quantity of coolant [L]
    WAc: Water distribution cost [JPY/L]
    WAQ: Initial quantity of water [L]
    AWAQ: Additional quantity of water [L]

Lubricant oil (*LOc*)

Lubricant oil is mainly used for spindle and slide way, so two equations are introduced. Here, oil-air lubricant is treated for spindle lubricant. The following equations are adapted to calculate the cost due to lubricant oil. Grease lubricant is not mentioned, but almost same equations can be adapted to calculate the cost.

$$LOc = Sc + Lc$$
(5)

where

        Sc: Cost per an NC program due to Spindle lubricant oil [JPY]
        Lc: Cost per an NC program due to slide way lubricant oil [JPY]

$$Sc = \frac{SRT}{SI} \times SV \times \left(SPc + SDc\right)$$
(6)

where

        SRT: Spindle runtime in an NC program [s]
        SV: Discharge rate of spindle lubricant oil [L]
        SI: Mean interval between discharges [s]
        SPc: Purchase cost of spindle lubricant oil [JPY/L]
        SDc: Disposal cost of spindle lubricant oil [JPY/L]

$$Lc = \frac{LUT}{LI} \times LV \times \left(LPc + LDc\right)$$
(7)

where

    LUT: Slide way runtime in an NC program [s]

    LI: Mean interval between supplies [s]

    LV: Lubricant oil quantity supplied to slide way [L]

    LPc: Purchase cost of slide way lubricant oil [JPY/L]

    LDc: Disposal cost of slide way lubricant oil [JPY/L]

Cutting tool (*Tc*)

Cutting tools are managed from the view point of tool life. So, tool life is compared with machining time to calculate the production cost in one machining. Also, the cutting tools, especially for solid end mill, are made a recovery by re-grinding, so these points are considered to construct cost equation.

$$Tc = \frac{MT}{TL \times \left(RGN + 1\right)} \times \left(\left(TPc + TDc\right) \times TW + RGN \times RGc\right)$$
(8)

where

        MT: Machining time [s]
        TL: Tool life [s]
        TPc: Purchase cost of cutting tool [JPY/kg]
        TDc: Disposal cost of cutting tool [JPY/kg]
        TW: Tool weight [kg]
        RGN: Total number of re-grinding
        RGc: Cost of re-grinding [JPY]

Metal chip (*CHc*)

Metal chips are recycled to material by electric heating furnace. This materialization process has to be considered. This equation is supposed to consider material kind, but electrical intensity of this kind of electric heating furnace is represent by kWh/t, so equation constructed in this research is calculated from total metal chip weight.

$$CHc = (WPV - PV) \times MD \times WDc \qquad\qquad (9)$$

where
   WPV: Work piece volume [cm³]
   PV: Product volume [cm³]
   MD: Material density of work piece [kg/cm³]
   WDc: Processing cost of metal chip [JPY/kg]

So far, cutting simulation system called VMSim (Virtual Machining Simulator) has been developed (Narita, et. al., 2000, 2002). Cutting force, cutting torque, machining time and machine tool motion which are the parameters to calculate cost can be predicted from NC program. Hence, prediction system for production cost has been developed by embedding the proposed calculation algorithm to VMSim.

4. CASE STUDIES

In order to show the feasibility of developed system, two case studies are introduced. In these case studies, machine tool is MB-46VA (OKUMA Corp.), cutting tool is carbide square end mill with 12mm diameter, 2 flutes and 30 deg. helical angle and work piece is medium carbon steel (S50C). The cost data are obtained by searching the companies' web site and asking the manufacture's branch offices. Table 1 shows the parameters of machine tool, work piece and cutting tool.

Table 1   Parameters of machine tool, work piece and cutting tool

Initial coolant quantity [L]	8.75
Additional quantity of coolant [L]	4.3
Initial quantity of dilution fluid [L]	166.25
Additional quantity of dilution fluid [L]	81.7
Mean interval between replacements of coolant in pump [Month]	5
Discharge rate of spindle lubricant oil [mL]	0.03
Mean interval between discharges for spindle lubrication [s]	480
Lubricant oil supplied to slide way[mL]	228
Mean interval between supplies [hour]	2000
Tool life [s]	5400
Total number of re-grinding	2
Material density of cutting tool [g/cm³]	11.9
Material density of work piece [g/cm³]	7.1

Case study 1:

Conventional prediction system of production cost for machining operation can not compare different machining strategies which manufacture same product shape, so this kind of comparison is shown first. Figure 4. shows the product shape and tool path pattern of two NC programs termed Program 1 and Program 2. Feed rate and spindle speed of each program are summarized in Table 2. These machining operation are also carried out by dry machining.

Analysis results of two NC programs are shown in Figure 5. In the figure, metal chip become profit in Japan, so this indicates minus value. Total production of Program 1 is larger than one of Program 2. So, from the view point of production cost, Program 2 is better than Program 1, though same product is manufactured. This kind of evaluation, which can not be realized by conventional evaluation system, can be achieved by developed system easily. That is to say various machining strategies effectively before real manufacturing.

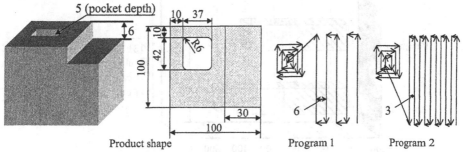

Figure 4. Product shape and tool path pattern of case studies

Table 2. Cutting conditions of two NC programs

	Program 1	Program 2
Spindle speed [rpm]	2500	5000
Feed rate [mm/min]	200	400

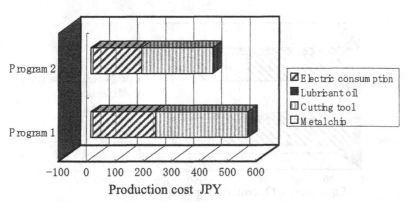

Figure 5: Analyzed production cost of two NC programs

Case study 2:

In order to verify the coolant effect on production cost, Program 1 and Program 2 with coolant usage is evaluated. Water-miscible cutting oil of A1 type (emulsion) is used in this case. Also, it is assumed that cutting tool life is extended to 1.5 times of original one due to coolant effect.

Analysis results are shown in Fig. 6. As shown, the total production costs of both NC programs are reduced from the ones of case study 1 (dry machining). This is the reason why cutting tool cost is reduced by the mitigation of tool wear due to the coolant effect. It is also found that the portions of coolant cost are very small, and the ones of peripheral devices run due to coolant usage like coolant pump, chip conveyer in machine tool is very small, too. Hence, the reduction of cutting tool cost is the most effective to realize the low cost machining in this case study. This kind comparison can be carried out quickly by developed system.

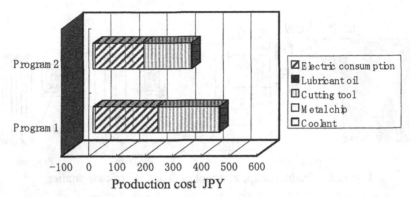

Figure 6. Analyzed production cost of two NC programs with coolant

Here, the environmental burden against global warming is evaluated using equivalent CO_2 emission intensity data (Narita, et. al., 2004). These emission intensities are obtained from environmental report, technical report, web page and industrial table. Environmental burden analysis can be realized that cost data in equations (2)-(9) is basically changed to equivalent CO_2 emission intensity data.

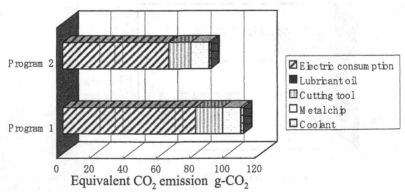

Figure 7. Analyzed equivalent CO_2 emission of two NC programs with coolant

Analyzed results are shown in Figure 7. As shown in the figure, Program 2 is better than Program 1 from the view point of equivalent CO_2 emission. This tendency is same to cost results. From the case study 2, it is found that the reduction of electric consumption of machine tool peripheral device is effective from the view point of equivalent CO_2 emission. In general, it is said that CO_2 emission has the proportional relationship to the cost, but CO_2 emission is not always correlate well with the cost in machining operation from the results of cost and CO_2 emission of electric consumption and cutting tool. Hence, in order to realize the low cost and low environmental burden machining, we have to evaluate them precisely and decide the improvement strategies depending on the situation. Using calculation model proposed in this research, production cost and environmental burden are compared easily before the real machining operation, so developed system will contribute enormously to the future manufacturing system.

5. CONCLUSIONS AND FURTHER WORK

Conclusions are summarized by the followings

- The cost calculation methods for machine tool operation using an activity-based model have been proposed and a cost prediction system has been developed;
- The feasibility of the developed system has been demonstrated through case studies.

Future work is how to take into account indirect labor cost, maintenance cost and fixturing cost. We hope this system will play an important role to contribute the cost down of manufacturing processes and improvement of manufacturing technologies.

Acknowledgments

We would like to express our sincere appreciation to OKUMA Corp. on thoughtful support.

REFERENCES

Brimson, J. A. (1997). Activity Accounting: An Activity-Based Costing Approach. John Wiley & Sons Inc.

Fujishima, M., Liu, J., Murata, I., Yamazaki, K., Xhang, X. (2002) Development of Autonomous Production Cost Accounting (APCA) and Its Implementation into CNC Machine Tool System, in Proc JUSFA2002, Hiroshima, Japan, 11-14

Narita, H., Shirase, K., Wakamatsu, H. and Arai, E. (2000) Pre-Process Evaluation and Modification of NC Program Using Virtual Machining Simulator, in Proc JUSFA2000, Ann Arbor, Michigan, U.S.A., 593-598

Narita, H., Shirase, K., Wakamatsu, H., Tsumaya, A. and Arai, E. (2002) Real-Time Cutting Simulation System of a Milling Operation for Autonomous and Intelligent Machine Tools, *International Journal of Production Research*, 40(15), 3791-3805

Narita, H., Kawamura, H., Norihisa, T., Chen,L.Y., Fujimoto, H., Hasebe, T. (2004) Prediction System of Environmental Burden for Machining Operation, in Proc JUSFA2004, Denver, Colorado, U.S.A., JL_016.pdf, CD-ROM

Ohashi, T., Arimoto, S., Miyakawa, S., Matsumoto, Y. (2000) Development of Extended Machining–Producibility Evaluation Method (Extended Quantitative Machining–Producibility Evaluation for Various Processes), *Trans. J. Soc. Mech. Eng.*, 2429-2436 (in Japanese)

Sashio, K., Fujii, S., Kaihara, T. (2004) A Basic Study on Cost Based Scheduling, in Proc ICMA2004, Suita, Osaka, Japan, 353-356

35. Information Modeling for Product Lifecycle Management

Q.Shu and Ch.Wang

College of Mechanical Engineering & Automation, Northeastern University, 110004, Shenyang, PR China
E-mail: shu_qilin@mail.neu.edu.cn

Product lifecycle modelling is to define and represent product lifecycle data and to maintain data interdependencies. This paper presents a framework of product lifecycle model that comprises three parts: product information model, process model based on product life cycle, and extended enterprise resource model. Further, the relationship and formation of product models at different stages are described. Finally, an integrated information architecture is proposed to support interoperability of distributed product data sources.

1. INTRODUCTION

In today's distributed manufacturing environment, how to manage the product data and distribute it to the right people who need it is critical to the success of an enterprise. At present, the existing systems such as CAx, ERP, PDM, SCM, CRM, eBusiness etc. are just solutions for some stages of a product lifecycle, which is difficult to support an enterprise to operate efficiently. Developing a complex product requires not only all collaboration among departments of an enterprise but also cooperation of other enterprises in different regions. To effectively allocate enterprise's resources and harmonize business activities, enterprises must integrate all information relevant to various stages of a product lifecycle and all the information throughout a product lifecycle can be accessed by everyone associated with its design, creation, sale, distribution, and maintenance. Now enterprises press for an information architecture to implement product lifecycle management. The product lifecycle modelling technology supports description, transmitting and sharing of data in distributed environment. Therefore, the product lifecycle modelling technology and related information management systems are taken seriously by academia and industries.

Many researchers have conducted study on product models at different stages of product lifecycle. Bidarra and Brosvoort described product structures and its function information by building engineering design model based on semantic feature method (Biddarra and Brosvoort, 2000). Jiao *et al.* proposed a generic bill-of-materials-and-operations for high-variety production management to integrate data of product structure and manufacturing information, which support engineering changes and transactions of customers' orders (Jiao *et al.*, 2000). Simon *et al.* presented a modelling technology of the life cycle of products with data acquisition features for washing machines based on a microcontroller and non-volatile memory (Simon *et al.*, 2001). This life cycle model with data acquisition features supported activities of design, marketing and servicing as well as end-of-life. Shehab *et al.* proposed a manufacturing cost model for concurrent product development to help inexperienced designers evaluate the manufacturing cost of products at the conceptual design stage (Shehab *et al.*, 2001).

To remain competitive and to respond the market rapidly, enterprises have to cooperate effectively as the form of extended enterprise to develop complex products. An extended enterprise comprises an Original Equipment Manufacturer (OEM), its supply chain, subsidiaries, consultants, and partners affiliated with the life cycle of a particular family of products. Rezayat identified a majority of the components needed to implement the E-Web and showed how E-Web can provide support for everyone associated with a product during its life cycle (Rezayat, 2000).

Manufacturing paradigms such as agile virtual enterprise, Internet-based manufacturing, and collaborative manufacturing require creating a distributed environment that enables integrated product, process, and protocols development in order to manage and maintain the distributed product data as a whole. Product lifecycle model provides a conceptual mapping mechanism to associate with data of product design, manufacturing, quality, cost, sales, operation, and service. The model also supports the access and operation of distributed data. Xue thinks that product lifecycle model should comprise a series of models at different stages and these models can be generated automatically by a method (Xue, 1999). Zhang and Xue proposed product lifecycle modelling method based on distributed databases, which defined the relationship between data located in different regions and activities by using and/or graph (Zhang and Xue, 2001). Tianfield built an advanced lifecycle model for complex product development (Tianfield, 2001). The model comprised concurrent engineering, production schedule control, virtual prototype, and enterprise information integration. Now it is commonly accepted that the Web is a major enabler in this regard due to its open standards, ease of use, and ubiquity, so some researchers proposed or implemented product lifecycle information infrastructure based on the Internet (Rezayat, 2000; Wang, 2001).

2. FRAMEWORK OF PRODUCT LIFECYCLE MODEL

Now 50–80% of all components in products from Original Equipment Manufacturers (OEMs) are fabricated by outside suppliers, and this trend is expected to continue (Rezayat, 2000). During a product lifecycle, the product data are generated and operated in multiple business processes among different departments in the extended enterprise. So, how to integrate these data and processes to support the product lifecycle activities is becoming very important. In this paper we present an information framework to describe all processes and activities during the lifecycle of a product by building integrated product information model, process model based on product life cycle, and extended enterprise resource model as shown in Figure 1.

It is impossible to describe all activities and processes during the product lifecycle only using a single model because of complexity of product developing and using process. Therefore, it is necessary to build a group of models according to different stages and aspects. We divide the product lifecycle into five stages, which are requirement analysis stage, conceptual design stage, engineering design stage, manufacturing stage, and service & support stage. Consequently, we propose that the product lifecycle model comprises requirement analysis model, conceptual design model, engineering design model, manufacturing model, and service & support model, and all related data and documents which stored in the distributed databases form those models through logical mapping. The product requirement analysis model represents customer's needs and interests in customer's language, and this model relates to subjective origins of a product. The product conceptualization model is a conceptual definition of a product including its functions, implementing theory, basic structure information, rough costing, and some key components' design etc. The product engineering design model is the core of product lifecycle model, which depicts both geometric and engineering semantic information by using CAD files and related data. The product manufacturing model describes the information relevant to the processes of fabrication and assembly such as material, manufacturing process planning, and assembly sequences etc. The service & support model is about the information of products' delivery state, installation, operation, training, maintenance, diagnosis, etc.

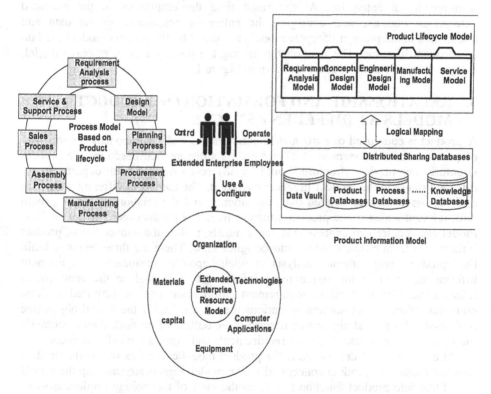

Figure 1 Framework of Product Lifecycle Model

The process model based on product lifecycle enhances and extends the functions of workflow in PDM systems, and it describes the relationship among data, applications, enterprise resources, organizations, and employees in various stages during the product formation. This model supports dynamical process definitions of product data, projects management, and process scheduling & planning. The process model comprises macro processes and various micro processes. The macro processes during a product lifecycle include the processes of requirement analysis, engineering design, process planning, procurement, manufacturing, assembly, sale, and service & support. However, the micro processes exits in those macro processes, e.g. the micro process of the examining and approving in the macro processes of both design and process planning. The process model guarantees the security and effectiveness of all product data by controlling the right operation of data by right people, and coordinating the relationship of various departments in the extended enterprise.

The extended enterprise resource model provides resource configuration for various stages of the product lifecycle. It describes the resource structures in the extended enterprise and relationship among the different resources. The extended enterprise resource model primarily comprises organizations, technologies, materials, capital, equipment, and various computer applications etc. In a manufacturing enterprise, employees are organized to perform corresponding tasks by a strict organizing structure and a set of rules, so they are the most important resource in an enterprise. At the same time the employees in the extended enterprises who use and configure the enterprise resources, operate data and documents in the product lifecycle model are guided by the process model based on the product lifecycle. The relationship among the process model, resource model, and product information model is shown in Figure 1.

3.   RELATIONSHIP AND FORMATION OF PRODUCT MODELS AT DIFFERENT STAGES

A product is composed of parts and/or components, and a component is composed of parts and/or sub-assemblies. Therefore, its structure tree describes the product structure. The product structure should be different when different departments at different stages in the product lifecycle operate it. The models in different stages are built by associating all related data, documents, and the relationship among them with relevant nodes in the product structure tree. The product requirement analysis model can acquire and express customers' needs, realize the conversion of product information from customer view into designer view. There are three levels to build the product requirement analysis model: acquire customers' requirement information, express the requirement information, and analyze the requirement information. The customers' requirement information can be acquired through communication with customers by various means, and finally the primal object tree is formed. This primal object tree describes the outline of product from customer's view, so it only expresses the basic requirements and expectations of customers.

At the conceptual design stage, the product model comprises two parts: product function model and product conceptualization model. First designers map the primal object tree into product function tree from the view of technology implementation. For example, customers need to acquire information from outside through their computers, so designers can map this requirement into the function of access to the Internet. Then designers may further map every node in the product function tree into one or more physical modules (parts or components), e.g. mapping the function of access to the Internet into physical module of network card or modem. Therefore, the product requirement analysis model can be converted into the product conceptualization model in this way.

At the engineering design stage, designers form a generic product structure through analyzing, computing, simulating, and testing by making full use of the product conceptualization model and considering product varieties, design constraints, and resources limitations etc.

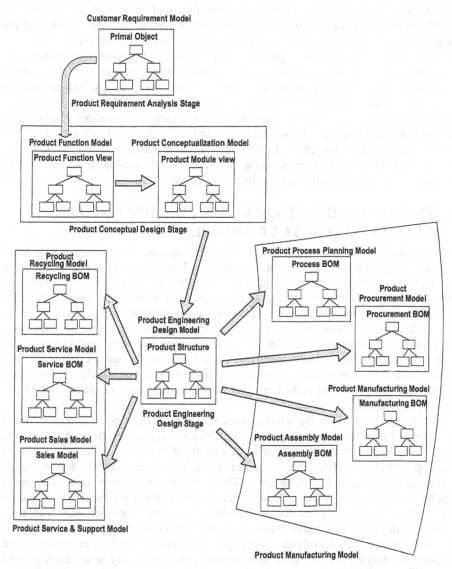

Figure 2 Relationships and Formation of Product Models at Different Stages

Every node in the generic product structure tree should be associated with related documents and data such as CAD files, design reports, analysis reports, test data, using materials, working space, kinematic parameters, and working precision etc. With these, we form the product engineering design model based on the product structure. The product engineering design model is the core staged model of the product lifecycle model, from which the manufacturing model and the service &

support model can be derived through reconfiguring the engineering BOM and adding additional related information by corresponding business department of the extended enterprise. For example, the process planning model at the manufacturing stage is built by reconfiguring the engineering BOM into process BOM and associating relevant process planning information with each node in the process BOM. In the same way the procurement model, the manufacturing model, and the assembly model can be formed based on the engineering design model.

The product service & support model collects all information about sale, transportation, installation, operation, maintenance, and recycling, which comprises three sub-models: the sale model, the service model, and the recycling model. All these information including various digital documents, CAD files (2D /3D), multimedia files (videos/audios/ animation) is associated with related nodes in the BOMs (sale BOM, service BOM, and recycling BOM). All these BOMs (sale BOM, service BOM, and recycling BOM) are reconfigured based on the product engineering design model (engineering BOM). The relationship and formation of product models at different stages is shown in Figure 2.

4. INTEGRATED INFORMATION FRAMEWORK OF PRODUCT LIFECYCLE MODEL

The product lifecycle management (PLM) system is a network-oriented manufacturing system, which comprises many sub-systems. These sub-systems are located in a distributed heterogeneous environment. Therefore, we must find effective methods for communication and sharing of information, especially those related to design and manufacturing, throughout the entire enterprise and the supply chain. The technologies that support such methods must be able to deal with distributed environments and databases, must ensure reliability and security, and must be practical. We present an integrated information framework of the product lifecycle model based on J2EE to realize the goal of providing the right information to the right person at the right time and in the right format anywhere within the extended enterprise.

As shown in Figure 3, the information framework of the product lifecycle model is composed of three tiers: application tier, service tier, and data tier. There are five application agents according to five stage models of the product lifecycle model in the application tier. They are: requirement agent, concept agent, design agent, manufacturing agent, and service agent. The five agents run in two containers, one is Applet container and another is application container. There are two ways to access information for every agent, that is Web-based way and non-Web way. For non-Web way, there is a client application to run on local computers, and for Web-based way the browser can download Web pages and Applets to client computers. The requirement agent, concept agent, and service agent are usually based on Web and they are implemented by Applets. However, the design agent and manufacturing agent are always run on Intranet within an enterprise, so they are implemented by client application instead of basing on Web.

The application agents access the basic service agents such as graphics agent, coordinating agent, security agent, query agent, and resource agent etc. in the service tier through various protocols (e.g. TCP/IP, SOAP, and IIOP etc.). The graphics agent displays the geometries (2D/3D) of products. The security agent guarantees the security of the whole product lifecycle management system by verifying identification and authorizing operation etc. Among these basic service agents, the query agent is a very important one, which responds to the querying requirements of users and provides the desired results to them by automatically searching all information sources in the product lifecycle model. The resource agent manages all system resources according to the request of other agents in order to avoid conflict between agents. These service agents are implemented based on Web container and EJB container. Web components provide Web services, which can be JSP pages or Servlets. Servlets are the classes of Java, which can dynamically respond to the request. However, JSP pages are based on text files that contain static texts and a fragment of Java program. When JSP pages are loaded, the Servlet executes the fragment of Java program and returns the results. EJB components are enterprise business components, which perform service functions of enterprise business such as security agent, resource agent, and coordinating agent etc.

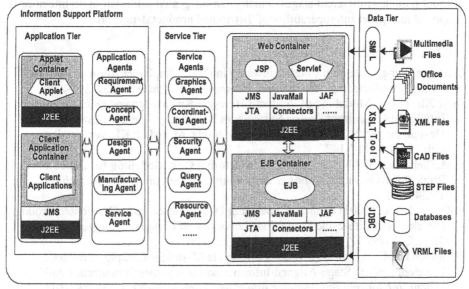

Figure 3 A Support Framework of Product Lifecycle Model

In the data tier, all the data are stored in physical media as various forms. To effectively manage and browse these data through the Internet, we convert the non-structured data with different formats into XML files by using XSLT tools. These files with different formats include office documents, CAD data, STEP files, and multimedia files etc. The structured data in the relational databases can be accessed through JDBC. The product lifecycle information framework based on the above three-tier architecture has the advantages of interoperation, reusability, and platform independence. The application agents in the application tier and the service agents in the service tier can be deployed in any servers compatible with J2EE, so in this way the PLM system can be established quickly and conveniently to meet enterprises' requirement for product lifecycle management. All data sources in the data tier can be stored in different locations and they can be managed through Web by various agents in the service tier. There is a great advantage to integrate and manage all partners' information during the product lifecycle. All PLM system components based on J2EE are reusable and extendable. They can be reused not only in the same system but in different systems (e.g. ERP, CRM etc.) as well.

5. CONCLUSION AND FURTHER WORK

Nowadays, information is the most important factor for a manufacturing enterprise. Product lifecycle information model helps a manufacturer to make decisions about management, design, production, operation, maintenance, and repair. The product lifecycle can be divided into five stages: requirement analysis, conceptual design, engineering design, manufacturing, and service & support. To integrate all information of a product lifecycle and support networked manufacturing mode, we present a framework of product lifecycle model that comprises three parts: product information model, process model based on product life cycle, and extended enterprise resource model. Further, we describe the relationship and evolvement of product models at different stages. Finally, an integrated information architecture is proposed to support interoperability of distributed product data sources.

REFERENCES

Chengen Wang, Chengbin Chu, Chaowan Yin (2001) Implementation of Remote Robot Manufacturing over Internet. *Computers In Industry* 45(3), 215-229

Deyi Xue (1999) Modelling of Product Life-cycle Knowledge and Data for an Intelligent Concurrent Design Systems, in Knowledge Intensive Computer Aided Design, Finger, S., Tomiyama, T., and Mantyla. M. (eds.): Kluwer Academic Publishers. pp117-142

E.M.Shehab, H. S. Abdalla (2001) Manufacturing Cost Modelling for Concurrent Product Development, *Robotics and Computer Integrated Manufacturing* 17:341-353

F.Zhang, D.Xue (2001) Optimal Concurrent Design Based upon Distributed Product Development Life-cycle Modelling. *Robotics and Computer Integrated Manufacturing* 17, 469-486

Huaglory Tianfield (2001) Advanced Life-cycle Model for Complex Product Development via Stage-Aligned Information-substitutive Concurrency and Detour. *Int Journal of Computer Integrated Manufacturing* 14(3), 281-303

Jianxin Jiao, Mitchell M.Tseng☐Qinhai Ma and Yi Zou (2000) Generic Bill-of-Materials-and-Operations for High-Variety Production Management, *Concurrent Engineering: Research and Applications* 8(4), 297-321

Mattew Simon, Graham Bee☐Philip Moore, Jun-Sheng Pu, Changwen Xie (2001) Modelling of the Life Cycle of Products with Data Acquisition Features, *Computers in Industry* 45:111-122

M.Rezayat (2000) The enterprise-web portal for life-cycle support, *Computer-Aided Design* 32:85-96

M. Rezayat (2000) Knowledge-based Product Development Using XML and KCs, *Computer-aided Design* 32, 299-309

R.Bidarra, W.F.Brosvoort (2000) Semantic Feature Modelling. *Computer-Aided Design* 32:201-225

36. Generic Core Life Cycle and Conceptual Architecture for the Development of Collaborative Systems

Tad Gonsalves and Kiyoshi Itoh

Information Systems Engineering Laboratory, Faculty of Science & Technology,
Sophia University, Tokyo, Japan
Email: {t-gonsal, itohkiyo}@sophia.ac.jp

In the conventional system development life cycle (SDLC), the system performance evaluation phase comes after the implementation phase. Our strategy is to project system performance estimate at the requirement analysis and design phase itself, much before the implementation phase. To achieve this objective, we propose a technology-neutral integrated environment for the core life cycle of system development. This core life cycle consists of three phases: system modelling, performance evaluation and performance improvement.

1. INTRODUCTION

Testing and evaluating system performance is an important stage in the development of a system. However, it is often ignored due to lack of time, tools or both. While designing systems, "designers are (blindly) optimistic that performance problems – if they arise – can be easily overcome" (Cooling, J, 2003). The designers test the performance of the system after the completion of design and implementation stages and then try to remedy the problems. The main difficulty with this 'reactive approach', Cooling states, is that problems are not predicted, only discovered. It is the concern of this study to *predict* the operational problems and to suggest a viable solution while designing the systems. The significance of performance design lies in the fact that the performance requirements and performance characteristics are already incorporated in the system at the design stage of the system. The system designers project an estimate of the system performance, as it were, at the requirement analysis stage itself of the system development life cycle.

In this paper we propose a core life cycle and an integrated environment for system development. The core life cycle consists of three phases - modelling, performance evaluation and performance improvement. System performance is evaluated by simulation and an initial improvement plan is suggested by the expert system. This improvement plan is incorporated in the system model and the performance evaluation simulation cycle is repeated. Thus, the originally planned system and its operation could, in principle, be repeated through a number of cycles in the integrated environment, till a refined system is obtained. Further, the core life cycle can be incorporated in the total system development life cycle.

Our target systems are collaborative systems. When multiple organizations work on a joint project or come together to share resources to enhance their business prospects, they constitute a collaborative system. The conceptual model of the system depicts the interrelationships between actors, tasks and collaborative activity in the system. Corresponding to these, actor sufficiency, task mobility and activity efficiency are chosen as performance indicators of the system. Performance improvement is by an expert system driven by qualitative rules. We have developed a systematic methodology for the conceptual core life cycle and indicated an implementation scheme for practical collaborative systems.

This paper is organized as follows. Section 2 introduces the salient features of collaborative systems which may be exploited in modelling the system and in evaluating and improving its performance. Section 3 describes the three integrated phases of the core life cycle. Section 4 deals with the conceptual architecture for the core life cycle and Section 5 describes in brief our attempts in implementing the core life cycle for real-life collaborative systems.

2. SALIENT FEATURES OF COLLABORATIVE SYSTEMS

When multiple organizations work on a joint project or come together to share resources to enhance their business prospects, they constitute a collaborative system. Business firms collaborating to enhance their business prospects, doctors and nurses in a clinic offering medical service to patients, teachers and staff members in a school offering educational service to students, manager and tellers in a bank offering financial service to customers, etc., are all examples of collaborative systems. Another special class of systems engaged in collaboration is collaborative engineering systems. The goal of these systems is to do engineering (or to provide 'engineering services') through collaboration.

Collaborative systems have several distinct features that may be exploited in the modelling, performance evaluation and performance improvement of the systems. In this section, we describe three salient features of collaborative systems that may be used in the different phases of system development.

2.1 Request-Perform Activity

Each collaborative activity in a collaborative system may be looked upon as a request-perform activity. Collaborative activity begins when the requesting collaborator requests the performing collaborator to perform the activity. The overall workflow in the collaborative system can then be viewed as a series of interrelated request-perform activities. Some request-perform activities are sequentially linked while others are concurrent; still others need to synchronize with one another.

The request-perform activity property can be used to adequately represent the collaborative system under consideration. The topology of all request-perform activities with their inter-relationships (sequential, merging, diverging, concurrent, synchronizing, etc.) can serve as a convenient system model.

2.2 Client-Server Systems

A client is anything that places a request for service and a server is anything that offers service in response to the request. In collaborative systems, the collaborators act as servers, offering service to the clients. Clients, in general, could be customers, orders, material for production, etc., that enter the system seeking service. Sometimes a given server may perform some part of the collaborative activity and request another server for the remaining part of the service; in other words, servers could become clients of other servers.

Waiting-line analysis, also known as queueing theory, deals with the computational aspects of the server in the client-server setup (Figure 1). Each collaborative activity may be represented by the queueing server and real-life collaborative systems may be modelled as a combination of several appropriate servers.

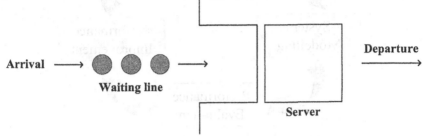

Figure 1: Model of a single queueing system

2.3 Discrete-Event Systems

A system is defined to be discrete-event if the phenomenon of interest changes value or state at discrete moments of time, as opposed to the continuous systems in which the phenomenon of interest changes value or state continuously with time (Banks, J, Carson, II, J.S, 1984). Collaborative systems are discrete-event systems because the events take place in discrete steps of time. An event is an occurrence that changes the state of the system. The beginning of service and the end of service are typical events in collaborative systems. In a clinic, for instance, a patient arriving for service, joining in the queue if the physician is busy, starting treatment and ending treatment are significant events that change the state of the system.

Both stochastic and deterministic systems can be simulated by using the discrete-event simulation approach (Ross, S.M, 1990). Discrete-event simulation is centred around the concept of event. This is because the only significant thing that contributes to the time-evolution of the system is the series of events. Without the series of events occurring in the course of the history of the system, the system would be static. Discrete-event simulation is just the right approach to simulate the performance of the discrete-event collaborative system.

3. CORE LIFE CYCLE

Our ultimate aim is to estimate the performance of the system at the requirement phase of the SDLC. To achieve this aim, we create a generic core life cycle consisting of just three phases - system modelling, performance evaluation and performance improvement. Each of these phases is described in detail in the ensuing sub-sections. From the planned system, a rough model is made in the beginning. The modelled system is simulated to yield its performance. The performance results are evaluated and suggestions for improvement are provided by an expert system. The operation improvements are incorporated into the model of the system and the entire cycle is repeated, yielding a refined system after each cycle. Furthermore, we make this approach integrated so that each phase, if not automatically, at least semi-automatically leads to the next phase.

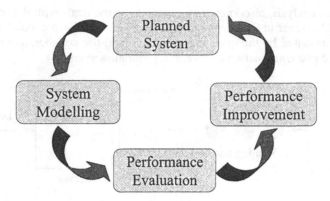

Figure 2: Core life cycle

3.1 Modelling

Model is an abstract concept that stands for the system. It is not the system itself, but something that represents the real system. Models should be made simple enough so that they can be simulated and improved. The model designer should not try to include too many details in the model, because that would make the model cumbersome and unwieldy. On the other hand, the designer should not make oversimplifications, because then the model will be far from the real-system it is supposed to represent.

The type of model to be chosen will depend on the system at hand. The model of a collaborative system should have the ability to account for the prominent features of collaboration. Any form of collaboration necessarily implies the presence of 'actors' (collaborators), 'activities' and 'objects upon which the action is performed' (tasks). Each of these constituent elements of collaborative systems is briefly described below.

Actors

In collaborative systems the collaborators are the main actors. They keep the system going by carrying on the collaborative work. They receive tasks and process them and make the system realize its ultimate goal. Actors, in practice, could be personnel or machines or a combination of both. The interaction among the actors is what essentially determines the nature of the collaborative system. The system model should have the capacity to grasp the prominent interactions among the collaborators.

Tasks

Tasks are processed by collaborators in the collaborative system. There could be a variety of inter-related tasks flowing in a variety of ways in the system. In designing the system model, the modeler should be able to decipher the different tasks and elaborately analyze their flow in the system. The flow analysis should lead to a well-defined flow-control mechanism in the model.

Activity

The work done on the task, (or the processing of the tasks) by the actors, is called activity. Collaboration consists of a series of activities that are linked to one another in a logical way. The collaborative activities present in a collaborative system are numerous and of a diverse nature. A sound model should be able to enumerate all the activities systematically, group them and arrange them in the model layout so as to give a comprehensive view of the entire system.

3.2 Performance Evaluation

Performance evaluation is more an art than a science. There is no hard and fast rule for performance analysis. It varies from developer to developer. Further, it may vary from system to system. Performance evaluation, to a great extent, will depend on the features of the system chosen by the manager/designer. In this section, we present some of the basic performance indicators that are of interest when faced with the task of performance evaluation of a collaborative system.

Since we have singled out actors, tasks and activities as the basic constituent elements of our collaborative system model, it follows that the performance indicators of the system operation should be constructed around these three basic elements. We choose the following three key performance indicators.

Actor sufficiency

The collaborators in a collaborative system are the actors in that system. They receive requests for service, process the requests and forward the requests (tasks) to subsequent servers in the system. When the actors at a given server are insufficient, service time tends to prolong and the utilization of the server increases. With the increase in service time, the queues in front of the servers grow in size, making the system operation unstable and leading to greater customer dissatisfaction. However, increasing the number of actors in nearly all cases leads to an increase in cost. Actor sufficiency, therefore, is a trade-off between operation cost and customer satisfaction.

Task mobility

Task mobility refers to the frequency of tasks' entry into the system and their flow from server to sever in the system. Another term for task mobility is workload and consists of service requests to the system. In the analytic approach, task mobility is usually represented by different probability distributions. We define task mobility as

$$\lambda = \text{Number of tasks / Time}$$

Task mobility is an important factor influencing the performance of the system. If the task mobility is low, system utilization is low, implying that the system resources are underutilized. However, system utilization rapidly increases with the increase in task mobility and soon approaches unity. Therefore, task mobility should be maintained at an appropriate level so as to maintain the system utilization in the optimum range of operation.

Activity efficiency

The utilization of the server is given by

$$\rho = \lambda / \mu$$

where μ is the service rate.

The utilization of the server, ρ, represents the fraction of total operation time the server is kept busy. This ρ, sometimes referred to as the traffic intensity, is also a measure of the efficiency of collaborative activity. System managers generally tend to drive the system to its maximum utilization. However, it is a fact of observation that the response time tends to infinity as utilization approaches unity. Storage is another problem that presents itself in the face of rapidly rising queues because of high utilization. From the failure-to-safety aspect, experts' heuristics suggest that the server utilization may not exceed 0.7 (Itoh, K, Honiden, S, *et al.*, 1990).

3.3 Performance Improvement

3.3.1 Qualitative Rules for Performance Improvement

Collaborative systems are usually large complex systems. It may not be possible to grasp all the variables that are in the system and the diverse interactions among them. Thus, quite a few complex collaborative systems could be systems with incomplete knowledge. Modelling, performance evaluation and performance improvement of such systems could be done by way of Qualitative Reasoning (QR). QR is an AI discipline that bases its reasoning on the nature and behaviour of the components that make up the system (Kuipers, J, 1994). A set of qualitative rules presented below can be established by accumulating knowledge of experts in the domain.

Table 1- Qualitative Rules for Performance Improvement

Performance Indicators	State	Number of actors (N)	Rate of activity (μ)	Rate of task mobility (λ)
Actor Sufficiency	Scarce	↑	↑	↓
	Normal	o	o	o
	Excess	↓	↓	↑
Task Mobility	Scarce	↓	↓	↑
	Non-uniform	o	o	↑/↓
	Excess	↑	↑	↓
Activity Efficiency	Minimal	↓	↓	↑
	Normal	o	o	o
	Critical	↑	↑	↓

(Notation ↑: Increase parameter; ↓: Decrease parameter; o: Do not change parameter)

The above principles could be further elaborated and structured as IF-THEN production rules, which can then be used in the construction of the performance-improvement knowledge-based system.

3.3.2 Meta-Rules for Performance Improvement

In addition to the above practical rules, one could envisage the following meta-rules that would be needed for the application of the above rules. The meta-rules act as guiding principles for the application of the performance improvement rules.

Availability

The parameters (number of actors in a given activity, duration of the given activity, task mobility, activity efficiency, etc.) that could, in principle, be changed so as to improve the performance of the system, satisfy the desirability condition or the availability condition. All these parameters are included in the available set. This is the largest set shown in Figure 3.

Feasibility

However, the values of some but not all the members of the available set can be changed. The system operation requirements and constraints are such that certain parameters cannot be changed. In other words, changes in these parameters are not feasible under the given conditions. Only the parameters that may be changed are included in the feasible set. Further, there may be a limit to the range of changes that are feasible for a given parameter. The feasibility conditions are normally laid down at the requirement analysis stage of the system development. The feasibility knowledge is therefore another important factor that needs to be included in the performance improvement knowledge-based system.

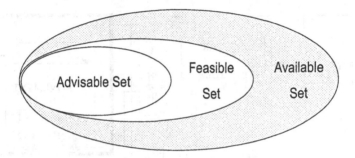

Figure 3. Available, Feasible & Advisable Sets

Advisability

In resolving bottlenecks, more important than the feasibility condition is the advisability condition. At times, it may be feasible to change a parameter, but the knowledge-based system (KBS) may judge that it is not advisable to change it for fear of exerting bad influence on other sections of the system. The advisable set, therefore, is only a subset of the feasible set. It contains only a handful of parameters which satisfy all the three conditions of availability, feasibility and advisability in resolving bottlenecks locally, while maintaining the global stability in system operation. If the changes made in one section of the system will end up in creating fresh bottlenecks or worsen the existing ones in the other section, then the *advisability* condition will prompt the KBS to issue a warning to the user. Advisability means coming up with a sound strategy for resolving bottlenecks such that changes made in the network are minimum.

4. CONCEPTUAL ARCHITECTURE FOR THE CORE LIFE CYCLE

The core life cycle of system development seeks to integrate the three phases of modelling, performance evaluation and performance improvement. The model of the collaborative system is typically a network of request-perform activities inter-related according to the logic of the collaborative system workflow. The model could be multi-layered, with each layer depicting in detail some aspect of collaboration. At the deepest level, there should be a representation of the network structure of the system.

Through an inter-conversion scheme, the network structure in the model could be imported into the Performance Evaluator (PE) and KBS (Figure 4). The qualitative rules and meta-rules form the backbone of the KBS. The KBS further relies on the PE for performance data of the system. It diagnoses the bottlenecks in system operation from this data and suggests a tuning plan for performance improvement to the user. The KBS and the user interact through the user interface. The integration of the three phases is illustrated in the diagram below.

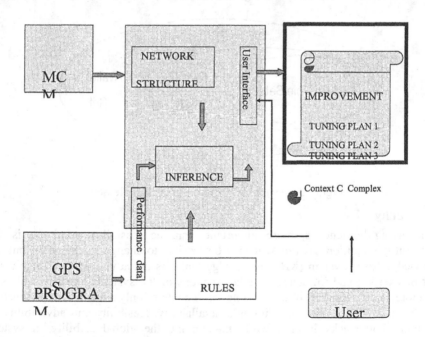

Figure 4. Conceptual Architecture for the Core Life Cycle

5. IMPLEMENTATION OF THE CORE LIFE CYCLE

We have attempted to create an integrated environment and to implement the three phases of modelling, performance evaluation and performance improvement (Gonsalves, T, Itoh, K, *et al.*, 2003, 2004a, 2004b). Microsoft Visio is used to create the descriptive model of the collaborative system. Arena, Simula, Simscript, GPSS, etc., are some of the well-known discrete-event simulation languages. We have chosen GPSS (General Purpose Simulation System) for developing the system performance evaluating program, because of its versatility and ease-of-use. Finally, the bottleneck diagnosis and parameter-tuning plan are by the expert system constructed by using Prolog-based Flex toolkit. Each of the three steps in system analysis and design and the use of the respective tools are discussed in the following sections.

5.1 Modelling

Our model is a descriptive model of the collaborative system. The basic unit of collaborative activity is represented by a 'Context'. The actors in the system are known as 'Perspectives'. The requestor of activity is the 'Left-hand Perspective' and the performer of the activity is the 'Right-hand Perspective. There exists an interface between the two Perspectives through which Token, Material and Information (TMI) pass. These three represent three types of (related) tasks in the system. The topology of inter-connected Contexts gives rise to the 'Multi-Context Map' (MCM) model of the collaborative system (Hasegawa, A, Kumagai, S, *et al.*, 2000).

There are two semi-automatic software converters included as macros in the MCM drafter – the Prolog converter and the GPSS converter. The former converts the MCM network visual information into Prolog facts, while the latter converts the TMI flow and MCM contexts information into GPSS skeleton program. The GPSS converter is semi-automatic in the sense that the user has to supply the given parameters required to simulate the performance of the system.

5.2 Performance evaluation

Performance evaluation of the system operation is carried out by GPSS simulation. GPSS is built around abstract objects known as entities. A GPSS simulation program is a collection of a number of entities. The most prominent entity types are transactions and blocks. GPSS simulations consist of transactions and a series of blocks. Transactions move from block to block in a simulation in a manner which represents the real-world system that the designer is modelling.

The initial values of parameters such as service times, average inter-arrival time, the capacity of each server, the respective distributions for inter-arrival times and service times and the sequence of random number generators are provided. The system operation is simulated for the desired length of time and the required steady-state performance measures are recorded. The simulation is performed a large number of times to reduce the random fluctuations. The final simulation data sheet gives the performance measures averaged over a large number of simulation runs.

5.3 Performance improvement

Performance improvement is by the Flex expert system. Flex is a Prolog-based software system, specifically designed to facilitate the development and delivery of expert systems. A very useful and appealing feature of Flex is the use of Knowledge Specification Language (KSL) in programming.

The expert system consults the simulation data sheet obtained from GPSS simulation and diagnoses the bottleneck parameters. The bottlenecks that are detected are listed and displayed to the user. The user then selects desired number of bottlenecks for resolving. The inference engine of the expert system fires the appropriate knowledge-based qualitative rules to resolve the bottlenecks. It interacts with the user to obtain information regarding feasibility. It makes use of the structural knowledge of the system obtained from MCM drafter to judge the propagation effects (advisability) of resolving the chosen bottlenecks. The output of the expert system is a set of performance parameters carefully selected for tuning.

6. CONCLUSION

Improving the performance of an established system is a cumbersome activity fraught with high risks and unjustifiable cost. Through the system development core life cycle we have indicated a way of projecting system performance estimate at the requirement analysis and design phase itself, much before the implementation phase of collaborative systems. The core life cycle consists of three phases: viz., system modelling, performance evaluation and performance improvement. The conceptual model of the system depicts the interrelationships between actors, tasks and collaborative activity in the system. Corresponding to these, actor sufficiency, task mobility and activity efficiency are chosen as performance measures of the system. Performance improvement is by a qualitative knowledge-based system. We have developed a systematic methodology for the conceptual core life cycle and indicated an implementation scheme for practical collaborative systems.

REFERENCES

Banks,J, Carson,II,J.S (1984) Discrete-Event System Simulation. New Jersey : Prentice-Hall.

Cooling,J (2003) Software Engineering for Real-time Systems. Addison-Wesley.

Gonsalves,T, Itoh,K,, Kawabata,R (2003) Performance Design and Improvement of Collaborative Engineering Systems by the application of Knowledge-Based Qualitative Reasoning, Knowledge Based Design Series, The ATLAS(1)

Gonsalves,T, Itoh,K,, Kawabata,R (2004) Perspective Allocating Qualitative Function for Performance Design and Improvement of Collaborative Engineering Systems (Proc ECEC2004, Hasselt, Belgium).

Gonsalves,T, Itoh,K,, Kawabata,R (2004) Use of Petri Nets in the Performance Design and Improvement of Collaborative Systems (Proc IDPT2004, Izmir).

Hasegawa,A, Kumagai,S,, Itoh,K (2000) Collaboration Task Analysis by Identifying Multi-Context and Collaborative Linkage. *CERA Journal* 8(1), pp61-71.

Itoh,K, Honiden,S, Sawamura,J,, Shida,K (1990) A Method for Diagnosis and Improvement on Bottleneck of Queuing Network by Qualitative and Quantitative Reasoning. *Journal of Artificial Intelligence* (Japanese) 5(1), pp92-105.

Kuipers,J (1994) Qualitative Reasoning Modeling and Simulation with Incomplete Knowledge. Cambridge : MIT Press.

Ross,S.M, (1990) A Course in Simulation. New York : Macmillan Publishing Company.

37. Integrated Process Management System and RFID Directory Services

Masayuki Takata[1], Eiji Arai[2] and Junichi Yagi[3]

1 The University of Electro-Communications Email: takata@cc.uec.ac.jp
2 Osaka University Email: arai@mapse.eng.osaka-u.ac.jp
3 Shimizu Corporation Email: junichi.yagi@shimz.co.jp

This paper describes an implementation of the Integrated Process Management System, which includes manufacturing process management for building parts, and also construction process management at construction site. To observe the flow of the building parts, RFIDs are stuck to all parts to be managed, and several checkpoints, which we named "gates", are introduced within the coherent process through part-manufacturing and building construction. The requirements of the RFID directory services are also discussed.

1. INTRODUCTION

This paper describes an implementation of the integrated construction process management system, which includes both manufacturing process management features for building materials and also construction process management features at construction site, and a proposal of an RFID directory service agent.

Recently, RFIDs are getting popular in logistics industries and manufacturing industries. The process management system for building construction and building materials manufacturing must cover these two aspects, and the use of RFIDs in construction industries will make the trace-ablity of the building materials more accurate.

When the implementations described here are realized, in a case of some troubles found in building materials, other building materials are re-allocated for the order, and if requested in advance, other new orders are submitted to the materials manufacturers. Those systems enable efficient project management, by means of providing all information of the both material manufacturing and building construction processes to all of material designing, material manufacturing, building designing, and building construction sites.

Through this implementation project, we found that the presense of an RFID directory, which translates RFID identifiers into WIP identifiers and vise versa, is the very point within the manufacturing line control, because it should change its translation tables as WIPs are assembled and disassembled within a manufacturing line.

2. APPROACH

2.1 Paragraphs

In this implementation, we aimed to confirm that the system operates properly on the whole. In order to make its information processing simple, the process management engine uses only typical durations to process each step in the manufacturing materials or installing them, and the bills of materials.

In order to trace WIPs(Work In Processes), we installed several checkpoints, which we named "gates", within the process through material-manufacturing and building-construction. On WIPs passing these gates, RFIDs are read and progress reports are collected to the process management system.

As the due time for passing the final inspection process of the installation to the building is deduced from the overall schedule of the building construction, the due time for passing each gates can be calculated from the given final due time and the typical durations from one gate to the next. In the other hand, when WIPs pass each gates, the estimated time for passing following gates can be calculated from the actual achievement time and the typical durations.

In these way, for each building materials types, we can obtain both due time for all demands passing all gates, and actual or estimated time for all WIPs. By associating each demands and each WIPs in the order of time passing a predefined gate for each building materials type, we carry out the allocation of demands and WIPs.

In the case of due time of allocated demands is earlier than estimated time of associated WIP passing by, we assume that tardiness is expected and some action is required, at the moment.

The due time for demands are re-calculated every time when the due for final inspection changes, and the estimated time of WIP passing by are recalculated every time when the WIP passes new gate. The reallocation of the demands and WIPs takes place, when either the list of demands or the list of WIPs sorted in order of time change.

3. THE IMPLEMENTATION

3.1 Gates

In order to trace WIPs, we have set up nine gates within both material-manufacturing and building-construction processes, as follows.

- Design approved,
- Ordering raw material,
- Start processing,
- Assembling,
- Shipping out from the manufacturing plant,
- Carrying into the construction site,
- Distributing within the construction site,
- Installing building material in the building,
- Final inspection.

It is easy for the system to change the total number of gates, to change typical durations from gate to the next. It is also possible to differ the typical durations for calculating due time with the due for the final inspection gate from the durations for calculating estimated time to pass gates with action achievements information.

When the WIPs pass gates, following processes are took place.

- Reading RFID on the WIP,
- Converting to the WIP identifier,
- Logging the time passing the gate,
- Logging the physical position of WIP,
- Logging the result of the post-process testing (optional).

These data are accumulated within an actual achievement database resides in the shared data space, described later.

3.2 Tracking Works with RFIDs

At each gate, the system gathers actual achievement information by means of RFIDs. In this implementation, we assumed to use read-only type RFIDs with 128 bit length identifiers.

Generally speaking, as each building materials consists of multiple parts which are manufactured independently in the manufacturing line, single building material may contain multiple RFIDs in it. On reading RFIDs of a building material, some of multiple RFIDs may respond and some may not, but the tracking engine should handle these information properly in any case. In some cases, an assembled WIP may be dis-assembled to find much more matching combination.

So, the tracking engine should have following features.

- Identifying the WIP from partial RFIDs information.
- Keeping RFID identifiers of all parts consisting the WIP.
- Keeping tree-structured information including assembling order and part structure, for the case of dis-assembling and re-assembling.

Furthermore, it is expected that reading some particular RFIDs instead of entering some information manually, such as operator's name, physical location of the work-cell, and others. So, the tracking system can judge whether given identifier represents some WIP or not.

3.3 Allocation

In this implementation, we use simple algorithm described in the Section 2 to allocate demand to corresponding WIP.

The basic data, which are the typical durations from one gate to the next or previous gate and the bills of materials, are given by initializing agent and stored in the shared data space among multiple processing agents, described later. In this implementation, the typical durations for due date deduction may differ from the typical durations for calculating estimated times passing following gates. Both durations are defined for all demand and WIP types as default values, but the users can define other values to override them for respective types.

We named the lists of same type WIPs, which are arranged in the order of actual expected time passing a predefined gate, as ``preceding list.'' In other hand, we named the list of same type demand, which are arranged in the order of due time at a predefined gate, as ``priority list.'' Their priority is defined only by the due time order, and we ignore any other value-related information which may affect on priority of demands.

All demands and WIPs are processed in building material type by type, and in the case of the material type made from multiple parts, the demands of those parts is newly created according to the bills of materials at the gate at which those parts are assembled together. The due time of the parts is set as same as the due time of the combined WIP at the same gate. The estimated time of one WIP to pass the assembling gate is calculated as the latest one of the estimated time of its parts to pass the gate.

In order of simplify the allocation algorithms, we assumed that there is no part types which are assembled into multiple building materials, in order to omit the idea of priority.

3.4 User Interface

In this subsection, we describe the user interface screens implemented. The user interface of this system was developed as the application of World Wide Web system, in order to make them accessible from not only desk-top computer systems but also portable data terminals.

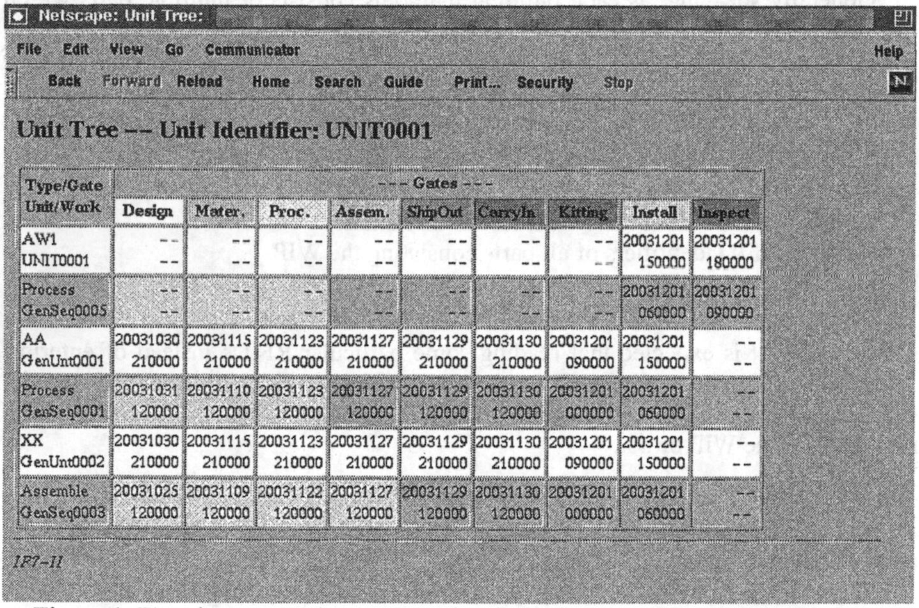

Netscape: Unit Tree:

File Edit View Go Communicator   Help

Back Forward Reload Home Search Guide Print... Security Stop

Unit Tree — Unit Identifier: UNIT0001

Type/Gate Unit/Work	Design	Mater.	Proc.	Assen.	ShipOut	CarryIn	Kitting	Install	Inspect
AW1 UNIT0001	-- --	-- --	-- --	-- --	-- --	-- --	-- --	20031201 150000	20031201 180000
Process GenSeq0005	-- --	-- --	-- --	-- --	-- --	-- --	-- --	20031201 060000	20031201 090000
AA GenUnt0001	20031030 210000	20031115 210000	20031123 210000	20031127 210000	20031129 210000	20031130 210000	20031201 090000	20031201 150000	-- --
Process GenSeq0001	20031031 120000	20031110 120000	20031123 120000	20031127 120000	20031129 120000	20031130 120000	20031201 000000	20031201 060000	-- --
XX GenUnt0002	20031030 210000	20031115 210000	20031123 210000	20031127 210000	20031129 210000	20031130 210000	20031201 090000	20031201 150000	-- --
Assemble GenSeq0003	20031025 120000	20031109 120000	20031122 120000	20031127 120000	20031129 120000	20031130 120000	20031201 000000	20031201 060000	-- --

IF7-11

Figure 1: Due time to passing gates and estimated time passing following gates

Figure 1 shows the status display for one particular association of the demand and the WIP, including the parts which demands and WIPs consists of. In each table entry, the upper row contains date information in the format of YYYYMMDD, and the lower one contains time in the format of HHMMSS.

The table entries with while background color show the due time of the demand for that gate, those with green background color show the actual achievement time of the WIP to pass the gate, and those with blue background color show the estimated time of the WIP to pass the gate calculated from the actual achievement time for the last gate passed and typical duration time.

In this table, the demands and the WIPs are shown in the preceding list or the priority list order.

The neighboring entries consisting upper demand line and lower WIP line show the allocation of demand and WIP. This allocation is subject to change, and the indications changes when due dates are changed or WIP passes new gate. In the case of tardiness expected in some allocated pairs, their actual achievement date and time are shown in red characters.

Figure 2 shows the status display for one particular type of building materials, including the parts which demands and WIPs consists of.

Type List -- Type Identifier: AW1

Type/Gate Unit/Work	Design	Mater.	Proc.	Assem.	ShipOut	CarryIn	Kitting	Install	Inspect
AW1 UNIT0001	-- / --	-- / --	-- / --	-- / --	-- / --	-- / --	-- / --	20031201 / 150000	20031201 / 180000
Process GenSeq0005	-- / --	-- / --	-- / --	-- / --	-- / --	-- / --	-- / --	20031201 / 060000	20031201 / 090000
AA GenUnit0001	20031030 / 210000	20031115 / 210000	20031123 / 210000	20031127 / 210000	20031129 / 210000	20031130 / 210000	20031201 / 090000	20031201 / 150000	-- / --
Process GenSeq0001	20031031 / 120000	20031110 / 120000	20031123 / 120000	20031127 / 120000	20031129 / 120000	20031130 / 120000	20031201 / 000000	20031201 / 060000	-- / --
XX GenUnit0002	20031030 / 210000	20031115 / 210000	20031123 / 210000	20031127 / 210000	20031129 / 210000	20031130 / 210000	20031201 / 090000	20031201 / 150000	-- / --
Assemble GenSeq0003	20031025 / 120000	20031109 / 120000	20031122 / 120000	20031127 / 120000	20031129 / 120000	20031130 / 120000	20031201 / 000000	20031201 / 060000	-- / --
AW1 UNIT0002	-- / --	-- / --	-- / --	-- / --	-- / --	-- / --	-- / --	20031203 / 150000	20031203 / 180000
Process GenSeq0006	-- / --	-- / --	-- / --	-- / --	-- / --	-- / --	-- / --	20031204 / 060000	20031204 / 090000
AA GenUnit0003	20031101 / 210000	20031117 / 210000	20031125 / 210000	20031129 / 210000	20031201 / 210000	20031202 / 210000	20031203 / 090000	20031203 / 150000	-- / --
Process GenSeq0002	20031031 / 120000	20031113 / 120000	20031125 / 120000	20031130 / 120000	20031202 / 120000	20031203 / 120000	20031204 / 000000	20031204 / 060000	-- / --
XX GenUnit0004	20031101 / 210000	20031117 / 210000	20031125 / 210000	20031129 / 210000	20031201 / 210000	20031202 / 210000	20031203 / 090000	20031203 / 150000	-- / --
Process GenSeq0004	20031025 / 120000	20031112 / 120000	20031125 / 120000	20031129 / 120000	20031201 / 120000	20031202 / 120000	20031203 / 000000	20031203 / 060000	-- / --

1F7-11

Figure 2: Status display for all works of specified building material type

This example shows that two building materials are currently under processing, and one WIP recovered its tardiness at the first gate, by shorten the duration for processing from the first gate to the second, but another WIP has delayed at the last gate it passed and tardiness at the gate for the final inspection is expected.

In order to recover such situations, one of following behavior is feasible.

- To shorten the duration from the gate to the next, as the first part has been done.
- To postpone the due time for the final inspection of the building materials installation.

Figure 3 shows the case of postponing the due time of the final inspection of the unit with unit-ID UNIT0001 for four days. As the result, the due time of the unit UNIT0001 and that of the unit UNIT0002 are reversed, and the parts set which are going to be used for those units are exchanged.

Type List -- Type Identifier: AW1

Type/Gate Unit/Work	Design	Mater.	Proc.	Assem.	ShipOut	CarryIn	Kitting	Install	Inspect
AW1 UNIT0001	--	--	--	--	--	--	--	20031205 150000	20031205 180000
Process GenSeq0006	--	--	--	--	--	--	--	20031204 060000	20031204 090000
AA GenUnt0001	20031103 210000	20031119 210000	20031127 210000	20031201 210000	20031203 210000	20031204 210000	20031205 090000	20031205 150000	--
Process GenSeq0002	20031031 120000	20031113 120000	20031126 120000	20031130 120000	20031202 120000	20031203 120000	20031204 000000	20031204 060000	--
XX GenUnt0002	20031103 210000	20031119 210000	20031127 210000	20031201 210000	20031203 210000	20031204 210000	20031205 090000	20031205 150000	--
Process GenSeq0004	20031025 120000	20031112 120000	20031125 120000	20031129 120000	20031201 120000	20031202 120000	20031203 000000	20031203 060000	--
AW1 UNIT0002	--	--	--	--	--	--	--	20031203 150000	20031203 180000
Process GenSeq0005	--	--	--	--	--	--	--	20031201 060000	20031201 090000
AA GenUnt0003	20031101 210000	20031117 210000	20031125 210000	20031129 210000	20031201 210000	20031202 210000	20031203 090000	20031203 150000	--
Process GenSeq0001	20031031 120000	20031110 120000	20031123 120000	20031127 120000	20031129 120000	20031130 120000	20031201 000000	20031201 060000	--
XX GenUnt0004	20031101 210000	20031117 210000	20031125 210000	20031129 210000	20031201 210000	20031202 210000	20031203 090000	20031203 150000	--
Assemble GenSeq0003	20031025 120000	20031109 120000	20031122 120000	20031127 120000	20031129 120000	20031130 120000	20031201 000000	20031201 060000	--

Figure 3: Status after changing due time

4. RFID DIRECTORY SERVICES

As we mentioned in the preceding section, in the manufacturing line control systems using RFIDs, we have to translate RFID identifiers into WIP identifiers and vice versa. This information consists of both associations of WIP identifier and RFID identifiers and tree-structured information including part structures.

As the parts are assembled and rarely but also disassembled in their manufacturing process, the set of RFIDs each WIPs are containing varies as their process progress, and the translation table should be updated as WIPs are processed. Basically, the translation table contains set of associations of a WIP identifier and a set of RFID identifiers. But once some parts with RFIDs are assembled and joined together, the biggest WIP containing the RFID should be answered to a request translating RFID identifier into WIP identifier.

As we can not change association information preparing cases of disassembly, the inclusion relations among WIP identifiers are maintained separately from the associations of WIP and RFID identifiers. The structure of this relation is similar to the structure of the bill of materials, and it is rather easy to trace the changes of an WIP through assembly process. But in the cases of the disassembling WIPs to change some of their components to re-assemble and to pass the inspection, it is very difficult to enumerate all the way to disassembling each WIP.

4.1 Basic Functionalities

The RFID directory service should have following functionalities:

- Keeping set of RFID identifiers, which are NOT used in order to identify WIPs. For example, RFIDs to identify operators, physical locations, gates in the manufacturing process, and others.
- Keeping associations between a WIP identifier and a set of RFID identifiers.
- Keeping the information of including relations among WIP structure, which can be obtained from the bill of materials.
- Tracking each WIP's current structure and its consisting parts.

4.2 Very Requied Functionalities

As this RFID system is used within manufacturing processes, the objects identified by RFIDs are assembled and disassembled. So the directory service should have following functionalities in order to handle joining and separating parts.

- Operation on joining some WIPs, which includes modification of WIP's structure information or creating a new WIP identifier and initialize its structure information.
- Operation on disassembling an WIP, which includes not only the modification of WIP's structure, but also includes finding the description of separation by scanning RFID identifiers. In many cases, as the disassemble process is not regular one in order to recover some difficulties introduced within preceding assemble processes, the set of the parts to be removed is not predictable and may varies in cases. This is because the RFID directory service should have abilities to guess members of some resulting sets themselves and their structures.

In the area of the life-cycle engineering, the reusing of the used parts will play a vital role in the parts management. In such application, the management of the parts should be based on technology of RFIDs, or something alike which can identify each parts from other same type parts, in order to trace back parts history and the past operation conditions, diagnostics results and maintenance logs.

For these purpose, further research is required on the RFID directory service agent, especially in the area of handling disassembling processes and old history management.

5. THE INFRASTRUCTURE SYSTEM

In this study, we used the system named ``Glue Logic'' (Takata and Arai, 1997) (Takata and Arai, 2001) (Takata and Arai, 2004) as the infrastructure to support multiple-agent processing system, in order to implement a kernel of MES (Manufacturing Execution System) controlling manufacturing processes.

The Glue Logic, which is designed and developed by Takata Laboratory of the University of Electro-Communications in Japan, includes an implementation of the active database and the network transparent programming environment, and supports data processing in the event driven programming paradigm.

Figure 4 shows the overall structure of this implementation.

5.1 Active Database

The active database is a subclass of the database systems, of which databases have an abilities to behave when it finds some changes of its contents, without waiting for external actions.

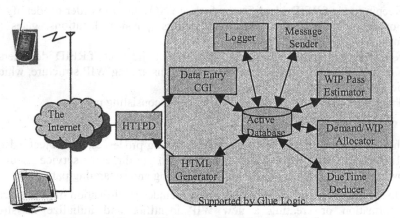

Figure 4: Overall Structure of the Implementation

The change of the contents includes

- when data is changed,
- when some relations are formed,
- when some new data become available.

and the behavior executed on these incidents includes

- changing contents of the database,
- calculating certain expression and assign the result into some variable,
- sending message to some client agents.

5.2 Aims and Functionalities of the Glue Logic

The Glue Logic is designed to make building manufacturing work-cell control systems easy and flexible, and also coordinates agents by means of followings;

- Providing field of coordination
- Implementing shared data space among agents
- Virtualizing agents within the name space of shared data
- Controlling message passing among agents
- Implementing mutual execution primitives
- Prompting agents to start processing
- Adapting control systems to real-time and network processing environment

As the Glue Logic supports event notification and condition monitoring features based on active database scheme, users can easily build real-time and event-driven application agents, only waiting for notification messages from the Glue Logic.

Each agents in an application system can be developed concurrently, and can be added, deleted or changed freely without modifying other existing agents. As the result of these, the Glue Logic compliant agents are easy to re-use, and the users can build large libraries of application agents.

5.3 Role of the Glue Logic

In this implementation, the flow of its data processing is as follows.

- When a WIP reaches new gate, or when a due time of a demand changes, the corresponding agent is activated via a Common Gateway Interface (CGI) for the

World Wide Web user interface. These agents updates the actual achievement records for WIPs or the due time requirement records for demands. These records are kept in the shared data space in the Glue Logic.

- When data within the shared data space is changed, some messages are sent from Glue Logic to agents which have already subscribed to the data items. This time, the agents keeping preceding lists or priority lists are informed, and update those contents.
- When the contents of preceding lists or priority lists are changed, the notification message is sent from Glue Logic to the allocation agent. The allocation agent reads preceding lists and priority lists, and then makeup associations between demands and WIPs.

As described above, the Glue Logic plays role of the conductor or alike of the whole system.

6. EVALUATION

6.1 Evaluation on the performance

We implement the system on Sun Netra T1 processor running on Solaris 8 OS. Through this implementation, we found following performance considerations.

- Re-allocation of the demand hardly took place unless due time of final inspection changed or some WIPs pass other preceding WIPSs.
- In many cases, as there are less than tens of WIPs concurrently being processed in both material-manufacturing and building-construction sites, re-allocation of the demand takes only a few seconds. But in the case of building materials with complex BoM (bill of material), as the re-allocation process occurs in multiple stages, it may takes more than ten seconds.
- It may takes a few minutes for operators to input data for gate passing process or changing due time. So, some features to inhibit processing during data entry may be required to prevent needless data processing.

6.2 Evaluation on the extend-ability

We used multiple agent support system in the implementation, in order to ease future extension of functionality. From this point of view, we found followings.

- As all information on the actual process achievements are kept within a database, any other agents can utilize these data for processing and user interface purposes, as follows:
  - displaying status of building materials within building floor plan chart
  - sending e-mail to notify shortage or tardiness of some materials, within a few seconds.
- It seems to be appropriate that the conversion from the identifier of RFID to building material identifier should be done by specialized subsystem in the management system. There may be many RFID classes representing objects other than WIPs.

6.3 Evaluation on the limitation

In this implementation, we introduced some limitations to simplify the system. They are as follows.

- There is no problem solving engine to minimize cost. In order to find best solution of re-allocation, it is required to minimize cost to re-distribute building materials, or it is required to determine which WIP to be scraped. Solving this problem may need massive computational power, because there may be combinational explosion. In the other hand, we can not define the function to evaluate the cost itself clearly in the practical world. For example, in many cases, two types of building materials are not equivalent in their priority.
- There is no clear decision rules to be embedded within the system. Some incidents can be processed automatically without human interventions, but some require human approvals. There is no clear border and the best way depends on its environment.
- In the calculation of the due time and estimated time to pass, we use ``typical durations." These values are constants and have less accuracy. In the following research, we have to link with APS (Advanced Planning and Scheduling) systems and MRP (Manufacturing Resource Planning) systems.

7. CONCLUSION

Through this implementation described above, we found that the integrated process management system including both part-manufacturing and building construction is feasible enough.

Especially, related to the use of RFIDs, we recognized the importance of the management agent, which we called RFID directory service agent in this paper. This agent keeps not only set of association between WIP itselves and the RFIDs included in the WIP, but also the inclusion relations among WIPs.

In coming years, we would like to test next implementation at the actual manufacturing and construction sites.

Acknowledgments

This research activity has been carried out as a part of the Intelligent Manufacturing Systems (IMS) international research program: "Innovative and Intelligent Parts-oriented Construction (IF7-II)". We appreciate the kind guidance of each members of this project.

REFERENCES

Takata,M., Arai,E. (1997) The Glue Logic: An Information Infrastructure System for Distributed Manufacturing Control. Proc. of the Int'l Conf. on Manufacturing Milestones toward the 21st Century, pp.549 - 554, Tokyo, Japan, July 23-25, 1997.

Takata,M., Arai,E. (2001) Implementation of a Layer Structured Control System on the "Glue Logic". In Mo, J., Nemes, L. (eds.) Global Engineering, Manufacturing and Enterprise Networks, Kluwer Academic Publishers, pp 488-496

Takata, M., Arai,E. (2004) Implementation of a Data Gathering System with Scalable Intelligent Control Architecture. Arai, E., J. Goossenaerts, F. Kimura and K. Shirase (eds.) (2005) Knowledge and Skill Chains in Engineering and Manufacturing: Information Infrastructure in the Era of Global Communications, Springer, pp 261-268

38. Seven Tools of Operation Standardization for Mass Production of a New Product

Kosei Sugitani, Hiroshi Morita and Hiroaki Ishii

Department of Information and Physical Sciences, Osaka University, Japan

Email:[k-sugitani, morita, ishii]@ist.osaka-u.ac.jp

We propose the effective tools of operation standardization for mass production of a new product. The cycle of operation standard consists of three stages of design, improvement and evaluation, and divided into seven steps, that is, decision, communication and understanding, observance, supervision, notice, decision again, and evaluation. The proposed seven tools of operation standardization (OS7) correspond to these steps. These tools enable us to realize mass production of a new product and to stabilize a product quality much earlier.

1. INTRODUCTION

Recently optical fiber communication services and devices have been wrestling with the problem of mass-production in the field of key material and devices. The key to dominant position in competition is the prompt supply of a product to customers with a steady production rate in quality. We often come across an important factor which brings great influences on a product quality. We should discover it and stabilize it in production line. It is important for us to standardize the operations.

The cycle of operation standardization consists of three stages of design, improvement and evaluation. Before the implementation of mass-production, we have defined the process of operation standardization as "design cycle". We have often implemented mass-production without clearly deciding an operation standard, which may have caused some serious quality problems. It is required to develop the effective tools for operation standardization to reduce these kinds of problems.

The process of operation standardization is divided into seven steps, that is, decision, communication and understanding, observance, supervision, notice, decision again, and evaluation. We propose the seven tools of operation standardization (OS7) corresponding to these steps to prevent the quality problems at the implementation of mass production of a new product. Table 1 summarizes the proposed OS7 together with the cycle and steps for operation standardization. We define the operation rank by using the color-coded card, and it enables us to communicate the information of operations with each other. OS7 is especially effective in the situation of mass production on which we can see dispersion of operations in a short term.

There are many problems that we have to try to implement mass-production so far. To begin with, we consider the cycle of operation standardization as the cycle of decision, observance, decision again, and observance. We often used each of these steps separately and not systematically moreover. Secondly, we could not exactly catch up the information of the importance of operations such as dispersion of operations, effect on quality and so on, because there is no cycle of communication to operators and understanding of operators. As a result, it took us long time to stabilize the quality of a new product in mass production.

OS7 is shown in Table 1, each of which is denoted in the following section.

Table 1. The seven tools of operation standardization

Cycle	Method	Content
Design	S1: Decision	Decision of standard of elemental operations
	S2: Communication and understanding	Analysis of the gap in recognition between operators and technicians
	S3: Observance	Analysis of the degree of observing operation standard briefly in starting operation
	S4: Supervision	Checking the percentages of observance operation standard in operation
Improvement	S5: Notice	Extraction an operation know-how operator detected in operation
	S6: Decision again	Taking priority to standard of elemental operations in worrying about quality problems
Evaluation	S7: Evaluation	Evaluation seven steps of operation standardization

2. SEVEN TOOLS OF OPERATION STADARDIZATION

2.1 S1: Tool for "Decision"

Before implementing mass-production, we check the operations if there are any lacks of standardization that have influence on the production. In many cases, some operations are not standardized. If operator declines the reduction of production which is caused by violating the operation standard, we have to decide a new operation standard at once. S1 is the tool for decision of standard of elemental operations, and is used as follows.

- Confirm the flow of elemental operations composed of unit operation.
- Watch elemental operations of operators and estimate dispersion of quality optimistically and pessimistically.
- Consider operators' skill, we give priority to operations which will be dispersed and improve their standard of elemental operations.
- Check the effectiveness of improvement.

If we can grasp important operations which have influence on the production by using S1, operations dispersion will be decreased and quality will be stabilized in manufacturing process. After we decide the operation standard again, we have to inform it to all operators, including the information about important operations, such as dispersion of operations.

2.2 S2: Tool for "Communication and Understanding"

We check the degree of understanding of operators so as to standardize each operation. We need to grasp the gap of recognition to each operation between operators and technicians. It is important to check the gap that may cause dispersion of quality. It could happen that some operators do not put emphasis on what technicians think important. Table 2 shows how to visualize the gap by using card. We have to make up for the gap using the card briefly. S2 is the tool for analysis of the gap in recognition between them, and is used as follows.

- Suppose a unit operation composed of some operations.
- Confirm the information of operation importance important have influence on the production.
- Confirm by handing to a technician the three cards, the A rank card, the B rank card, and the C rank card, and let them select one of the cards. One operation is shown on one card. If a technician selects the A rank card which is red, he thinks that the operation is the most important one. Here we define the A rank card as the most important card which have influence on the production. B rank card of which is yellow is less important than the A rank card, but operation of B rank is reasonably important. The C rank card which is blue is a normal operation where dispersing is acceptable.
- Let operators choose a card as we do to technicians.

We will be surprised at the difference of their chosen color-coded cards. An operator will show the C rank card of the operation from his operation experience, even if technician shows the A rank card as to the same operation. The problem is that there are gaps of the recognition between operators and technicians. We have to solve this problem.

Table 2. Gap in recognition

Elemental operation	Recognition to operation		Gap
	Technician	Operator	
Put two materials into the box	Red card	Yellow card	Yes
Carry box	Blue card	Blue card	No
Close box	Yellow card	Blue card	Yes

2.3 S3: Tool for "Observance"

The operators are informed the determined operation standard. So it is important to provide an appropriate operation environment to observe operation standard. We need to examine S3, the method of brief analysis of the degree of observing operation standard, from the viewpoint of prevention against occurrence and outflow. If we use S3 effectively, we can catch up the cause that operators have missed to observe the operation standard. S3 is used as follows.

- Check the degree of observing an operation standard from the point of occurrence prevention, and estimate it. (Table 3)
- Check the degree of observing an operation standard from the point of outflow prevention, and estimate it. (Table 4) The rank is determined according to the process capability index Cp.
- Evaluate each elemental operation by using the rank of occurrence and outflow against prevention given in Table 3 and Table 4. (Table 5)

- After finishing the total evaluation, we take measures against each high score greater than one with lower scores and confirm the effectiveness of theirs.

Table 3. The rank of occurrence against prevention

Rank of occurrence against prevention	Content
1	Equipped, Automatic
2	Tooled
3	Manual Inspected, Exist of Limit sample
4	Exist of operation standard
5	Exist of operation standard, No control

Table 4. The rank of outflow against prevention

Rank of outflow against prevention	Content
1	Equipped, Automatic (Cp>1.33)
2	Half Automatic (Cp>1.0)
3	Control of a fixed quantity (0.8<Cp<1.0)
4	Check, No record (Cp<0.8)
5	No check

Table 5. The method of analysis of the degree of observing operation standard briefly in starting operation

Element operation trouble	Occurrence prevention	Rank	Outflow prevention	Rank	Total Evaluation	Goal
More than three materials	Exist of one-point standard	3	Check No record	4	E	A
Short of hydrochloric acid	Exist of standard No control	4	No check	5	F	B
Open cover	No check	5	Warning	2	E	A

2.4 S4: Tool for "Supervision"

At the end of the design cycle, we confirm that operators observe the operation standard in manufacturing process. S4 is the tool for checking the percentages of observation of operation standard. If we find the operators who do not observe the operation standard, we give precedence to the A or B rank card. We inquire of the operator why it is difficult to observe the operation standard. Then it makes clear which step has a problem in the seven steps. S4 is used as follows.

- Make the sheet arranged in the elemental operations. (Table 6)
- Superintendents check mainly the percentages of observance of operation standard through observance of watching an operator operating in manufacturing line. And they estimate the point of each of elemental operations, writing check sheets.
- If they find the operator who does not observe the operation standard, they will ask him the reason. Sometimes, he may say that it is easy for him to maintain the operation standard because of his inexperienced skill.

Table 6. The method of checking the percentages of observed operation standard

Operation point	Operation content	Evaluation	Cause of not observation		Measure to meet
			Factor	Operation standardization step	
Seize the box	One time	X	Use of old operation standard	Communication	□
Put powder into box	Five times	△	Not decided of putting time into box	Decision	□
Mix powder	□	O	□	□	□

Next we define the improvement cycle including notice and decision again step; the process we extract factors that cause quality problems by the elemental operations, and stabilize quality of product. We will decrease defects of the design cycle through this cycle by deciding operation standard again.

2.5 S5: Tool for "Notice"

In the improvement cycle, we will decide operation standard again to find remaining some important factors. We extract an operation know-how that is not stipulated in the operation standard and have influence on the production. We define the operation know-how that is found by operators and is not known to the technicians as the operation standard again. We will try to make the measures for each of the know-how, and confirm the effectiveness. If we will find out an important elemental operation, we define the rank of operation again, and append them to the operation standard. S5 is used as follows.

- Prepare the big size paper on the wall.
- We adopt the most important quality characteristic, and let operator paper cards written in the cause have influence on the production. In papering cards, operators distinguish the A or B rank card from the C rank it. (See Figure 1.) After papering the cards, we give a priority to the card to make measures from the technical viewpoint.
- Add a useful know-how which will be effective to operation standard in it.
- Confirm the effectiveness of it by time series analysis.

Notice that S5 resembles to cause and effect diagram, but the difference is that the operator uses the color-coded cards and confirms the effectiveness by time series analysis.

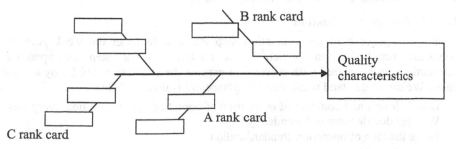

Figure 1. Extraction of an operation know-how detected in operation by operators

We have had experienced in meeting with the quality problems caused by that the operators do not observe the operation standard before starting mass-production. We seldom find out the cause at once. Then it is necessary to stabilize the dispersion of each of operations in the manufacturing line at first. Through the improvement cycle, we can reduce the dispersion of quality operations that have influenced on.

Table 7. The matrix of given priority to elemental operations

		Importance of elemental operations	
		High	Low
Operation Standard	No observance	Do observation	Simplify the operation standard
	Observance	Reduce dispersion	Permission of dispersion
No operation standard		Decide operation standard and reduce dispersion	Permission of dispersion

2.6 S6: Tool for "Decision again"

In worrying about the quality problems, we will take the priority to the standard of elemental operations, and choose an operation that should be standardized again from some elemental operations by using the systematic chart of sub-material items. At first, we select the most important process by checking inspection items by using QC flow chart. If we find uninspected items or unobserved items though the observation by the operators, we will give a higher priority to them. If not, we should select the process composed of most dispersed one. Secondly, we select the most important elemental operation from selected process by constructing the matrix. S6 is used as follow.

- Make the factor chart for the quality characteristics composed of unit operations.
- Press the most important process by checking inspection items in QC flow chart or observation of operator in operation.
- Select the most important process by checking inspection items by using QC flow chart.
- Select the most important elemental operation from selected process by making the matrix.

Suppose that there are many factors have influence on the production. S6 is the systematically method of giving priority to decision operation standard again. We should take measures about meet them as Table 7.

2.7 S7: Tool for "Evaluation"

Finally, we proceed to the evaluation step that can find out the weak point of operation standardization, which is a relatively weak step of operation standardization compared with others. We express the evaluation tool S7 by a radar chart. We can understand weak step at a glance. S7 is used as follows

- Decide levels and contents of evaluation of operation standardization every step. We may decide them independently.
- Judge the step of operation standardization.

- Paper the result of evaluation on the wall in manufacturing line, and operators understand the weak point of the operation standardization. Then we can connect with the improvement of a constitution in each section.

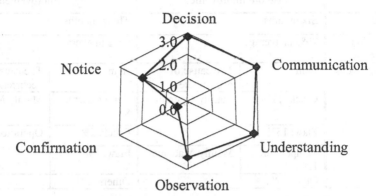

Figure 2. Evaluation the steps of operation standardization

3. CASE STUDY

We have used the proposed OS7 to the case of a semiconductor communication company. We have observed the time to the steady rate in quality to confirm the effectiveness of OS7 by comparing before and after the improvement. Many technicians noticed the cause of quality problem slightly, and they knew the three inferior modes (Table8), but they did not notice the cause of occurrence. So they tried to change the condition of machines many times, not reducing dispersion of operations in manufacturing line, however, there is no sign of improvement even after six months. It was a problem a group of technicians where the leader of the group was not interested in operation dispersion.

In the design cycle, we subjected to more fluently inferior modes of crack and flaw, caused by operation. The other technicians thought that it was caused by the condition of machines. We decided 33 operation standards again to reduce dispersion of these modes, using S1, and it took eight days, and to communicate them to all operators how dispersion of every elemental operations is with their rank, using S2. And we checked the observance of the operators to keep operation with a steady rate, using S3. For example, we prepared the visual operation standard and the tool for cutting to easy operation. Next, superintendents checked the observance of the operators in manufacturing line once a week, using S4. The rate of observance of operation standard rose up to 84%.

In the improvement cycle, we tried to reduce the dispersion of elemental operations many times, using S5 and S6. We could extract 23 operator's know-how beyond expression in visual operation standard in this process.

In total, in the efforts of using OS7, we could reduce dispersion elemental operations and improved the production from 29% to 63% in average. Table 9 shows the comparison of before and after improvement in operation standard.

Table 8. Comparison of before and after improvement of manufacturing index

	Before Improvement		After Improvement	
Period	Six months		Three months	
The yield of Product	29% in average		63% in average	
Inferior goods mode	Item	The cause of occurrence	Item	The cause of occurrence
	Crack: 17%	Mainly Operation	Temperature: 8%	Mainly Machine
	Flaw: 15%		Crack: 2%	Operation
	Temperature: 12%	Mainly Machine	Flaw: 1%	
	Others: 2%		Others: 1%	

Table 9. Comparison of before and after improvement in operation standard

Step	Before Improvement		After Improvement	
	Content		Content	
Decision	Element operations: 23 Unit operations: 6 (Stuffs decided operation standard only)	28 days	Element operations: 56 Unit operation: 6 (Stuffs and operator)	8 days
Communication and understanding	From two technicians to only one group leader	1 day	From two technicians to all operators	12 days
Observance	A Little operation standard only	A little check	Prepared tooled in addition to operation standard	Total 72 days
Supervision	No check→60%	Total 4 days	60%→84%	Total 27 days
Notice	Stuff only	2 know-how	All operator and some technicians	Total 23 know-how
Decision again	Not Try	□	Focus the subject of improvement of elemental operation standard	Total 14 days
Evaluation	Not Try	□	Go to Supervision step	Total 5days

4. CONCLUSION

We have proposed the seven tools of operation standardization (OS7) to effectively establish the mass-production of a new product. We have suggested that we should divide process of operation standard into seven steps, that is, decision, communication and understanding, observance, supervision, notice, decision again, and evaluation. The proposed systematic seven tools of operation standardization (OS7) corresponding to these steps effectively prevent the quality problems at the implementation of mass production of a new product. OS7 enable us to realize mass-production of a new product and stabilize a product quality much more quickly. Note that OS7 is not a completed tool and should be reformed as operators can use it easily in production line.

REFERENCES

Fukuda, R (1997) Building Organizational Fitness: Management Methodology for Transformation and Strategic Advantage. Productivity Press Inc.

Sugitani, K (2000) The study of operation standardization for mass production of a new product (I). Proceedings of the Japanese Society For Quality Control (in Japanese)

Sugitani, K (2001) The study of operation standardization for mass production of a new product (II). Proceedings of the Japanese Society For Quality Control (in Japanese)

Sugitani, K (2002) The study of operation standardization for mass production of a new product (III). Proceedings of the Japanese Society For Quality Control (in Japanese)

Sugitani, K (2004) The study of operation standardization for mass production of a new product (IV). Proceedings of the Japanese Society For Quality Control (in Japanese)

Sumitomo Electric Industry (1993) Practical use of cause and effect diagram by card in standardization and quality control. No. 5-4200 (in Japanese)

39. Workbench: A Planning Tool for Faster Factory Optimization

Georg F. Wiesinger,
Chair of Factory Organization,
University Dortmund,
Email:wiesinger@lfo.uni-dortmund.de

A key requirement for a successful facility planning and design is to become suitable for new use as well as to be adaptable for new products, technologies or capacities. The software solution "Workbench" is designed for an Information Infrastructure System for the planning of large logistics networks as well as for the network structures of the facilities in an enterprise. The "Workbench" ensures a better information flow and basis for Factory planning and enables planners without great expert planning knowledge.

Keywords: Logistics and Production Network Structures, Knowledge Management, Facility Planning, Supply Chain Management, Business Process Reengineering

1. INTRODUCTION

In today's competitive global markets, facility planning has taken on a completely new meaning. In the past, facility planning was primarily considered to be a science, today it is a strategy. Companies, institutions, and businesses no longer compete against each other individually. These entities now align themselves into cooperatives, organizations, logistics networks, associations, and ultimately synthesized supply chains to remain competitive by bringing the customer into the process. In future real competition will happen between large logistics networks and not between enterprises anymore.

The method project M6 "Body of Construction Rules ", Collaborative Research Center 559[84] Modelling of Large Logistics Networks, which is financed by the German Research Foundation (Deutsche Forschungsgemeinschaft DFG), designed an Information Infrastructure System called "Workbench" for the planning of big logistics networks. The fundamentals of such networks are the plants and stores with the material flow and intralogistics. For six years now scientists from the fields of Logistics, Transport Engineering, Warehousing, Industrial Management, Factory Organization, Mathematical Statistics, Theoretical Informatics, Applied Computer Science and Systems Analysis have been working on joint research projects that ensure a wide range of interdisciplinary competencies.

---

[84] Research project SFB 559 funded by the University of Dortmund, the Fraunhofer Institute for Material Flow and Logistics and the German Research Foundation (DFG)

»Large Logistics Networks« exist where different goods or products are transformed and transported over several stages by different cooperating partners. They do not only include material and information flow, but also the organizational framework, the required resources, policies for planning and control, as well as the people acting in this environment (Beckmann, 1996). All partners in these networks provide multiple, but also complementary services or competencies. The components of these large logistics networks (organizations, resources, goods, information, knowledge etc.) are connected by numerous different relations (Kuhn, 1995). The number of connections is growing permanently, because there is an increasing demand for more competitive products or services that can only be satisfied by cooperating with powerful partners within these logistics networks (Beckmann, 2000).

Factory or facility planning can only be successful, if the structures are able to handle the logistics and manufacturing processes. To guarantee the business processes and to achieve the necessary performance of the factory the layout structures of facilities have to be designed in different alternatives and tested to get the solution closest to optimum. It should be mentioned, that factory or facility planning, as addressed in this paper, includes broad fields of application (production plant, warehouse, retail store). The already developed information Infrastructure System and knowledge management application "Workbench" is very useful for the planning of facilities or any portion of these. Factory planning uses a great deal of methods, empirical and analytical approaches. It is important to recognize, that contemporary concepts consider the facilities of a factory as dynamic entities. It is a key requirement for a successful facility plan to become suitable for new use as well as to be adaptable for new products, technologies or capacities.

The major benefit of the new construction method combined with an Information and Communication Technology ICT support, introduced in this paper, is the reduction of complexity, which is achieved by construction catalogues combined with best practices for organizational and technical design questions. The presented catalogues lead to reuseable construction elements in configurable networks which depend on changing frame conditions

2. CHALLENGES OF LOGISTICS NETWORKS DESIGN AND FACTORY PLANNING

The facilities we plan today must help an organization to achieve Supply Chain Excellence by a plant on demand. The plants on demand have to be designed and controlled in correlation to the logistics network (supply chain), the material flow, and the production processes. The high tech production processes, the permanent changing and improvement of technologies, and the complex logistics system are drivers of change for the innovation targets in factory planning and logistics network design.

These are the reasons for Factory planning and logistics to steadily grow more complex to be faster nowadays. Facility Planning is no longer a unique process at the start-up of an enterprise. Rather facility management is changing to a permanent planning based on a successful strategy to survive in the fast changes of competitive markets. Whereas it becomes more and more difficult for planners and operating companies to keep the overall views over these difficult and complex planning tasks.

Today it is impossible to plan a big factory or a large logistics network by only one general planner. Instead we need a network of specialists to solve these high complex planning tasks in logistics and production networks.

At the present time, we lack the experience of methodical construction processes that link organizational aspects, like network policies and value chain design, with technical aspects of information exchange and material flow. Most companies are in search of new instruments and technologies to extend their knowledge of setting up new processes, facilities and of designing organizational aspects like network strategies and policies, organizational structure, information flow, and performance measurement metrics.

Knowledge has become a main guarantor for market excellence, especially in high tech and service industry. Far from that, business partners face a growing lack of guidance through all the information avail-able. The challenge is to determine which information is relevant for whom and how to provide them with this information in different situations and planning phases of given projects.

The ultimate goal of the study of the structures of network structures is to understand and explain the workings of systems built upon these networks.

The progress in the state of the art of research for a better understanding of network structures is slowly growing in different faculties, like in chemistry (Fox, 2001), biology (Dune, 2002) and the information technology (world wide net) for example (Ebel, 2002), improving also the understanding of the effects of those structures.

3. THE GENERIC ARCHITECTURE OF WORKBENCH

This paper proposes a new approach to engineering logistics and network oriented facilities. The engineering methodology for planning tasks and aspects of facilities is achieved by using process models (Kuhn, 1995), project reference models (Kühling, 2000), and construction catalogues (Wiesinger, 2002) in combination with best practices for organizational and technical design questions. The facilities are not only planned as a physical object, but rather as process oriented and process fulfilling dynamic, adaptable and reusable modules for the following product generations.

The paper will also show how using the software tool "Workbench", which provides alternative construction objects for different network structures, with changing capacities of processes in construction catalogues, can accelerate and improve the design process in facility planning. The technical concept of "Workbench" is a content management system which has been qualified for the specific needs of logistics engineers and facility planners by using UML specifications. The presented solution will show how new Internet based technologies will support the network design process.

This new approach to engineering logistics network oriented facilities will lead to the reduction of complexity and to a change from planning based on personal experience to a knowledge based information system for a planning on a new level.

3.1 The Process Chain Model of Dortmund University

The elementary model of logistics systems as the "germ cell" of logistics networks is the Process Chain Model from Prof. Kuhn (Kuhn,1995).

In recent years the Process Chain Model has been proven to be successful in more than 50 business-reengineering projects in the automotive, manufacturing, process, and food industry. The components of the Process Chain Model reflect the design parameters like policy, process, structure, and flow. The paper shows how the elementary model has been extended by a construction method for logistics systems that guides the engineer through the design steps in order to structure design questions and to link them with the construction catalogues.

Designing facilities in logistics systems requires both the modelling of static and dynamic aspects. The concept of the »Process Chain Model« (compare Figure 1) is based on the system theory and basic principles of cybernetics.

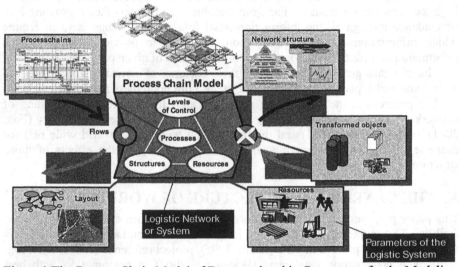

Figure 1 The Process Chain Model of Dortmund and its Parameters for the Modeling of Logistics Systems

Therefore it is suitable for the process oriented modelling of logistics systems that they can be characterized. Besides other modelling concepts the main goal of the Process Chain Model is to integrate both, technical as well as organizational aspects. The core object of the Process Chain Model is the »Process Chain Element« and its five different parameters »Levels of Control«, »Processes«, »Structures«, »Resources« and »Flows«. The Process Chain Element represents a logistics system on different levels, e.g. a whole company, a warehouse, a department, a distribution network, or a business service covering all aspects which are relevant for logistics. Each parameter of a Process Chain Element can be visualized and documented by a number of methods. The most important parameter of a Process Chain Element is »Processes«, which can be modelled by a generic business process diagram suitable for the modelling of material and information flows between different organizations. Exemplary processes may be Material Delivery, Order Scheduling or Component Assembly. The other parameters, as described above, can also be modelled following rules based on experience. All the modelled objects of different parameters are linked with each other.

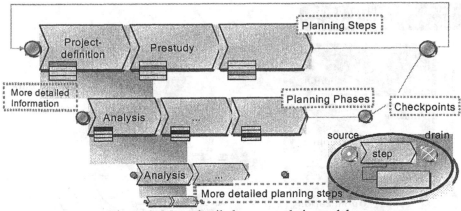

Figure 2. More detailed process chain models

3.2 The Process Reference Model of Dortmund University

Logistics and Facility planners use more or less complex or specified project templates or reference models for their design tasks. Examples are methods for facilities planning, software engineering, or business process reengineering. All of them use equal concepts and most of them can be transformed easily from one to the other. However, existing reference models lack important features like flexible structure, flexible checkpoints, or quality gates between different steps, as well as links to other objects like planning data, methods, or construction elements.

For a more extensive use of systematic concepts in logistics or facility design projects the existing models have been extended to this reference model. It is structured into the following phases »Problem Identification«, »Preparation«, »Master Planning«, »Detailed Planning«, »Prototyping«, »Implementation«, »Improvement und Controlling« (Fang, 1996) (see Figure 3).

The existing research, particularly in the domain of engineering, economics and informatics, has shown the need for the development of an integrated reference model for designing processes in logistics as starting point for a model-based planning approach. The reference model presented fulfills the following requirements:

- Generic model of a standardized documentation of design processes in logistics
- Guidelines for specific design tasks in all steps and phases
- Allows individual iterations and leave outs following the »Evolutionary Prototyping Process«
- Modular (phases, steps, activities; checkpoints etc.) and self similar
- Applicable for all kinds of logistics and facility design projects including technical, organizational aspects etc.
- Links to other components of the framework: data, methods, construction elements etc.

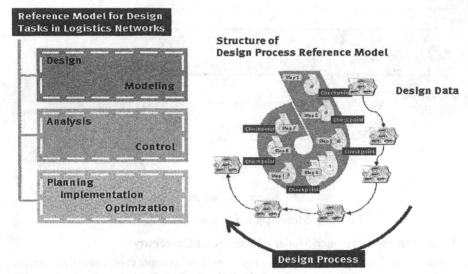

Figure 3. Design Process Reference Model

During the different phases the designer has to solve different planning tasks, which are rougher for example in the first planning phases and more detailed in the final realization planning. The modularity and the self similarity of the "Design Process-" and the "Reference- Model" results from the structuring of multiple design processes especially in the detailed planning phase. To solve a complex design task (for example the planning of a distribution concept for a specific sales channel) a huge number of detailed problems have to be solved, like warehousing, transportation management, structure of distribution network, communication standards etc.. These problems can be divided into independent, but related projects. They also can be described by using the same reference model, taking into account all relevant interfaces between the different levels. In other contexts (different superordinated project, different project focus, other partners etc.) all projects stand for themselves and can be started independently. Therefore, logistics planners gain profit in both situations.

Every phase or step of a project modeled using the reference model represents a complete unit with fix requirements and expected results, so that every participating planner knows its input data, course of activities, and out-put. Within a project the defined checkpoints or milestones have to be fulfilled.

An additional result after performing a project in that way is a newly modelled and documented prototype of a logistics or facility planning project. Examples for such projects are: Development and Implementation of a Postponement Strategy, Direct Delivery, Collaborative Demand Planning, Planning of a Warehouse, Customizing and Implementation of a Shop Floor Control System, Planning of the material flow in a layout, or Implementation of a B2B procurement scenario. All these project examples follow a generic structure placing emphasis on specific project steps and requirements.

3.3 Method Toolbox

A further benefit of the developed project reference model is found in the integrated concept of the method toolbox. It makes a specific adjustment of the design methods possible by taking into account specific project goals and the given situation. The overall structure of a project can be documented by using the project reference model. In detail, the solutions of partial problems can be achieved effectively by using manageable but systematic methods. Usually, the application of a single method is not sufficient in order to solve a common problem in logistics. Therefore on a lower level the project reference model has interfaces to specific methods to describe the solution of certain problem solving tasks. This microstructure allows a flexible composition of different types of existing methods and other related project steps.

Workbench contains helpful techniques for structuring and sorting the knowledge contents. The two most important techniques are categorization and cataloguing.

Categorization: The category system helps to ontologically classify comprised knowledge objects. The system is equipped with main categories as so-called roots, which orientate by the knowledge objects of the work-bench. Therefore, the roots methods, model elements, planning tasks, and project steps do already exist. The category system allows associations between the individual categories.

Cataloguing: In addition to that construction catalogues can be de-fined. Functions and elements of a catalogue are frequently used in systems, which are to be designed differently. The catalogues can be subdivided into a pattern part with systematizing classification characteristics respectively and further attributes.

Successfully proven modeling elements like methods, construction elements (Wiesinger, Laakmann and Hieber, 2002) and design process patterns (Kühling, 2000) will extent the knowledge base of the Workbench.

3.4 Construction Catalogs and Construction Rules: Level of Modeling and Examples

Construction catalogues can also solve certain organizational problems, which are selected, sorted, and structured problem-specific for the documentation of exemplary solutions in the context of organizational architecture. Construction catalogues are manually manageable information memories, which are adjusted to the parameters of the Process Chain Model, the project reference model, and the method toolbox. They are structured systematically and each element has additional characterizing attributes. By using this structure, the planner has access to the content of the construction catalog, represented by the construction elements. Every decision for selecting a construction element follows a systematic procedure, so it can be reconstructed from every following project step.

With a catalog as documentation form, research has a suitable tool for documenting results and experiences for practical use. This meets the requirement for an extendable repository for the different fields in logistics. Examples of converted catalogs within the range of camp planning are de-scribed by Fang (Fang, 1996).

3.5 Collaborative Knowledge Workbench

The practical use of reference models during the logistics planning shows that their application and the quick and straight access to the knowledge is a sophisticated and time-consuming activity. Today, the support of the model-based project execution by information and communication technology (ICT) is a critical factor for the success of this concept. The efficient co-operation of all planners in a project for the design of certain fields of Large Logistics Networks or a complex factory planning needs an integration plat-form. Important features of ICT are the administration and management of the information objects and their links, the workflow management, project management and documentation. A collaborative Workbench allows the connection of all components of the modeling framework presented in Figure 4.

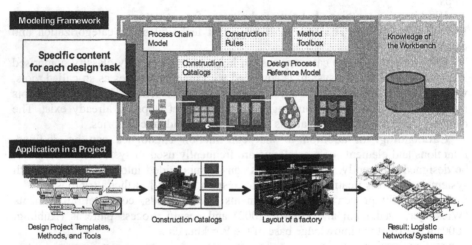

Figure 4. Modeling Framework and Application in a project by the "Workbench"

The technical background of the collaborative Workbench is based on a content management system, which is qualified for the specific needs of logistics design projects. With the widespread use of the Internet and the opportunities of e-Business and web services the technical concept has already been outlined. A substantial component of Workbench is the modeling and configuration of individual design projects based on the predefined patterns. Other modules of Workbench allow the planners to identify relevant planning data for certain steps, the documentation of results and the availability check of all required project outcomes. By using this structured and unified modeling framework, the process modeling language and the approved concept of construction catalogs will simplify the design of components of logistics networks.

The general procedure for logging on knowledge has proven to be successful in the past. It is structured into several steps and does not exclude iterations. Because it was presented at the DIISM conference in 2002 (Wiesinger, 2002)(Wiesinger, 2002a) this reference procedure will not be explained in detail again.

4. ARCHITECTURE OF THE "WORKBENCH"

Based on the described design architecture, the Dortmund LFO implemented collaborative Information- and Knowledge- Management system "Workbench", a computer aided application. The planning tool "Workbench" is able to pre-process the planning knowledge with the most important planning aspects by using the planning knowledge of already realized projects. By using these experiences in a structured knowledge management system for future planning projects, the planning is getting on a to-tally new dimension. The modular description of building, construction, manufacturing, and planning steps, which are the aspects of the service pack-ages, enables the planner to manage the current planning task by using the experience of projects already realized by the explicit knowledge of the Workbench. "Workbench" is an intelligently linked database, subdivided into knowledge categories, which are important for the respective project tasks. As there are for example strategies, processes, methods, design aspects, and so on (see Figure 6). This tool provides useful assistance during the planning and optimizing phase of logistical networks. As a particularly user-friendly and internet-based system it supports the execution of logistical planning projects. Furthermore it can be used for cross-linking the partial projects of SFB by correlating the main results within the platform.

The access to this form of knowledge database helps the planners on the one hand to coordinate their work and on the other hand to accelerate the exchange of data and information in order to improve and speed up the whole planning process. Furthermore it helps to support all processes and phases within the whole product life cycle as an integrated system and thus provides high flexibility, utmost cost saving potential, and a great deal of up-to-dateness to the manufacturer.

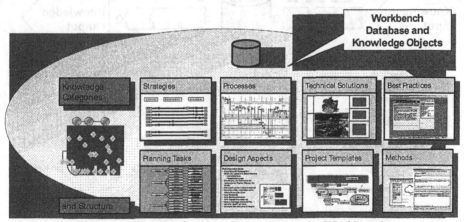

Figure 5. The planning aspects handled in the Workbench

The planning knowledge contained in "Workbench" is filed into so-called knowledge objects, which can be combined to knowledge object groups. The term "knowledge object" illustrates that planning knowledge can only be integrated into the data processing system when it is described in detail according to formal guidelines (compare figure 6). The following knowledge object groups are used in the Workbench system for user administration: "classifying knowledge objects" (category system, construction catalogues), "knowledge objects for description of logistical systems" (model elements, attributes, and planning parameters), "project related knowledge objects" (planning tasks, processing models, and project steps), and further data objects.

The planning tasks, as terms of reference for planning projects, were also included into the Workbench. Planning tasks are the starting point for further knowledge objects, which can be networked problem-specific, for example planning targets, planning aspects, planning data, or processing models.

The result of our efforts is a construction kit for planning large networks in logistics as well as network structures of a production layout plan. Standardised model elements were set up, which serve as generic term for several knowledge objects. As model elements real objects were set up (resources) as well as abstract facts (concepts, strategies, processes). The differentiation of model elements was done by means of the classifications according to the categories and catalogues mentioned above.

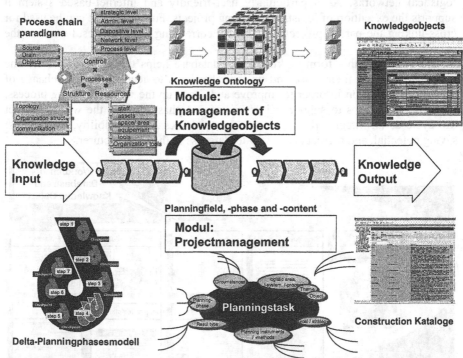

Figure 6. Input and Output of the Knowledge management system "Workbench"

In retrospect we succeeded in designing a platform for supporting planning, modelling, and optimisation of logistical networks and to fill it with basic contents, which will allow access to as well as networking of knowledge of processes, structures, and resources.

5. CONCLUSION AND OUTLOOK

Different industries require a common demand for a concept or tool for the big challenges of global markets to a permanent readiness for strategic innovation projects and derived facility planning activities. The author led a study of the BMBF "Faster ramp-up processes for production systems" in 2002 in Germany. This study examined and analyzed the methods, instruments, tools, simulation concepts, modelling concepts, and the general state of the technology of relevant industry partners in automotive, electronic and mechanical engineering branches. The result of that study was an urgent need for supporting tools, instruments and methods to accelerate and ascertain the ramp up phase in serial production industries (Kuhn, 2002).

The rapid product lifecycles and the efforts of a shorter time to market and time to customer (Wiesinger, 2001), needs in future more Ramp Ups of serial products than we have today (Wiesinger, 2002b). This greater amount of Ramp Ups should also be realized in a shorter time, because only in the starting phases especially high tech product can achieve highest prices in the market (Wiesinger, 2002c). The slower competitors will lose these valuable gains of the first phases because of the saturation of the market and the following product generation of the high tech product (Wiesinger, 2001). "Workbench" as a powerful Knowledge Management Tool can be used to manage a faster ramp up to achieve a shorter time to customer.

The Dortmund SFB 559 (SFB559, 2004) owns a unique pool of logistical and planning knowledge. The manifold findings and experiences of methods and application projects of SFB 559 resulting from interdisciplinary experience should be edited and documented for future research and utilisation by planners.

The Workbench of our partial project M6 allows documenting the findings and experiments in a knowledge management system in a structured and user-friendly way. The enormous amount of explicit and implicit planning knowledge acquired by SFB 559 can now be structured and documented by the design architecture of partial project M6 without a loss of their dependencies mutual influencing. The intelligence of the logistical net-work, so to speak - will not get lost in the knowledge management system.

The focus of the works of the just now appropriated third SFB phase till 2007 is supposed to be put from the development of a technical solution to the utilisation of the design environment. The existing registration system concerning integration and processing of real planning knowledge can be used by the generalisation of this knowledge, leading to standardised contents which are available for reuse.

The main target of the project is to continue the collection of knowledge, either already existing or still to be gathered, to create a complete and viable platform which can be used project-specifically. For this purpose the knowledge has to be documented, structured, and finally networked intelligently by means of the design architecture and Internet based data processing support developed by M6.

Acknowledgments

This research is part of the Collaborative Research Centre (SFB) 559: "Modeling Large Networks in Logistics" at the University of Dortmund together with the Fraunhofer Institute of Material Flow and Logistics founded by Deutsche Forschungsgemeinschaft (DFG). Internet: www.sfb559.uni-dortmund.de

REFERENCES

Beckmann, H. (1996) Theorie einer evolutionären Logistik-Planung. Praxiswissen, Dortmund

Beckmann, H. (2000) Supply Chain Method Handbook. Praxiswissen, Dortmund

Dunne, J.A., Williams, R.J., Martinez N.D. (2002) Network structure and bio-diversity loss in food webs: Robustness increases with connectenance, Ecology Letters 5, pp 558-567

Ebel, H., Mielsch, L.-I., Bornholdt, S. (2002) Scale-free topology of e-mail networks, Phys.Rev E 66, 035103

Fang, D. (1996) Entwicklung eines wissensbasierten Assistenzsystems für die Planung von Lagersystemen. Reihe Fabrikorganisation. Praxiswissen, Dortmund

Fox, J.J., Hill C.C. (2001) From topology to dynamics in biochiamical net-works, Chaos 11; pp 809-815

Kühling, M. (2000) Gestaltung der Produktionsorganisation mit Modell- und Methodenbausteinen. Dissertation Universität Dortmund

Kuhn A. et. al. (2002) Endbericht "fast ramp up" Schneller Produktionsanlauf von Serienprodukten BMBF Förderkennzeichen 02PD9001; Verlag Praxiswissen, Dortmund, ISBN Nr. 3-932-775-92-9

Kuhn, A. (1995) Prozessketten in der Logistik. Entwicklungstrends und Umsetzungsstrategien, Praxiswissen, Dortmund

Sonderforschungsbereich SFG 559 (2004) http://www.sfb559.uni-dortmund.de (25.09.2004)

Wicsinger G., Housein G. (2002c) Management von Wissen, Qualifikation und Beziehungen im Anlauf Wissensmanagement und Personalqualifikation als Garant für einen schnellen Produktionsanlauf von Serienprodukten Artikel zum Thema Ramp-Up aus der WT-Online Heft 10

Wiesinger G., Laakmann F., Guenther S. (2002a) Collaborative Knowledge Workbench for Supply Chain Design Projects. Proc. of the 6th International Conference on Engineering Design and Automation; Hawaii, USA

Wiesinger G., Laakmann F., Hieber R. (2002) Supply Chain Engineering and the use of a supporting knowledge Management Application. In Arai, E. (ed) Pre-proceedings of DIISM 2002, Osaka University Press, Osaka, Japan

Wiesinger G. (2002b) Leitartikel: Schneller Produktionsanlauf von Serienprodukten Wettbewerbsvorteile durch ein anforderungsgerechtes Anlaufmanagement; WT-Online Heft 10

Wiesinger, G., Giourai H. (2001) Potential for fast ramp-up Processes in Production Systems IML annual report.